Parrots of the World

Princeton Field Guides

Rooted in field experience and scientific study, Princeton's guides to animals and plants are the authority for professional scientists and amateur naturalists alike. *Princeton Field Guides* present this information in a compact format carefully designed for easy use in the field. The guides illustrate every species in color and provide detailed information on identification, distribution, and biology.

Albatrosses, Petrels, and Shearwaters of the World, by Derek Onley and Paul Scofield
Birds of Australia, Eighth Edition, by Ken Simpson and Nicolas Day
Birds of Borneo: Brunei, Sabah, Sarawak, and Kalimantan, by Susan Myers
Birds of Chile, by Alvaro Jaramillo
Birds of the Dominican Republic and Haiti, by Steven Latta, Christopher Rimmer, Allan Keith, James Wiley, Herbert Raffaele, Kent McFarland, and Eladio Fernandez
Birds of East Africa: Kenya, Tanzania, Uganda, Rwanda, and Burundi, by Terry Stevenson and John Fanshawe
Birds of Europe, Second Edition, by Lars Svensson, Dan Zetterström, and Killian Mullarney
Birds of India, Pakistan, Nepal, Bangladesh, Bhutan, Sri Lanka, and the Maldives, by Richard Grimmett, Carol Inskipp, and Tim Inskipp
Birds of Kenya and Northern Tanzania: Field Guide Edition, by Dale A. Zimmerman, Donald A. Turner, and David J. Pearson
Birds of the Middle East, by R. F. Porter, S. Christensen, and P. Schiermacker-Hansen
Birds of Nepal, by Richard Grimmett, Carol Inskipp, and Tim Inskipp
Birds of Northern India, by Richard Grimmett and Tim Inskipp
Birds of Peru, by Thomas S. Schulenberg, Douglas F. Stotz, Daniel F. Lane, John P. O'Neill, and Theodore A. Parker III
Birds of the Seychelles, by Adrian Skerrett and Ian Bullock
Birds of Southeast Asia, by Craig Robson
Birds of Southern Africa, by Ian Sinclair, Phil Hockey, and Warwick Tarboton
Birds of Thailand, by Craig Robson
Birds of the West Indies, by Herbert Raffaele, James Wiley, Orlando Garrido, Allan Keith, and Janis Raffaele
Birds of Western Africa, by Nik Borrow and Ron Demey
Caterpillars of Eastern North America: A Guide to Identification and Natural History, by David L. Wagner
Coral Reef Fishes, by Ewald Lieske and Robert Meyers
Dragonflies and Damselflies of the West, by Dennis Paulson
Mammals of Europe, by David W. Macdonald and Priscilla Barrett
Mammals of North America, Second Edition, by Roland W. Kays and Don E. Wilson
Minerals of the World, by Ole Johnsen
Nests, Eggs, and Nestlings of North American Birds, Second Edition, by Paul J. Baicich and Colin J. O. Harrison
Palms of Southern Asia, by Andrew Henderson
Parrots of the World, by Joseph M. Forshaw
The Princeton Field Guide to Dinosaurs, by Gregory S. Paul
Raptors of the World, by James Ferguson-Lees and David A. Christie
Seeds of Amazonian Plants, by Fernando Cornejo and John Janovec
Sharks of the World, by Leonard Compagno, Marc Dando, and Sarah Fowler
Stars and Planets: The Most Complete Guide to the Stars, Planets, Galaxies, and the Solar System, Fully Revised and Expanded Edition, by Ian Ridpath and Wil Tirion
Whales, Dolphins, and Other Marine Mammals of the World, by Hadoram Shirihai and Brett Jarrett

PARROTS

OF THE WORLD

JOSEPH M. FORSHAW
Illustrated by Frank Knight

Princeton University Press
Princeton and Oxford

Dedicated to our wives, Beth and Heather,
in appreciation of their support, understanding, and patience

Published in the United States, Canada, and the Philippine Islands in 2010 by
Princeton University Press, 41 William Street, Princeton, New Jersey 08540
nathist.princeton.edu

Published in the United Kingdom and European Union in 2010 by
Christopher Helm, an imprint of A&C Black Publishers Ltd., 36 Soho Square, London,
W1D 3QY
www.acblack.com

Published in Australia and New Zealand in 2010 by
CSIRO Publishing, 150 Oxford Street (PO Box 1139), Collingwood VIC 3066,
Australia
www.publish.csiro.au

Library of Congress Cataloging-in-Publication Data

Forshaw, Joseph Michael.
   Parrots of the world / Joseph M. Forshaw ; illustrated by Frank Knight.
      p. cm.—(Princeton field guides)
   Includes bibliographical references and index.
   ISBN 978-0-691-14285-2 (pbk. : alk. paper)
   1. Parrots—Identification. I. Title.
QL696.P7F64 2011
598.7'1—dc22   2010002333

Artwork was facilitated by the loan of museum specimens from the National Museum
of Natural History (USNM), Smithsonian Institution, Washington, D.C.

🌀 Smithsonian

This book has been composed in Goudy (main text) and Arial (headings, labels, and
captions)

Printed on acid-free paper

Edited and designed by D & N Publishing, Baydon, Wiltshire, UK

Printed in China

10 9 8 7 6 5 4 3 2

CONTENTS

List of Color Plates   6

Preface   8

Plan of the Book   10

Introduction   15

Parrots in the Australasian Distribution (Plates 1–62)   22

Parrots in the Afro-Asian Distribution (Plates 63–76)   146

Parrots in the Neotropical Distribution (Plates 77–144)   174

Extinct or Presumed Extinct Parrots (Plates 145–146)   310

Index of English Names   314

Index of Scientific Names   320

LIST OF COLOR PLATES

Parrots in the Australasian Distribution

Plate 1   Black Cockatoos (in part)
Plate 2   Black Cockatoos (in part)
Plate 3   Gray and Pink Cockatoos
Plate 4   Yellow-crested White Cockatoos
Plate 5   Fan-crested Cockatoos
Plate 6   Corellas (in part)
Plate 7   Corellas (in part)
Plate 8   *Chalcopsitta* Lories
Plate 9   *Pseudeos* and *Eos* Lories (in part)
Plate 10   *Eos* Lories (in part)
Plate 11   *Trichoglossus* Lories (in part)
Plate 12   *Trichoglossus* Lories (in part)
Plate 13   *Trichoglossus* Lories (in part)
Plate 14   *Trichoglossus* Lories (in part)
Plate 15   *Psitteuteles* Lorikeets
Plate 16   *Glossopsitta* Lorikeets and Swift Parrot
Plate 17   *Vini* and *Phigys* Lorikeets
Plate 18   *Lorius* Lories (in part)
Plate 19   *Lorius* Lories (in part)
Plate 20   *Charmosyna* Lorikeets (in part)
Plate 21   *Charmosyna* Lorikeets (in part)
Plate 22   *Charmosyna* Lorikeets (in part)
Plate 23   *Charmosyna* Lorikeets (in part)
Plate 24   *Charmosyna* Lorikeets (in part)
Plate 25   New Guinea Highlands Lorikeets
Plate 26   Hanging Parrots (in part)
Plate 27   Hanging Parrots (in part)
Plate 28   Hanging Parrots (in part)
Plate 29   Pygmy Parrots (in part)
Plate 30   Pygmy Parrots (in part)
Plate 31   Fig Parrots (in part)
Plate 32   Fig Parrots (in part)
Plate 33   Fig Parrots (in part)
Plate 34   Fig Parrots (in part) and Guaiabero
Plate 35   *Geoffroyus* Parrots (in part)
Plate 36   *Geoffroyus* Parrots (in part)
Plate 37   *Psittacella* Tiger Parrots (in part)
Plate 38   *Psittacella* Tiger Parrots (in part)
Plate 39   Racquet-tailed Parrots (in part)
Plate 40   Racquet-tailed Parrots (in part)
Plate 41   Racquet-tailed Parrots (in part)
Plate 42   *Tanygnathus* Parrots (in part)
Plate 43   *Tanygnathus* Parrots (in part)
Plate 44   Eclectus Parrots (in part)
Plate 45   Eclectus Parrots (in part) and Pesquet's Parrot
Plate 46   *Alisterus* King Parrots
Plate 47   *Aprosmictus* Parrots
Plate 48   *Polytelis* Parrots
Plate 49   *Barnardius* Parrots
Plate 50   Red-capped Parrot and Rosellas (in part)
Plate 51   Rosellas (in part)
Plate 52   Rosellas (in part)
Plate 53   Rosellas (in part)
Plate 54   *Psephotus* Parrots (in part)
Plate 55   *Psephotus* Parrots (in part) and Bluebonnet
Plate 56   *Neophema* Parrots (in part)
Plate 57   *Neophema* Parrots (in part) and Budgerigar
Plate 58   Ground and Night Parrots
Plate 59   *Prosopeia* and *Eunymphicus* Parrots
Plate 60   *Cyanoramphus* Parrots (in part)
Plate 61   *Cyanoramphus* Parrots (in part)
Plate 62   *Nestor* and *Strigops* Parrots

Parrots in the Afro-Asian Distribution

Plate 63   *Coracopsis* Parrots
Plate 64   Gray Parrot and *Poicephalus* Parrots (in part)
Plate 65   *Poicephalus* Parrots (in part)
Plate 66   *Poicephalus* Parrots (in part)
Plate 67   African Lovebirds (in part)
Plate 68   African Lovebirds (in part)
Plate 69   African Lovebirds (in part)
Plate 70   *Psittacula* Parrots (in part)
Plate 71   *Psittacula* Parrots (in part)
Plate 72   *Psittacula* Parrots (in part)
Plate 73   *Psittacula* Parrots (in part)
Plate 74   *Psittacula* Parrots (in part)
Plate 75   *Psittacula* (in part) and *Psittinus* Parrots
Plate 76   Hanging Parrots (in part)

Parrots in the Neotropical Distribution

Plate 77   Blue Macaws
Plate 78   Large *Ara* Macaws (in part)
Plate 79   Large *Ara* Macaws (in part)
Plate 80   Large *Ara* Macaws (in part)
Plate 81   Smaller Green Macaws (in part)
Plate 82   Smaller Green Macaws (in part)
Plate 83   Macaw Allies
Plate 84   *Aratinga* Conures (in part)
Plate 85   *Aratinga* Conures (in part)
Plate 86   *Aratinga* Conures (in part)

Plate 87   *Aratinga* Conures (in part)
Plate 88   *Aratinga* Conures (in part)
Plate 89   *Aratinga* Conures (in part)
Plate 90   *Aratinga* Conures (in part)
Plate 91   *Aratinga* Conures (in part)
Plate 92   *Aratinga* Conures (in part)
Plate 93   *Aratinga* Conures (in part)
Plate 94   *Aratinga* Conures (in part)
Plate 95   *Cyanoliseus* and *Myiopsitta* Parrots
Plate 96   *Enicognathus* Conures
Plate 97   *Pyrrhura* Conures (in part)
Plate 98   *Pyrrhura* Conures (in part)
Plate 99   *Pyrrhura* Conures (in part)
Plate 100 *Pyrrhura* Conures (in part)
Plate 101 *Pyrrhura* Conures (in part)
Plate 102 *Pyrrhura* Conures (in part)
Plate 103 *Pyrrhura* Conures (in part)
Plate 104 *Pyrrhura* Conures (in part)
Plate 105 *Pyrrhura* Conures (in part)
Plate 106 *Pyrrhura* Conures (in part)
Plate 107 *Pyrrhura* Conures (in part)
Plate 108 Mountain Parakeets (in part)
Plate 109 Mountain Parakeets (in part)
Plate 110 *Forpus* Parrotlets (in part)
Plate 111 *Forpus* Parrotlets (in part)
Plate 112 *Forpus* Parrotlets (in part)
Plate 113 *Forpus* (in part) and *Nannopsittaca*
   Parrotlets
Plate 114 *Touit* Parrotlets (in part)
Plate 115 *Touit* Parrotlets (in part)
Plate 116 *Touit* Parrotlets (in part)
Plate 117 *Brotogeris* Parakeets (in part)
Plate 118 *Brotogeris* Parakeets (in part)
Plate 119 *Brotogeris* Parakeets (in part)

Plate 120 *Pionites* Parrots
Plate 121 *Gypopsitta* Parrots (in part)
Plate 122 *Gypopsitta* Parrots (in part)
Plate 123 *Pionopsitta* and *Triclaria* Parrots
Plate 124 *Hapalopsittaca* Parrots (in part)
Plate 125 *Hapalopsittaca* (in part) and
   Amazonian Parrots
Plate 126 *Pionus* Parrots (in part)
Plate 127 *Pionus* Parrots (in part)
Plate 128 *Pionus* Parrots (in part)
Plate 129 *Amazona* Parrots (in part)
Plate 130 *Amazona* Parrots (in part)
Plate 131 *Amazona* Parrots (in part)
Plate 132 *Amazona* Parrots (in part)
Plate 133 *Amazona* Parrots (in part)
Plate 134 *Amazona* Parrots (in part)
Plate 135 *Amazona* Parrots (in part)
Plate 136 *Amazona* Parrots (in part)
Plate 137 *Amazona* Parrots (in part)
Plate 138 *Amazona* Parrots (in part)
Plate 139 *Amazona* Parrots (in part)
Plate 140 *Amazona* Parrots (in part)
Plate 141 *Amazona* Parrots (in part)
Plate 142 *Amazona* Parrots (in part)
Plate 143 *Amazona* Parrots (in part)
Plate 144 *Amazona* Parrots (in part)

Extinct Parrots

Plate 145 Extinct or Presumed Extinct Parrots
   (in part)
Plate 146 Extinct or Presumed Extinct Parrots
   (in part)

PREFACE

As pointed out by Dean Amadon in his Foreword to my *Parrots of the World* (Lansdowne Editions 1973, 1981, 1989), parrots always have been of particular interest to mankind, mainly because of their popularity as pets. This popularity is well documented and dates from early times. It is possible that the Rose-ringed Parakeet *Psittacula krameri* from northern Africa was known to the ancient Egyptians, though there appear to be no records in their writings or art. Ctesias, a Grecian slave who became court physician to Artaxerxes II in 401BC, gave a fairly accurate description of the Plum-headed Parakeet *Psittacula cyanocephala* and wrote romantically of the bird's ability to speak the language of its native India and the claim that it could be taught to speak Greek. It probably was Alexander the Great who introduced to Europe tame parrots from the Far East, and Alexandrine Parakeet, the English name for *Psittacula eupatria*, honors the warrior king. Aristotle almost certainly based his description of parrots on birds brought back by the triumphant armies of his pupil, Alexander. Parrots, presumably *Psittacula* species from northern Africa and the Middle East and possibly the Gray Parrot *Psittacus erithacus* from tropical Africa, were well known to the ancient Romans, and talking birds were status symbols among the noble classes. Voyages of discovery to Asia and the Americas during the fifteenth and sixteenth centuries resulted in new parrot species being brought back to Europe and trading in live birds soon commenced, eventually resulting in the domestication of some species, most notably the Budgerigar *Melopsittacus undulatus* and the Cockatiel *Nymphicus hollandicus* from inland Australia. Contrasting with this popularity of parrots as pets has been a longstanding indifference among ornithologists, researchers, and field observers towards the group, and only in recent decades has there been a change in attitude. Thankfully, the change has been quite dramatic, and it probably is true to say that at the present time parrots are one of the most intensely studied groups of birds. Interest among fieldworkers also is very strong, with the focus often being on the conservation needs of rare or endangered species. Among birdwatchers there is increasing attention being given to parrots, and "parrot-watching" tours are becoming more commonplace. I have accompanied these tours in Australia, while in South America the spectacular aggregations of parrots at traditional "clay licks" have generated regional ecotourism enterprises.

To meet identification needs associated with this upsurge of interest in parrots, *Parrots of the World: An Identification Guide* (Princeton University Press 2006) was produced. In that book, Frank Knight and I attempted to address all aspects of identification, both in the field and at close quarters, the latter to meet the needs of researchers, museum or zoo curators, aviculturists, and officials administering national or international conservation programs, including the Convention on International Trade in Endangered Species of Wild Fauna and Flora (CITES). Because of its broad scope in the coverage of identification, that volume was designed as a handbook, so restricting its practical usage as a field guide. This shortcoming we now are addressing in production of this *Parrots of the World*, which is designed primarily for use in the field, though of course it will find a place in both private and institutional libraries. Preparation of the text and plates has been very much a collaborative effort, and together we gratefully acknowledge generous assistance received from a number of sources.

First, we express our appreciation to authors of the published works consulted in the course of preparing the plates and text. In addition to the published references, we have used unpublished information kindly made available by Thomas Arndt, Jessica Eberhard, Leo Joseph, José Vicente Rodríguez-Maheca, Thomas Schulenberg, Luís Fábio Silveira, Louise Warburton, and Carlos Yamashita.

Assistance with selecting "localities" for observing species was kindly given by Thomas Arndt, D. Avinandan, Donald Brightsmith, Enrique Bucher, Robert Clay, Adrián S. Di Giacomo, Bennett Hennessey, Thomas Jenner, Lee Jones, Olivar Komar, Jeremy Minns, Salvadora Morales, Fábio Olmos, John O'Neill, John Penhallurick, Aasheesh Pittie, Craig Robson, José Vicente Rodríguez-Maheca, Thomas Schulenberg, Luís Fábio Silveira, Andréa Ulian, Barry Walker, Louise Warburton, and Carlos Yamashita.

Curators and staffs facilitated study by both author and illustrator of specimens in collections at the American Museum of Natural History, New York, U.S.A.; the Academy of Natural Sciences, Philadelphia, U.S.A.; the Australian Museum, Sydney, Australia; and the Australian National Wildlife Collection, Canberra, Australia. Additional studies of specimens were made by the author at the Western Foundation of Vertebrate Zoology, Camarillo, California, U.S.A., and the Museum of Victoria, Melbourne, Australia. Ken Mays assisted the author with the examination of specimens at the American Museum of Natural History, New York, and the Academy of Natural Sciences, Philadelphia, and Rae Anderson similarly assisted with the examination of specimens at the Western Foundation of Vertebrate Zoology, Camarillo. Data from specimens were kindly provided by Mary LeCroy, Paul Sweet, Shannon

Kenney, and Margaret Hart (American Museum of Natural History, New York), Leo Joseph and Nate Rice (Academy of Natural Sciences, Philadelphia), and Robin Panza (Carnegie Museum of Natural History, Pittsburgh, Pennsylvania). Interinstitutional loans of specimens were arranged through the Australian Museum, Sydney, and the Australian National Wildlife Collection, Canberra, and we are grateful to Walter Boles, Leo Joseph, and Robert Palmer for arranging these loans. For loans of specimens we thank directors, curators, and collection managers at the American Museum of Natural History, New York; Academy of Natural Sciences, Philadelphia; Australian Museum, Sydney; Australian National Wildlife Collection, Canberra; Field Museum of Natural History, Chicago; Macleay Museum at University of Sydney; Museum of Victoria, Melbourne; National Museum of Natural History, Smithsonian Institution, Washington, D.C.; Peabody Museum at Yale University, New Haven, Connecticut, U.S.A.; Queensland Museum, Brisbane, Australia; and Western Foundation of Vertebrate Zoology, Camarillo, California.

Particularly helpful in preparing the colored plates were digital images of specimens kindly supplied by Mark Adams (British Museum of Natural History, Tring, U.K.); Renato Gaban-Lima (Museu de Zoologia, Universidade de São Paulo, Brazil); Shannon Kenney and Margaret Hart (American Museum of Natural History, New York); Julie Reich (Academy of Natural Sciences, Philadelphia); Christopher Millensky (National Museum of Natural History, Smithsonian Institution, Washington, D.C.); and John O'Neill (Louisiana State University Museum of Zoology, Baton Rouge, Louisiana, U.S.A.). For the same purpose, photographs of living birds were provided by Cyril Laubscher, Alan Lieberman, Rainer Niemann (Arndt-Verlag, Germany), and Matthias Reinschmidt (Loro Parque, Tenerife, Spain). The illustrator prepared working sketches of living birds housed at the aviaries of Peter Gowland in Canberra.

Research associated with this project was undertaken by the author as Research Associate in the Department of Ornithology at the Australian Museum, Sydney, and support given by the trustees, director, and staff at the museum is gratefully acknowledged. Particularly helpful was the provision of library facilities, and copies of important references were supplied by Walter Boles. Literature searches were undertaken in the library at the Western Foundation of Vertebrate Zoology, Camarillo, and copies of references from publications held at that library were supplied by Jon Fisher and Linnea Hall. Special thanks go to Robert Kirk and colleagues at Princeton University Press for bringing the project to fruition.

Finally, we pay special tribute to our wives and families for the support, encouragement, and understanding given to us when so much of our time was being directed to this project.

Joseph M. Forshaw
Canberra, Australia
1 November 2009

PLAN OF THE BOOK

Because this field guide is intended primarily for use in the field, species are arranged geographically instead of taxonomically, so bringing together some species that are likely to be encountered in the same area. In addition to being more useful in the field, a geographical arrangement offers continuity, for currently there are ongoing investigations into the phylogeny of parrots, mostly involving biochemical analyses, and intraordinal arrangements probably will remain unresolved for some time.

Parrots occur mostly in the Southern Hemisphere, and are most prevalent in tropical regions. Following extinction of the Carolina Parakeet *Conuropsis carolinensis* in North America, the Slaty-headed Parakeet *Psittacula himalayana* is the most northerly distributed species, occurring at lat. 34°N in eastern Afghanistan, and now that the Red-fronted Parakeet *Cyanoramphus novaezelandiae* is no longer present on Macquarie Island the southern limit of distribution is on Tierra del Fuego, where the Austral Conure *Enicognathus ferrugineus* ranges south to lat. 55°S. For convenience, the worldwide range may be divided into three geographical components, which I have identified as the Australasian Distribution, the Afro-Asian Distribution, and the Neotropical Distribution, and these are shown in figure 1. The Neotropical Distribution is well set apart, but separation of the Australasian and Afro-Asian Distributions is to some extent arbitrary. I have adopted the modified Wallace's Line as the boundary, so incorporating Wallacea, west to the Lesser Sunda Islands, and the Philippine Islands. *Loriculus* hanging parrots occur on both sides of this boundary, so species are treated in both Distributions according to their ranges. Extinct or presumed extinct species from throughout the worldwide range are illustrated on plates 145 and 146.

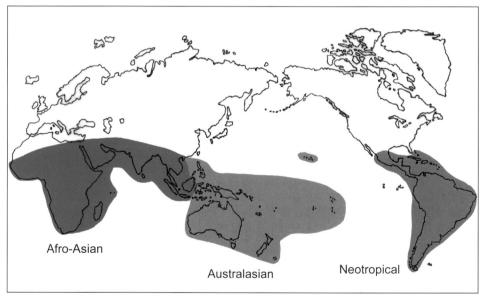

Figure 1. Units of geographical distribution of parrots adopted in this book.

Parrots are very well represented in the Pacific Distribution and show marked endemism centered on Australia, the predominant land mass, New Guinea, New Zealand, and neighboring islands. Insular forms restricted to one or few islands are prevalent in Polynesia and the Indonesian Archipelago. Lorikeets and cockatoos are restricted entirely to the Pacific Distribution, and there is strong diversity among other parrots. This diversity is in marked contrast to the uniformity found elsewhere in the worldwide range, and often is reflected in Australia being referred to as the "land of parrots."

In the Afro-Asian Distribution, which extends from Africa and Madagascar east to Indochina, the Malay Peninsula, and the Greater Sunda Islands, parrots are poorly represented and divergence above specific level is slight. Africa and the Indian subcontinent are the major land masses, and it is on these that most parrots occur. *Psittacula* is the only genus occurring in both Africa and Asia, with the Rose-ringed

Parakeet *P. krameri* being the most widely distributed of all parrots. We have evidence that in at least one region—the Mascarene Islands in the Indian Ocean—parrots formerly were more numerous and apparently comprised a dominant component of the avifauna. The critically endangered Mauritius Parakeet *Psittacula echo* is the only endemic species surviving on these islands, but Newton's Parakeet *P. exsul* from Rodrigues and the Mascarene Parrot *Mascarinus mascarinus* are known from specimens. Also, fossil remains from these islands indicate that the number of parrot species and the high endemism that formerly occurred there is remarkable for such a small archipelago.

Although characterized by a lack of diversity, parrots are very well represented in the Neotropical Distribution, where are found some of the most familiar groups such as the large, spectacular *Ara* macaws, the short-tailed, stolid *Amazona* parrots, and the slim, narrow-tailed *Aratinga* conures. The dominant geographical component is the South American continent, where most species occur, and here the distribution of parrots is influenced strongly by two major topographical features—the Andes and the Amazon River basin. The influence of the Andes is reflected in the association of some species with one or more of three forested zones: tropical (up to 1000m), subtropical (1000 to 2500m), or temperate (2500m up to the tree line at about 3500m), and altitudinal limits in the distribution of parrots are more pronounced in South America than elsewhere in the worldwide range. Rainforests in the Amazon River basin are frequented by many species, some of which do not occur elsewhere and may be restricted to certain types of forest.

Color Plates and Their Descriptive Texts

Descriptive texts and distribution maps accompany the color plates, on which are illustrated all extant species and the more divergent subspecies, with significant sex or age differences also being depicted. In many instances, sightings in the field are of parrots in flight, and often it is in flight that diagnostic plumage features, notably wing and tail patterns, are most conspicuous. Thus where relevant and helpful, upperside and underside flight images are included, though these are not to scale.

On the color plates all birds not identified as juveniles or subadults are adults, and the sex is indicated only if the species is sexually dimorphic. For example, "♂" indicates that the bird is an adult male, and "♀" indicates that it is an adult female. No symbol is given for adults without visual sex differences, so read "adult ♂ ♀" for all birds not otherwise identified. Similarly, "juv ♂" and "juv ♀" indicate that there are sexual differences in the plumages of juveniles, but "juv" denotes that juveniles of both sexes are alike.

English Names

For most species, I have retained English names used in *Parrots of the World: An Identification Guide* (Princeton University Press 2006), and again have adopted "conure" as a collective name for species in *Aratinga*, *Pyrrhura*, and allied genera. Together with "macaw" and "amazon," which are widely accepted, "conure" has a long history of almost universal usage in avicultural literature, and well identifies a distinctive assemblage of neotropical parrots.

Mention should be made of the terms "parrot" and "parakeet," and "lory" and "lorikeet," because they can cause some confusion. There is no biological basis for distinguishing "parrots" from "parakeets" or "lories" from "lorikeets." In general terms, "parrots" and "lories" are larger birds with short, squarish tails, and "parakeets" and "lorikeets" are smaller parrots with long, graduated tails, but in many instances these distinctions do not hold. Indeed, in Australia the terms "parakeet" and "lory" are not used. I have not been consistent with the use of these terms, being content to accept whichever is in common usage.

Scientific Nomenclature

Not unexpectedly, ongoing research into the phylogeny of parrots, mostly involving advanced investigatory techniques, brings about changes in nomenclature, especially at the genus and species levels. Contributing to this instability is the present trend of elevating distinctive isolates from subspecies to species and affording generic differentiation to "species groups" when subgeneric differentiation probably is more appropriate. I suspect that already we have too many species and genera of neotropical parrots, which really do constitute a most homogeneous assemblage of forms. While acknowledging these difficulties, I have attempted to adopt in this book a nomenclature that reflects current thinking on relationships up to genus level. For the most part, I have accepted the nomenclature adopted in *The Howard and Moore Complete Checklist of the Birds of the World* (Dickinson 2003), but have incorporated some changes proposed in recent studies, most of which have focused on neotropical species. References for these studies are listed at the end of this section.

Descriptive Texts

A standard format is followed in the brief descriptive texts, although slightly different sequences are employed for monotypic and polytypic species. Key distinguishing plumage features are stated, with sexual differences or differences in juveniles being noted. Morphological features referred to in these descriptions are shown in figure 2. Where relevant to field identification, vocalization, habitat preferences, and

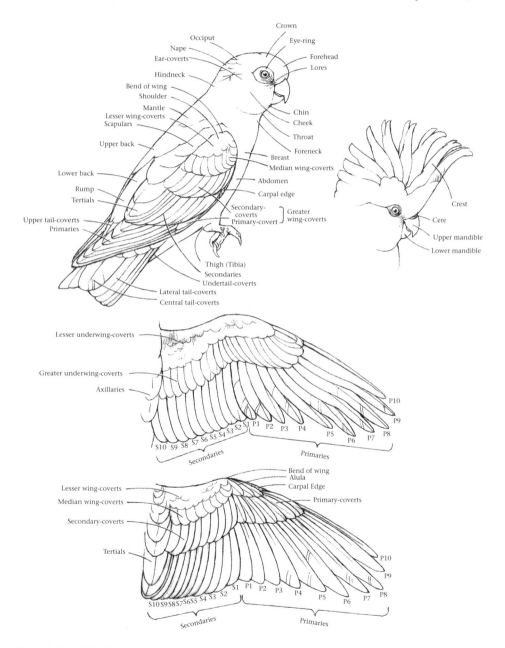

Figure 2. Descriptive features of a parrot with terms used in the text.
Top: external features of a Galah Eolophus reseicapilla (left); head of a Sulphur-crested Cockatoo Cacatua galerita (right).
Below: underside of wing (top) and upperside of wing (bottom).

behavioral characteristics are itemized. Distribution of the species then follows and altitudinal limits to the range are included. An absence of altitudinal data indicates either that the species occurs at all altitudes within its range (most likely for lowland species) or that altitudinal limits have not been determined. The status is given, with particular attention being focused on species that are endangered and those that are listed on CITES Appendix I. Criteria set out by BirdLife International (2004) are adopted for threatened or endangered species. For polytypic species there follows a statement on the number of subspecies together with a comment on the ease of field identification; poorly differentiated subspecies are separated by minor differences, mainly size, that cannot be detected by observation, discernible subspecies often can be differentiated only by close or persistent observation, and well-marked subspecies are readily distinguished in the field. Unless stated otherwise, the nominate subspecies is described first, and descriptions of other subspecies relate to that of the nominate or another named subspecies. The numbering of subspecies is merely to indicate the degree of geographical variation, which can be relevant when determining identification.

There are parrots of unmistakable appearance and conspicuous occurrence. Others, most notably on islands, are the sole parrot inhabitants of their ranges. In these situations little or no difficulty should be experienced with field identification. However, occurrences of two or more species of similar appearance in the same area are commonplace, so where relevant I have listed key distinguishing features of look-alike species.

When nominating localities that are especially promising for observing particular parrots, I have given preference to national or state parks, nature reserves or game parks, forest reserves, and commercially operated lodges or guesthouses catering to birdwatchers. A number of these lodges or guesthouses are featured on the World Wide Web, and some include bird lists on their websites. National or regional birding associations also can provide up-to-date information on prime localities for parrot-watching, and hiring professional guides often can make the difference between success or failure in searches for rare or elusive species. Localities in Australia, and some places in New Zealand and Papua New Guinea, are suggested on the basis of personal experience, but for localities elsewhere I have relied on literature records and recommendations from local fieldworkers or tour operators.

Distribution Maps

Ranges of all subspecies, and of all monotypic species, are shown on maps facing the color plates, and for some widely distributed polytypic species there is an additional map showing the total range of the species. There always is some valid basis for finding fault with distribution maps, especially where the only available data from which they can be prepared are meager or poorly documented. A species cannot be expected to occur uniformly throughout its range, for dispersal often is patchy and determined by the availability of suitable habitat. Also, there is a likelihood of a species or subspecies occurring irregularly outside the normal range, and even within its normal range local occurrences may be influenced by seasonal or long-term changes in food availability. Nevertheless, maps are quite helpful as indicators of likely areas of occurrence, and are particularly valuable for demonstrating sympatry or allopatry, especially in the ranges of "look-alike" species.

References

Arndt, T., 2006. A revision of the *Aratinga mitrata* complex, with the description of one new species, two new subspecies and species-level status of *Aratinga alticola*, J. Ornithol., 147: 73–86.

Arndt, T., 2008. Anmerkungen zu einigen *Pyrrhura*-Formen mit der Beschreibung einer neuen Art und zweier neuer Unterarten, *Papageien*, 21: 278–286.

Birdlife International, 2004. *Threatened Birds of the World*, CD-ROM.

Caparroz, R., and J. M. Barbanti Duarte, 2004. Chromosomal similarity between the Scaly-headed Parrot (*Pionus maximiliani*), the Short-tailed Parrot (*Graydidascalus brachyurus*) and the Yellow-faced Parrot (*Salvatoria xanthops*) (Psittaciformes: Aves): a cytotaxonomic analysis, *Genetics and Molecular Biology*, 27: 522–528.

Christidis, L., and W. E. Boles, 2008. *Systematics and Taxonomy of Australian Birds*, Collingwood: CSIRO Publishing.

Dickinson, E. C. (ed.), 2003. *The Howard and Moore Complete Checklist of the Birds of the World*, 3rd (rev.) edn, Princeton: Princeton University Press.

Eberhard, J. R., and E. Bermingham, 2005. Phylogeny and comparative biogeography of *Pionopsitta* parrots and *Pteroglossus* toucans, *Molecular Phylogenetics and Evolution*, 36: 288–304.

Forshaw, J. M., 2006. *Parrots of the World: An Identification Guide*, Princeton: Princeton University Press.

Joseph, L., 2000. Beginning an end to 63 years of uncertainty: the neotropical parakeets known as *Pyrrhura picta* and *P. leucotis* comprise more than two species, *Proc. Acad. Nat. Sci. Phil.*, 150: 279–292.

Joseph, L., 2002. Geographical variation, taxonomy and distribution of some *Pyrrhura* parakeets, *Ornith. Neotrop.*, 13: 337–363.

Joseph, L., and D. Stockwell, 2002. Climatic modeling of the distribution of some *Pyrrhura* parakeets of north-western South America with notes on their systematics and special reference to *Pyrrhura caeruleiceps* Todd, 1947, *Ornith. Neotrop.*, 13: 1–8.

Ribas, C. C., and C. Y. Miyaki, 2004. Molecular systematics in *Aratinga* parakeets: species limits and historical biogeography in the "*solstitialis*" group, and the systematic position of *Nandayus nenday*, *Molecular Phylogenetics and Evolution*, 30: 663–675.

Ribas, C. C., R. Gaban-Lima, C. Y. Miyaki and J. Cracraft, 2005. Historical biogeography and diversification within the neotropical parrot genus *Pionopsitta*, *J. Biogeogr.*, 32: 1409–1427.

Ribas, C. C., L. Joseph, and C. Y. Miyaki, 2006. Molecular systematics and patterns of diversification in *Pyrrhura* (Psittacidae) with special reference to the *picta-leucotis* complex, *Auk*, 123: 660–680.

Russello, M. A., and G. Amato, 2004. A molecular phylogeny of *Amazona*: implications for neotropical parrot biogeography, taxonomy, and conservation, *Molecular Phylogenetics and Evolution*, 30: 421–437.

Tavares, E. S., C. Yamashita, and C. Y. Miyaki, 2004. Phylogenetic relationships among some neotropical parrot genera (Psittacidae) based on mitochondrial sequences, *Auk*, 121: 230–242.

INTRODUCTION

Few groups of birds are more easily recognized by even the most casual observer than the parrots, and this is due largely to the universal popularity of some species as pets, most notably the Budgerigar *Melopsittacus undulatus* and the Cockatiel *Nymphicus hollandicus*. The most conspicuous external feature making all species easily recognizable as parrots is the short, blunt bill with a downcurved upper mandible fitting neatly over a broad, upturned lower mandible. Also prominent is the zygodactylous foot with two toes pointing forward and two turned backward. Other less obvious characteristics include the proportionately large, broad head and short neck, the thick and prehensile tongue, and nostrils set in a bare or feathered fleshy cere at the base of the upper mandible. Despite the strong homogeneity of morphological features separating parrots from other birds, it will be apparent from illustrations in this book that parrots come "in all shapes and sizes." This variation in external appearances is the key to field identification.

Size alone distinguishes the largest and smallest species. Being so much larger than other species, the *Anodorhynchus* and *Ara* macaws are instantly recognizable, the majestic Hyacinth Macaw *Anodorhynchus hyacinthinus* measuring approximately 100cm in total length. Weighing up to 3kg, the Kakapo *Strigops habroptila* from New Zealand is the largest parrot by weight, and its bulkiness is quite obvious in the field. At the other extreme, the diminutive *Micropsitta* pygmy parrots of New Guinea are less than 9cm in total length and thus easily overlooked or often mistaken for small passerines. Size can be the key factor in distinguishing parrots of similar appearance, and this is particularly evident with the two *Coracopsis* species on Madagascar, the Red-tailed Black Cockatoo *Calyptorhynchus banksii* and the Glossy Black Cockatoo *C. lathami* in eastern Australia, and with *Amazona* and *Pionus* species in the neotropics.

Obvious mostly in overhead flight, various characteristic shapes derive mainly from structure of the wings and tail. To illustrate these identifying features, diagrams of the flight silhouettes of some species are shown in fig. 3. As depicted in these diagrams, wings can be long, narrow, and pointed, as in the Cockatiel *Nymphicus hollandicus*, the Swift Parrot *Lathamus discolor*, and most *Psittacula* species, broad and pointed, as in many lories or lorikeets, or broad and rounded, as in *Eclectus* and *Amazona* parrots. Tails are particularly variable, being long or short, narrow or broad, and rounded, squarish, wedge-shaped, or pointed. Tails also may feature specially structured feathers, such as markedly elongated central feathers in the Papuan Lorikeet *Charmosyna papou* and Long-tailed Parakeet *Psittacula longicauda*, or the bare-shafted, spatulate-tipped central feathers in *Prioniturus* species. Long, pointed wings, often opened in a backward-swept formation, together with a long, narrow, pointed tail produce a characteristically streamlined appearance in flight, whereas broad, rounded wings and a short squarish or rounded tail produce a very different "top-heavy" flight silhouette, and these differences are important for field identification.

Of course, plumage coloration is the feature most relied upon for identification. Apart from the cockatoos, most parrots have a predominantly green plumage. Cockatoos stand apart because of the absence of green in their coloration, as also do a few, normally unmistakable "non-green" parrots. Red, yellow, and blue feature consistently in distinguishing markings, which tend to be concentrated on the head, uppersides of the wings, and rump to upper tail-coverts, thus producing color patterns that are of prime importance in identifying most species. Accurate identification often is determined by certain prominent features, such as frontal bands, cheek-patches, ear-coverts, nuchal collars, or "wing-patches," and this is particularly true when separating closely-allied species. Among peculiarly distinctive plumage features are the bare face or head of Pesquet's Parrot *Psittrichas fulgidus*, the Vulturine Parrot *Gypopsitta vulturina*, and the Orange-headed Parrot *G. aurantiocephala*. Extreme sexual dimorphism distinguishes the Eclectus Parrot *Eclectus roratus*, and erectile, elongated nuchal feathers are present only on the Hawk-headed Parrot *Deroptyus accipitrinus*. Colors of the bill, iris, and feet also can be important elements in field identification. Colored bills, notably red, orange, or yellow, are present in many species occurring in both the Australasian and Afro-Asian Distributions, but are decidedly uncommon in neotropical species.

Obvious modifications in bill structure can aid identification, but other anatomical differences normally are inconspicuous. An enormous, projecting bill is diagnostic of the Palm Cockatoo *Probosciger aterrimus*, and the peculiarly bulbous bill of the Glossy Black Cockatoo *Calyptorhynchus lathami* usually is discernible when birds are observed while feeding. Narrow, protruding bills with elongated, less decurved upper mandibles are conspicuous distinguishing features of the Slender-billed Corella *Cacatua tenuirostris*, the Western Corella *C. pastinator*, the Red-capped Parrot *Purpureicephalus spurius*, and the Slender-billed Conure *Enicognathus leptorhynchus*. Conversely, bill differences that separate the very similar Carnaby's Black Cockatoo *Calyptorhynchus latirostris* and Baudin's Black Cockatoo *C. baudinii* seldom are obvious to field observers. Likewise, the "brush-tipped" tongues of lories and lorikeets can be

1. Broad, pointed wings and short, pointed or wedge-shaped tail.
2. Broad, pointed wings and short, squarish or rounded tail.
3. Broad, pointed wings and long, graduated tail.
4. Broad, pointed wings and long, rounded tail.
5. Broad, rounded wings and short, squarish or rounded tail.
6. Broad, rounded wings and long, graduated tail.
7. Narrow, pointed wings and long, sharply graduated tail.

Figure 3. Some examples of principal flight silhouettes for parrots. (The drawings are not to scale.)

seen only when a feeding bird is observed at close quarters, and identification of these nectar-feeding parrots normally is made from their behavior.

Vocalization

The sharply metallic call-notes of parrots are distinctly harsh and unmelodic. Generally, they are based on a short, simple syllable or combination of simple syllables, and variation lies primarily in the timing of repetition. Apart from the shrill, sibilant notes given by *Micropsitta* pygmy parrots, *Cyclopsitta* fig parrots, *Loriculus* hanging parrots, and some small lorikeets, calls normally are disproportionately loud relative to the size of the birds, and can be heard from afar, often well before the parrots come into view. In general, larger species have lower-pitched calls, as evidenced by the hoarse, guttural cries of *Ara* macaws and the discordant screeches of most cockatoos, but there are interesting exceptions. Australian platycercine parrots have pleasant, whistle-like calls, and one species, the Red-rumped Parrot *Psephotus haematonotus*, often emits a prolonged series of trills or whistling notes, almost a song. Other species with unusually musical calls include the Singing Parrot *Geoffroyus heteroclitus* from the Solomon Islands and the Purple-bellied Parrot *Triclaria malachitacea* from southeastern Brazil, whereas the Amazonian Hawk-headed Parrot *Deroptyus accipitrinus* has a remarkably wide repertoire of piping notes, musical whistles, and chatterings or squawks. Such distinctive call-notes are excellent aids to field identification.

Descriptions of calls given in this book have been sourced mostly from published texts.

Habitats

Lowland, tropical rainforest is the habitat in which parrots are particularly prevalent, and they seem to be more common along the edges of forest where it borders a watercourse or track, or where it adjoins open clearings. Of course, this apparent preference for edge habitats could be due to a greater ease of observation in more open situations. Although generally showing a close association with trees, especially those lining watercourses, parrots that inhabit open country tend to have a broader habitat tolerance than forest-dwelling species, and sometimes become residents or regular visitors in urban parks and gardens. There are some distinctive highland forms, but parrots generally are less common at higher altitudes, and those species that do occur there normally are absent from, or scarce in neighboring lowlands.

In both forested and open country habitats, subtle differences in preferences can be shown by some, often closely-related species, and an awareness of these preferences can be very helpful in field identification. In Amazonia, some *Ara* macaws and *Amazona* amazons favor either wet, seasonally inundated (várzea) forests or drier (terra firme) forests on elevated ground. Similar preferences for humid, wet forests or drier broadleaved forests are shown by some *Psittacula* species in Southeast Asia. Likewise in open country habitats, more densely wooded areas are preferred by some parrots, while other species favor grasslands or shrublands with widely spaced stands of trees. Additionally, there are species that occur in very specialized habitats. One such species is the terrestrial Ground Parrot *Pezoporus wallicus*, which in southern Australia is found only in coastal and contiguous mountain heathlands, a very restricted habitat that is rapidly disappearing. Also in southern Australia, the Rock Parrot *Neophema petrophila* occurs along the seaboard and nests in crevices under overhanging rocks above high-water level. Dominance of particular plants or vegetation communities that serve as primary sources of food or of nesting sites is a habitat requirement for some species. In Amazonia, local occurrences of some macaws and *Aratinga* conures are dependent on the presence of *Mauritia* palms, while farther south Red-spectacled Amazons *Amazona pretrei* and Slender-billed Conures *Enicognathus leptorhynchus* are largely confined to *Araucaria* forests, and Tucumán Amazons *Amazona tucumana* occur mostly in *Alnus* forests or woodlands. Similarly in southeastern Australia, the Glossy Black Cockatoo *Calyptorhynchus lathami* is found only where *Allocasuarina* trees, its primary food source, are present, and in the southwest the range of the Red-capped Parrot *Purpureicephalus spurius* coincides with the distribution of marri *Corymbia calophylla*, an important food tree. Less obvious are reasons for the virtual restriction of Black-cheeked Lovebirds *Agapornis nigrigenis* to *Colophospermum* woodland in southern Africa.

For the most part, I have used self-explanatory, simplified terms, such as savanna woodland, cultivation, and urban parklands, in habitat descriptions, but explanations are needed for the following, sometimes peculiarly regional, descriptive terms:

caatinga (northeastern Brazil): semiarid to arid scrubland with sparse groundcover of few grasses and dominated by cacti and deciduous, often spiny trees and bushes remaining leafless for many months and with characteristically pale gray bare branches;

cerrado (inland Brazil): semiarid scrubland with sparsely scattered low trees and bushes having characteristically gnarled or twisted branches bearing thick, grooved bark and leathery leaves (arboreal
termitaria, which provide nesting sites for many parrots, are prevalent);

gallery forest or woodland: narrow strips of forest or woodland bordering watercourses, usually in grassland
or open woodland;

igapó forest (Amazonia): humid or wet lowland forest flooded by stationary water along riverbanks and
sometimes occurring along streams in terra firme forest;

mallee scrubland (Australia): semiarid to arid scrubland with woody shrubs of low to medium height
and widely spaced multi-stemmed or "mallee" eucalypts;

mallee woodland (Australia): low semiarid to arid woodland (closed or open) on red sandy soil and
dominated by multi-stemmed or "mallee" eucalypts;

pantanal (southern South America): a vast expanse of seasonally inundated grassland across floodplains
of mid reaches of the Rio Paraná;

tepuis (eastern Amazonia): spectacular vertical-walled, flat-topped mountains that remain heavily
shrouded in fog at higher elevations and bear stunted, mostly endemic vegetation;

terra firme forest (Amazonia): humid or wet lowland forest on elevated, dry ground and not subject to
inundation;

várzea forest (Amazonia): humid or wet lowland forest seasonally flooded for several months once or
twice each year and situated mainly on floodplains of major rivers.

Habits

Recovery data for marked birds (usually fitted with leg rings or bands) are available for very few parrots,
mostly Australian species, so information on movements has been compiled mostly from observational
records. Only two Australian species—the Swift Parrot *Lathamus discolor* and the Orange-bellied Parrot
Neophema chrysogaster—are known to be totally migratory; both of these species breed only in Tasmania
and move across Bass Strait to overwinter in southeastern mainland Australia. Blue-winged Parrots
Neophema chrysostoma also leave Tasmania to overwinter in southeastern mainland Australia, where there
is another resident breeding population. Conversely, only a few parrots restricted to small islands are
known to be totally sedentary, but even among these species there are some that regularly move between
islands. The vast majority of parrots undertake some movements, varying from seasonal shifts in altitudinal range to regular or irregular dispersals over large distances or merely local wandering in search of food.
This predilection for wandering causes some unpredictability in local occurrences of even resident species.

Most parrots are gregarious, associating during much of the year in flocks of varying sizes. Flocks tend
to be smaller during the breeding season, while pairs are occupying nesting territories, and reach peak
numbers with the return of those pairs and their offspring at the end of the breeding season.

Only the Kakapo *Strigops habroptila* from New Zealand and the Night Parrot *Pezoporus occidentalis* from
mainland Australia are known to be nocturnal, though other species have been observed flying about
and calling on moonlit nights. In inland Australia, Bourke's Parrot *Neopsephotus bourkii* regularly comes
to watering places well after nightfall and before sunrise. Daily activities generally follow a consistent
routine, commencing at sunrise with much loud calling accompanying departure from communal nighttime roosts, the birds often traveling high along regular flight paths to distant feeding grounds. Feeding
takes place during early to mid morning, the middle of the day then being spent sheltering amidst the
shading foliage of trees or bushes. Feeding resumes in the late afternoon, followed by visits to favored
watering places, and at dusk the birds return to the communal roosts, where preroosting aerobatics and
loud vocalizing often precede settling down for the night. These communal nighttime roosts often are
traditional, remaining in use for many years, and are ideal sites for observing the occupants.

The strong flight of most parrots often is undertaken with rapid, shallow wingbeats and, in addition
to distinguishing silhouettes of flying birds (see fig. 3), variable flight patterns, with differences in speed,
straight or weaving direction, undulation, gliding, and so on, can aid identification. For example, the
flight of *Eclectus*, *Pionus* and *Amazona* species is characterized by wingbeats entirely below body level,
and flying Red-tailed Black Cockatoos *Calyptorhynchus banksii* drift noticeably from side to side, as if
being blown alternatively off and then back on course. The strangely raptor-like flight of the
Hawk-headed Parrot *Deroptyus accipitrinus* from Amazonia is a particularly good characteristic for field
identification of that species.

Lorikeets and *Loriculus* hanging parrots feed on nectar and pollen, and *Micropsitta* pygmy parrots
apparently feed on lichen and fungus, but the diet of other parrots comprises mainly fruits, seeds, nuts,

and berries. Insects and their larvae are important food items for some species, especially some *Calyptorhynchus* cockatoos, and have been recorded even in stomach contents from species that normally feed on grass seeds. It is noteworthy that feeding on the ground is prevalent among Australian parrots, but elsewhere food is procured mostly in trees or shrubs. A local presence of parrots may be dependent on available food supplies, and concentrated food sources, such as profusely flowering or fruiting trees and expanses of seeding grasses, can attract large numbers of birds, so offering excellent observational opportunities. Similarly, large numbers of parrots regularly gather at traditional "clay-licks" to take mineral-rich soil.

Nesting usually takes place in hollows in trees or holes in arboreal or terrestrial termitaria, and occasionally in holes in earth-banks or cliff-faces, or in crevices among rocks. If in termitaria, the tunnel and nesting chamber are excavated by the parrots, while natural hollows in trees or old nesting holes of other birds, such as woodpeckers or barbets, frequently are enlarged and altered by chewing away at the walls or entrance. *Cyclopsitta* fig parrots, *Geoffroyus* parrots, and the Red-breasted Pygmy Parrot *Micropsitta bruijnii* are among the few parrots known to excavate nesting holes in rotting tree stumps. These excavated holes and the tunnels dug into termitaria are good telltale signs of the local presence of nesting parrots. More conspicuous than the noisy parrots themselves are the large communal nests of interwoven sticks built by Monk Parakeets *Myiopsitta monachus* in treetops scattered across the pantanal of South America.

Status and Conservation

Disappearance in 2000 of the last known Spix's Macaw *Cyanopsitta spixii* from the wild in northern Brazil and rediscovery in July 2002, in the mountains of western Colombia, of a small remnant population of the Indigo-winged Parrot *Hapalopsittaca fuertesi*, lost since its initial finding in 1913, epitomizes the parlous status of many parrots, especially in the neotropics. Of the 356 extant species recognized by Birdlife International, no fewer than 123 species, or 34.6 percent, are listed as being near-threatened to endangered, thus making parrots one of the most threatened groups of birds (Birdlife International 2004).

Parrots are not immune from the pressures affecting all wildlife, and already there are signs that some species, including the Gang Gang Cockatoo *Callocephalon fimbriatum* of southeastern Australia, are being affected adversely by global warming. Habitat interference is, by far, the most serious threat to parrots, with the great majority of endangered populations facing varying degrees of habitat loss, degradation, or fragmentation. Of particular concern is the widespread destruction of tropical and subtropical rainforests, which are preferred habitats for many species. In tropical regions, land clearance often is motivated by economic forces, with logging concessions, for example, featuring prominently in national and international commerce. Mining and the conversion of lands to pastoral or agricultural use can be primary factors in land clearance or can thwart any hopes of rehabilitating logged areas. These practices are prevalent in Southeast Asia and Equatorial Africa, where there is intense pressure for rapid commercial gain and greatly increased food production, but it would be tragic if in the process insufficient effort was made to preserve viable stands of the magnificent rainforest, which for centuries has been identified with these regions. Already rapid declines in parrot populations have taken place throughout the Indonesian Archipelago and in the Philippine Islands, where a number of species have become critically endangered. Similarly in the Congo River basin and in West Africa, the Gray Parrot *Psittacus erithacus* has disappeared from parts of its former range because of widespread land clearance. In the neotropics too, there is a major, ongoing transformation of the Amazonian landscape, where mining and agricultural or pastoral activities are bringing about the fragmentation of previously extensive tracts of tropical rainforest. Even more damaging has been extensive deforestation in coastal Brazil, where the long-term survival of a number of endemic species is dependent on the sustained viability of protected stands of forest in reserves and national parks.

Parrots with specialized habitat requirements are especially at risk from habitat loss. *Araucaria* forests in southeastern Brazil, home of the Red-spectacled Amazon *Amazona pretrei*, are being cleared at an alarming rate, and in the Andean highlands of northern Ecuador and western Colombia, where the spectacular Yellow-eared Conure *Ognorhynchus icterotis* is virtually restricted to stands of *Ceroxylon* palms, almost total destruction of the palms has brought the parrots to the brink of extinction. The terrestrial Ground Parrot *Pezoporus wallicus*, a specialized inhabitant of coastal heathlands in southern Australia, is threatened by expanding urbanization and agricultural development along the seaboard.

The particularly high vulnerability of parrots confined to small, isolated islands is well demonstrated by the fact that 16 of the 18 species listed as extinct in 1981 had been endemic to islands, and many currently endangered parrots are restricted to islands. Of concern is the threatened status of endemic

Amazona species in the Lesser Antilles, West Indies, the endangered status of some *Eos* lories and *Cacatua* cockatoos in the Indonesian Archipelago, and the disappearance of *Vini* lories from islands they formerly inhabited in the South Pacific Ocean. Similarly at risk are species found only within very restricted ranges on continental landmasses. A disproportionately high concentration of such species occurs in the highlands of Colombia and Ecuador, where the Yellow-eared Conure *Ognorhynchus icterotis*, the Golden-plumed Conure *Aratinga branickii*, several *Hapalopsittaca* species, and some *Pyrrhura* conures are among parrots seriously threatened by continued clearing of highly fragmented upland forests.

Much has been said of the love/hate relationship between humans and parrots, a relationship manifested primarily in the unequaled popularity of parrots as cagebirds and in the impact of some species on agriculture. Damage to crops by parrots has been reported from a number of countries, but to date there has been little evaluation of the problem. When looking at conflicts between parrots and agriculture, attention often is unduly focused on Australia, where crop damage is regularly cited as a reason for relaxing the prohibition on exports of live birds. I am amazed that any credence at all is given to the claim that trapping and export of so-called "pest species" would alleviate conflicts with agriculture, for the two issues are totally divorced from each other; export certainly would not reduce crop damage and could pose very serious environmental and economic risks in importing countries. It is unlikely that crop damage can be eliminated, but trials undertaken in Australia have demonstrated that damage levels can be reduced significantly by modifications of farming practices or by adopting crop-protection measures based on sound ecological principles.

Although not wishing to detract from the significance of capture for the live-bird trade as a pressure adversely affecting parrot populations, I must stress that little is to be gained by prohibiting capture if the effects of habitat loss are neglected—the two are complementary! Capture for the live-bird market seldom is a primary pressure threatening the survival of a species, but as a secondary pressure it takes on a much greater significance when the species is already affected adversely by a primary pressure such as habitat loss. Too often does rarity give rise to increased demand, with high prices being offered by "collectors" who want birds simply because they are rare, and will take any measures, legal or illegal, to acquire them.

Surveys and monitoring programs are being carried out on endangered parrots in a number of countries, particularly in the neotropics and the Indonesian Archipelago, and detailed studies of the breeding biology of *Ara* macaws in Costa Rica and Amazonia will provide vital information on potential recruitment levels in local populations. As essential prerequisites for effective conservation or management programs, these surveys and investigations are dependent on the compilation of reliable field data, which in turn relies on accurate identification. By assisting with identification, I trust that this field guide can contribute to the conservation of parrots.

COLOR PLATES

PLATE 1 BLACK COCKATOOS (in part)

22

Large black cockatoos with or without colored tail-bands; mostly arboreal; forests and woodlands; seeds, fruits, and insect larvae in diet; labored flight with deep wingbeats; harsh screeches and shrill whistles.

PALM COCKATOO *Probosciger aterrimus* 60cm

Unmistakable; no colored tail-band; prominent crest; enormous bill; red naked face; sexes alike, JUV with yellow barring on underside; loud *keer-eeeow* whistle, and harsh *raark*. Rainforest and adjacent woodland; pairs or small groups. **DISTRIBUTION** New Guinea and adjacent islands, and northernmost Queensland, Australia; up to 1350m; declining, CITES I. **SUBSPECIES** three poorly differentiated and one distinctive subspecies. 1. *P. a. aterrimus* smaller size. *Range* Aru Islands and Misool in western Papuan Islands, Indonesia. 2. *P. a. macgillivrayi* larger than *aterrimus*. *Range* southern New Guinea, between Fly and Balim Rivers, and Cape York Peninsula, northernmost Queensland, Australia. 3. *P. a. goliath* larger than *macgillivrayi*. *Range* western Papuan Islands, except Misool, and southern New Guinea. 4. *P. a. stenolophus* like *goliath*, but narrower crest feathers. *Range* northern New Guinea. **SIMILAR SPECIES** Pesquet's Parrot *Psittrichas fulgidus* (plate 45) red underparts, no crest. Red-tailed Black Cockatoo *Calyptorhynchus banksii* (plate 2) tail-band red (♂) or barred orange (♀), different call. **LOCALITIES** Iron Range National Park, Cape York Peninsula, northernmost Queensland, Australia. Crater Mountain Research Station, Chimbu Province, Papua New Guinea.

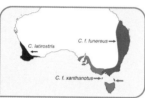

YELLOW-TAILED BLACK COCKATOO
Calyptorhynchus funereus 67cm

Unmistakable; yellow tail-band and ear-coverts; bill black (♂) or white (♀); eye-ring pink (♂) or gray (♀), JUV like ♀; loud *keee-ow* whistle. Pairs or flocks; attracted to *Pinus* plantations. **DISTRIBUTION** eastern Australia; up to 2000m; increasing. **SUBSPECIES** two subspecies differentiated mainly by size. 1. *C. f. funereus* larger size and longer tail with dark flecking. *Range* central Queensland south to eastern Victoria. 2. *C. f. xanthanotus* smaller size; shorter tail with little dark flecking. *Range* Tasmania and Bass Strait islands, where only black cockatoo, and eastern Victoria to southeastern South Australia. **SIMILAR SPECIES** Red-tailed *C. banksii* and Glossy *C. lathami* Black Cockatoos (plate 2) tail-band red (♂) or barred orange (♀); different calls. **LOCALITIES** Werrikimbe National Park, New South Wales. Cradle Mountain National Park, Tasmania.

CARNABY'S BLACK COCKATOO *Calyptorhynchus latirostris* 55cm

One of two almost indistinguishable black cockatoos with white tail-band and white ear-coverts; black (♂) or white (♀) bill without elongated upper mandible; eye-ring pink (♂) or gray (♀), JUV like ♀; loud *wy-lah* cry. Pairs or small flocks; attracted to *Pinus* plantations. **DISTRIBUTION** (see map above) southwestern Australia; endangered. **SIMILAR SPECIES** Baudin's Black Cockatoo *C. baudinii* (see below) with elongated upper mandible. Red-tailed Black Cockatoo *C. banksii* (plate 2) see above. **LOCALITIES** Yanchep, Moore River, and Lesueur National Parks, Western Australia.

BAUDIN'S BLACK COCKATOO
Calyptorhynchus baudinii 56cm

Distinguishable from Carnaby's Black Cockatoo only at close quarters; bill with elongated upper mandible; more prolonged *wy-lah* call. Not attracted to *Pinus* plantations. **DISTRIBUTION** extreme southwestern Australia; near-threatened. **SIMILAR SPECIES** Carnaby's Black Cockatoo *C. latirostris* (see above) without elongated upper mandible. Red-tailed Black Cockatoo *C. banksii* (plate 2) see above. **LOCALITIES** Leeuwin-Naturaliste and D'Entrecasteaux National Parks, Western Australia.

PALM COCKATOO

P. a. aterrimus

Juv

♂

P. a. goliath

P. a. stenolophus

YELLOW-TAILED BLACK COCKATOO

C. f. funereus

♀

♂

C. f. xanthanotus

C. f. funereus

♀

♂

CARNABY'S BLACK COCKATOO

♀

♂

BAUDIN'S BLACK COCKATOO

PLATE 2 BLACK COCKATOOS (in part)

24

RED-TAILED BLACK COCKATOO
Calyptorhynchus banksii 60cm

Larger of two similar black cockatoos with tail-band red (♂) or barred orange (♀ & JUV); prominent recumbent crest; far-carrying, screeching *kree-ee* or *krur-rr*. Noisy and conspicuous; often in flocks feeding on ground. **DISTRIBUTION** mainland Australia; common in north, scarce in south. **SUBSPECIES** five subspecies differ in size and female plumage. 1. *C. b. banksii* large size; ♂ uniformly black; bill dark gray; ♀ & JUV head and upper wing-coverts spotted pale yellow, underparts barred yellow, tail-band barred orange, and bill white. *Range* north Queensland to northeastern New South Wales, northeastern Australia. 2. *C. b. macrorhynchus* large size; ♂ like *banksii*; ♀ & JUV like *banksii*, but pale barred tail-band, head and upper wing-coverts spotted palest yellow, and pale barring on underparts. *Range* Kimberleys to Gulf of Carpentaria, northern Australia. 3. *C. b. samueli* smaller than *banksii*. *Range* four isolated populations across inland eastern, central, and western Australia. 4. *C. b. graptogyne* like *samueli*, but ♀ & JUV with head, upper wing-coverts and underparts more heavily marked with darker yellow. *Range* isolated in southwestern Victoria and southeastern South Australia; endangered. 5. *C. b. naso* like *samueli*, but larger bill; ♀ & JUV with head, upper wing-coverts, and underparts more heavily marked yellow. *Range* extreme southwestern Australia; near-threatened. **SIMILAR SPECIES** Glossy Black Cockatoo *C. lathami* (see below) smaller with inconspicuous, short crest; bulbous bill; ♀ with yellow patches, not spots on head, no yellow barring on underparts; quiet, inconspicuous; different call. Yellow-tailed Black Cockatoo *C. funereus* (plate 1) yellow tail-band and ear-coverts; different call. Carnaby's *C. latirostris* and Baudin's *C. baudinii* Black Cockatoos (plate 1) white tail-band and ear-coverts; different call. **LOCALITIES** Kakadu National Park, Northern Territory. Lakefield National Park, Queensland.

GLOSSY BLACK COCKATOO
Calyptorhynchus lathami 48cm

Smaller black cockatoo with tail-band red (♂) or barred orange (♀ & JUV); very short, inconspicuous crest; peculiarly bulbous bill; soft, wheezy *taar-red*. Feeds exclusively on seeds of casuarinas, so always in or near these trees; pairs or family trios; arboreal; quiet and unobtrusive. **DISTRIBUTION** coastal and subcoastal eastern Australia, including Kangaroo Island. **SUBSPECIES** three doubtful subspecies differ by bill size. 1. *C. l. lathami* head and underparts sooty brown; ♀ with yellow feathers on head; JUV with yellow spots on upper wing-coverts and sides of head, and lower underparts barred pale yellow. *Range* southeastern Queensland to eastern Victoria; uncommon. 2. *C. l. halmaturinus* like *lathami*, but larger bill. *Range* Kangaroo Island, South Australia; endangered. 3. *C. l. erebus* like *lathami*, but smaller bill. *Range* Paluma Range to Dawson-Mackenzie-Isaac Rivers basin, northeastern Queensland; scarce. **SIMILAR SPECIES** Red-tailed Black Cockatoo *C. banksii* (see above) larger with prominent crest; ♀ & JUV yellow spotted; noisy and conspicuous; different call. Yellow-tailed Black Cockatoo *C. funereus* (plate 1) yellow tail-band and ear-coverts; different call. **LOCALITIES** Werrikimbe National Park, New South Wales. Western River Conservation Park, Kangaroo Island, South Australia.

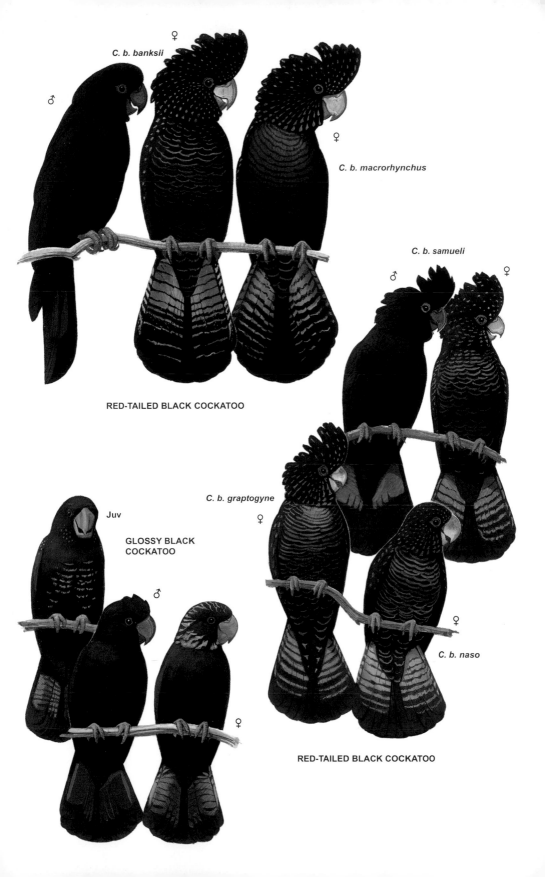

C. b. banksii

♂

♀

C. b. macrorhynchus

♀

C. b. samueli

♂ ♀

RED-TAILED BLACK COCKATOO

C. b. graptogyne

♀

Juv

GLOSSY BLACK COCKATOO

♂

♀

♀

C. b. naso

RED-TAILED BLACK COCKATOO

PLATE 3 GRAY AND PINK COCKATOOS

26

GANG GANG COCKATOO
Callocephalon fimbriatum 35cm

Unmistakable; midsized pale-barred gray cockatoo with forward-curving filamentary crest; head and crest red (♂) or gray (♀ & JUV); distinctive rasping call like rusty hinge. Cooler mountain and lowland forests, visiting urban gardens; pairs or small groups; arboreal; confiding and easily approached; "owl-like" flight with deep wingbeats. **DISTRIBUTION** southeastern mainland Australia from lat. 32°S south to southeastern South Australia; up to 2000m; locally common, generally uncommon; introduced to Kangaroo Island, South Australia. **LOCALITIES** National Botanic Gardens, Canberra. Kosciuszko National Park, New South Wales.

GALAH *Eolophus roseicapilla* 35cm

Unmistakable; midsized pink and gray cockatoo with short, recumbent crest; carunculated bare eye-ring; iris dark brown (♂), pink-red (♀), or pale brown (JUV); distinctive *chet* and *tit-ew*. All wooded areas except dense forests; common in urban parklands; ubiquitous family groups to large flocks; feeds mostly on ground; strong flight with full, rhythmic wingbeats. **DISTRIBUTION** Australia generally, chiefly inland; mostly below 1300m; abundant in expanding range. **SUBSPECIES** two distinctive and one poorly differentiated subspecies. 1. *E. r. roseicapilla* crown and crest pale pink; eye-ring gray-white. *Range* western and central Australia, south of Great Sandy Desert and east to southern Northern Territory. 2. *E. r. albiceps* crown and crest white; eye-ring dull red. *Range* eastern Tasmania and eastern mainland Australia, north to about lat. 20°S and west to Simpson Desert. 3. *E. r. kuhli* generally paler than *albiceps*; shorter crest. *Range* northern Australia, from Kimberley division of Western Australia east to north Queensland. **LOCALITIES** easily seen throughout range.

MAJOR MITCHELL'S COCKATOO
Lophocroa leadbeateri (formerly *Cacatua leadbeateri*) 35cm

Unmistakable; midsized pink and white cockatoo with forward-curving multicolored crest; yellow band in crest narrow (♂) or broad (♀); iris dark brown (♂), pink-red (♀), or pale brown (JUV); quavering *creek-ery-cree*. Arid and semiarid woodlands; pairs or small flocks; feeds in trees or on ground; wary; flapping wingbeats with brief gliding. **DISTRIBUTION** inland Australia, except northeast; up to 300m; uncommon. **SUBSPECIES** two poorly differentiated subspecies. 1. *L. l. leadbeateri* crest scarlet with prominent yellow band. *Range* inland southeastern Australia, from southwestern Queensland and western New South Wales to northwestern Victoria and central-eastern South Australia. 2. *L. l. mollis* crest darker red with little or no yellow band (♂) or ill-defined yellow band (♀). *Range* inland western and central Australia, east to Eyre Peninsula and Lake Eyre basin, South Australia. **LOCALITIES** Currawinya National Park, Queensland. Hattah-Kulkyne National Park, Victoria. Eyre Bird Observatory, Western Australia.

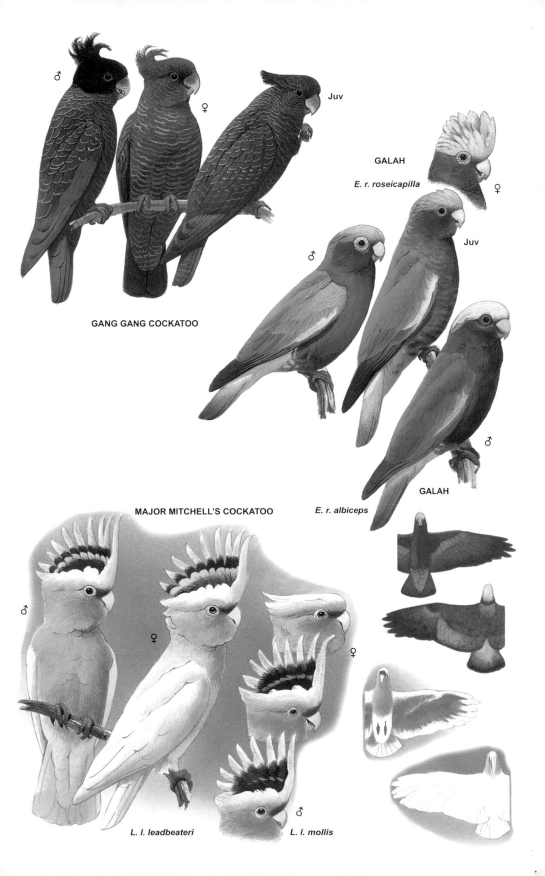

GANG GANG COCKATOO

GALAH

E. r. roseicapilla

Juv

♂

♀

Juv

GALAH

E. r. albiceps

♂

MAJOR MITCHELL'S COCKATOO

♂

♀

♀

♂

L. l. leadbeateri

L. l. mollis

PLATE 4 **YELLOW-CRESTED WHITE COCKATOOS**

28

YELLOW-CRESTED COCKATOO
Cacatua sulphurea 33cm

Smaller of two white cockatoos with yellow forward-curving crest and black bill; yellow undersides of wings and tail; bare eye-ring cream-white; sexes alike, and JUV resembles adults; loud, harsh screeches. Forests, woodlands, plantations; pairs or small groups; mostly arboreal; wary; shallow wingbeats with gliding.

DISTRIBUTION Sulawesi to Lesser Sunda Islands and islands in Flores and Java Seas, Indonesia; up to 1200m; critically endangered; introduced to Hong Kong and Singapore. **SUBSPECIES** three poorly differentiated and one distinctive subspecies. 1. *C. s. sulphurea* yellow ear-coverts and yellow bases to feathers of head and underparts. *Range* Sulawesi and adjacent islands and islands in Flores Sea; introduced to Hong Kong and Singapore. 2. *C. s. parvula* paler yellow ear-coverts and less yellow on head and underparts. *Range* Nusa Penida and Lesser Sunda Islands, except Sumba. 3. *C. s. abbotti* larger than *parvula*. *Range* Masalembu Island in Java Sea; near extinction. 4. *C. s. citrinocristata* (Citron-crested Cockatoo) crest and ear-coverts orange. *Range* Sumba, Lesser Sunda Islands. **LOCALITIES** Komodo National Park, Komodo Island. Langgaliru-Manipeu, Sumba Island.

SULPHUR-CRESTED COCKATOO
Cacatua galerita 50cm

Larger than Yellow-crested Cockatoo; little yellow on ear-coverts or bases to feathers of head and underparts; raucous *raa-aach*. All wooded lands and many urban areas; mostly arboreal in north, ground-feeding in south; common cagebird; large flocks in south, mostly pairs and small groups in north; very noisy and conspicuous; shallow wingbeats with gliding. **DISTRIBUTION** northern and eastern Australia, New Guinea, and western Papuan and Aru Islands, Indonesia; up to 2000m; abundant in south, less numerous in north; introduced to southwestern Australia, New Zealand, Palau Islands in Micronesia, Kai Islands and some islands of East Moluccas, Indonesia, and Taiwan. **SUBSPECIES** five poorly differentiated subspecies. 1. *C. g. galerita* yellow suffusion on ear-coverts; white eye-ring. *Range* eastern and southeastern Australia, from Tasmania and southeastern South Australia north to north Queensland. 2. *C. s. queenslandica* smaller size; broader, ridged bill. *Range* Cape York Peninsula and southern Torres Strait islands, northernmost Queensland. 3. *C. g. fitzroyi* larger than *queenslandica*; eye-ring blue. *Range* northern Australia, from Kimberley division of Western Australia east to Gulf of Carpentaria, north Queensland. 4. *C. g. triton* like *fitzroyi*, but broader crest feathers. *Range* western Papuan Islands, Indonesia, and New Guinea; introduced to Palau Islands, some islands of East Moluccas, Indonesia, and probably the subspecies in Taiwan. 5. *C. g. eleonora* smaller than *triton*. *Range* Aru Islands; introduced to Kai Islands. **SIMILAR SPECIES** Corellas (plate 6) no yellow crest; white bill; different calls. **LOCALITIES** easily seen in range in Australia.

BLUE-EYED COCKATOO
Cacatua ophthalmica 50cm

Large white cockatoo with backward-curving yellow crest, and only cockatoo in range; eye-ring blue; sexes alike, JUV resembles adults; nasal *aa-aah* or *naa-aa*. Forest canopy; arboreal; pairs or small groups; noisy and conspicuous; fluttering wingbeats with gliding.
DISTRIBUTION New Britain, eastern Papua New Guinea.
LOCALITY Wide Bay, eastern New Britain.

YELLOW-CRESTED COCKATOO

C. s. sulphurea

C. s. citrinocristata

C. s. parvula

SULPHUR-CRESTED COCKATOO

C. g. galerita

C. g. triton

BLUE-EYED COCKATOO

PLATE 5   FAN-CRESTED COCKATOOS

30

SALMON-CRESTED COCKATOO
Cacatua moluccensis 52cm

Unmistakable, and only cockatoo in range; large salmon-pink cockatoo with backward-curving, deeper salmon-pink fan-like crest; black bill; eye-ring cream-white; sexes alike, JUV resembles adults; discordant, harsh screech. Forests and tall secondary growth; pairs and small groups; arboreal, wary; noisy and conspicuous; rapid, shallow wingbeats with short glides. **DISTRIBUTION** Seram, Haruku, and Saparua, in South Moluccas, Indonesia; also nearby Ambon, where possibly introduced; up to 900m; vulnerable and declining; CITES I. **LOCALITIES** Manusela National Park, Seram, and Hitu Peninsula, Ambon.

WHITE-CRESTED COCKATOO
Cacatua alba 46cm

Unmistakable, and only cockatoo in range; large white cockatoo with backward-curving, all-white fan-like crest; black bill; eye-ring cream-white; sexes alike, JUV resembles adults; very nasal, high-pitched screech. Forests, woodlands, plantations; pairs, small groups or roosting flocks; arboreal; fairly tame, but wary where hunted; noisy and conspicuous; rapid, shallow wingbeats with short glides. **DISTRIBUTION** Halmahera and adjacent islands, North Moluccas, Indonesia; birds on Obi and Bisa, Central Moluccas, probably escaped pets; up to 900m, mostly below 600m; vulnerable and declining; introduced to Taiwan. **SIMILAR SPECIES** Sulphur-crested Cockatoo C. *galerita* (plate 4) sympatric feral population in Taiwan; prominent forward-curving yellow crest. **LOCALITY** Kali Batu Putih, Halmahera, North Moluccas.

SALMON-CRESTED COCKATOO

WHITE-CRESTED COCKATOO

PLATE 6 CORELLAS (in part)

32 Midsized to large white cockatoos with short, recumbent crest and white bill; yellow underwings and undertail; prominent bare eye-ring; sexes alike, JUV resembles adults. Noisy and conspicuous; often in large flocks, especially at roosts near water; fluttering wingbeats with short glides.

SLENDER-BILLED CORELLA
Cacatua tenuirostris 37cm

Smaller of two corellas with elongated, sharply-pointed bill; frontal band, lores, band across foreneck, and bases to feathers of head and breast scarlet; blue eye-ring more extensive beneath eye; quavering *curr-ur-rup*. Woodlands and farmlands; feeds on ground, digging up seeds, roots, and bulbs. **DISTRIBUTION** southeastern mainland Australia from southeastern South Australia to central Victoria and southwestern New South Wales; up to 400m; feral populations in or near many urban centers outside natural range; very common. **SIMILAR SPECIES** Little Corella C. *sanguinea* (see below) no scarlet band on foreneck; short, blunt bill. **LOCALITIES** Deniliquin, New South Wales. Grampians National Park, Victoria.

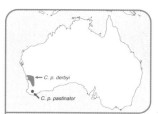

WESTERN CORELLA *Cacatua pastinator* 45cm

Larger than Slender-billed Corella and without scarlet band across foreneck; trisyllabic chuckling. Forests, woodlands and farmlands; feeds on ground. **DISTRIBUTION** southwestern Australia; up to 400m. **SUBSPECIES** two subspecies differentiated by size.
1. C. *p. pastinator* larger size. *Range* extreme southwest, mostly near Lake Muir; endangered. 2. C. *p. derbyi* smaller size. *Range* wheatbelt from Geraldton south to about lat. 32°S; common and increasing.
SIMILAR SPECIES Little Corella C. *sanguinea* (see below) distinguishable only at close quarters when short, blunt bill visible. **LOCALITIES** Lake Muir Reserve and Watheroo National Park, Western Australia.

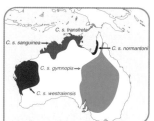

LITTLE CORELLA *Cacatua sanguinea* 38cm

One of four smaller corellas with short, blunt bill, but only one with dark blue eye-ring extending underneath eye; chuckling *curr-ur-rup*. Open woodlands, farmlands, urban parklands; feeds mostly on ground. **DISTRIBUTION** southern New Guinea and inland Australia; up to 400m; introduced to Tasmania; feral populations in coastal Australia; abundant in expanding range. **SUBSPECIES** five slightly differentiated subspecies. 1. C. *s. sanguinea* pale pink-orange lores and bases to feathers of head. *Range* Kimberley division of Western Australia east to Gulf of Carpentaria, Queensland, northern Australia. 2. C. *s. normantoni* smaller than *sanguinea*. *Range* western Cape York Peninsula, north Queensland. 3. C. *s. transfreta* brownish-yellow underwings and undertail. *Range* southern New Guinea between Kumbe and lower Fly Rivers. 4. C. *s. gymnopis* more extensive darker pink-orange on lores and bases to feathers of head and breast. *Range* inland eastern Australia, west to about long. 133°E. 5. C. *s. westralensis* orange-red lores and bases to feathers of head and underparts; deeper yellow underwings and undertail. *Range* coastal and inland central-western Australia, east to about long. 123°E. **SIMILAR SPECIES** Slender-billed Corella C. *tenuirostris* (see above) scarlet band across foreneck; elongated, sharply-pointed bill. Western Corella C. *pastinator* (see above) elongated, sharply-pointed bill. **LOCALITIES** Kakadu National Park, Northern Territory, Kinchega National Park, western New South Wales, and Hattah-Kulkyne National Park, northwestern Victoria, Australia.

SLENDER-BILLED CORELLA

WESTERN CORELLA

LITTLE CORELLA

C. s. sanguinea

C. s. westralensis

C. s. gymnopis

PLATE 7 CORELLAS (in part) AND COCKATIEL

34

DUCORPS'S CORELLA *Cacatua ducorpsii* 31cm

Only white cockatoo in range; smaller corella with short, blunt bill, white lores and prominent blue eye-ring; harsh *eerk-eerk*. Primary forest, tall secondary growth, village gardens; pairs and small flocks; mostly arboreal; wary; erratic flight with jerky wingbeats and gliding. **DISTRIBUTION** Bougainville and Buka Islands, easternmost Papua New Guinea, and Solomon Islands east to Malaita, but apparently absent from San Cristobal group; up to 1700m, mostly below 700m; common. **LOCALITIES** Loloru Crater, Bougainville. Komarindi Catchment Conservation Area, Guadalcanal, Solomon Islands.

GOFFIN'S CORELLA *Cacatua goffiniana* 32cm

Another smaller corella with short, blunt bill; lores and bases to feathers of head salmon-pink; palest blue, almost white eye-ring; harsh monosyllabic screeches and nasal, quavering cry. Primary and secondary forest; also urban parklands and gardens in Singapore and Taiwan; pairs or flocks; rapid, shallow wingbeats with gliding. **DISTRIBUTION** Yamdena and Larat, Tanimbar Islands; Indonesia; near-threatened, CITES I; introduced to Singapore, where common, and to Taiwan. **SIMILAR SPECIES** Yellow-crested Cockatoo C. *sulphurea* (plate 4) sympatric feral population in Singapore; prominent forward-curving yellow crest; black bill; different calls. Sulphur-crested Cockatoo (plate 4) sympatric feral population in Taiwan; larger with forward-curving yellow crest; black bill; different calls. **LOCALITIES** Yamdena, Tanimbar Islands. Sentosa and St. John's Islands, Singapore (feral population).

RED-VENTED CORELLA
Cacatua haematuropygia 31cm

Unmistakable; only white cockatoo in range; red undertail-coverts; undersides of wings and tail suffused yellow; circle of bare white eye-skin; sexes alike, JUV like adults; raucous *eeeek* to *owwwk* or *rouuuk*. Forests, secondary growth, mangroves, cultivation; singly, pairs, small flocks, no longer large flocks; noisy and conspicuous; swift, direct flight with rapid wingbeats. **DISTRIBUTION** formerly widespread in Philippine Islands, but now extirpated in much of range; mainly lowlands; critically endangered, CITES I. **LOCALITIES** Rajah Sikatuna National Park, Bohol, and Mount Apo National Park, Mindanao, Philippines. St. Paul's Subterranean National Park and Calauit Wildlife Sanctuary, Palawan, Philippines.

COCKATIEL *Nymphicus hollandicus* 32cm

Unmistakable; small gray cockatoo with prominent white "wing-patches"; fine, tapering crest; long, strongly graduated tail uniformly gray (♂) or finely barred white with outermost feathers yellow barred gray (♀ & JUV); prolonged *queel-queel*. Open woodlands, farmlands, urban parks, and gardens; small to large flocks; noisy and conspicuous in flight, but can pass undetected while feeding on ground; swift, direct flight with distinctive "streamlined" silhouette; domesticated cagebird with many color mutations. **DISTRIBUTION** mainland Australia, chiefly the interior; up to 400m. **LOCALITIES** Macquarie Marshes Nature Reserve, central New South Wales. Diamantina National Park, western Queensland. Uluru National Park, southern Northern Territory.

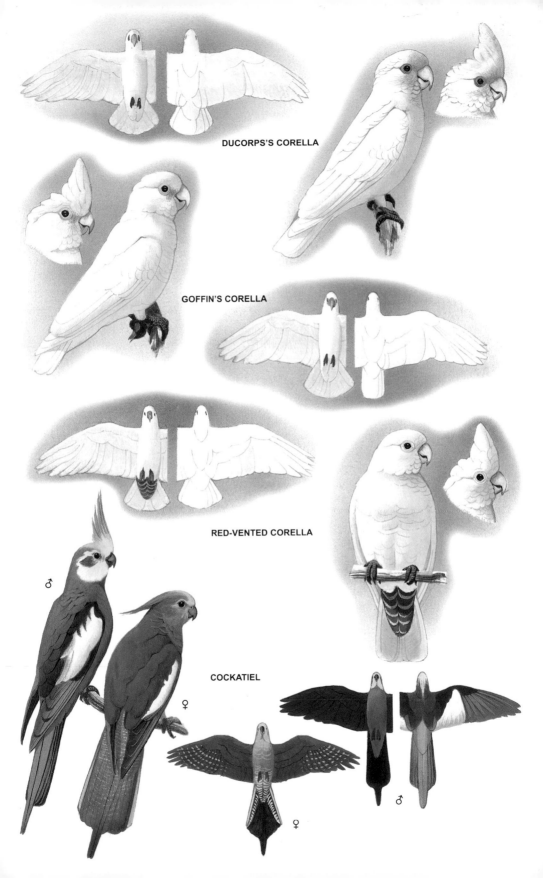

DUCORPS'S CORELLA

GOFFIN'S CORELLA

RED-VENTED CORELLA

COCKATIEL

PLATE 8 *CHALCOPSITTA* LORIES

36 Adapted for feeding on nectar, pollen, and fruits, lories and lorikeets are arboreal and move about, often in noisy, swift-flying flocks, searching for flowering or fruiting trees and shrubs. *Chalcopsitta* species are midsized lories with long, rounded tails; black eye-ring and black bare skin at base of bill; sexes alike, JUV duller with pale eye-ring and skin at base of bill. Forests, secondary growth, woodlands, plantations; screeching and twittering calls. Species replace each other geographically.

BLACK LORY *Chalcopsitta atra* 32cm

Unmistakable; only black parrot with blue rump and black bill. **DISTRIBUTION** western New Guinea and adjacent islands, Indonesia; up to 200m; common. **SUBSPECIES** three well-marked subspecies. 1. *C. a. atra* all-black with blue rump and olive-yellow undertail. *Range* Batanta and Salawati, western Papuan Islands, and western Vogelkop Peninsula. 2. *C. a. bernsteini* reddish purple on forehead and thighs. *Range* Misool, western Papuan Islands. 3. *C. a. insignis* (Rajah Lory) face, thighs, and underwing-coverts red. *Range* eastern Vogelkop, Rumberpon Island, and Onin and Bombrai Peninsulas, West Papua. **LOCALITIES** Beachfront on Salawati, and Sorong district, Vogelkop.

YELLOW-STREAKED LORY
Chalcopsitta sintillata 31cm

Green prominently streaked yellowish; dark face and red forecrown; black bill. **DISTRIBUTION** Aru Islands, Indonesia, and southern New Guinea; up to 800m; common. **SUBSPECIES** two well-marked and one slightly differentiated subspecies. 1. *C. s. sintillata* underwing-coverts red. *Range* southern New Guinea east to lower Fly River, western Papua New Guinea. 2. *C. s. chloroptera* underwing-coverts green. *Range* southern Papua New Guinea east from lower Fly River. 3. *C. s. rubrifrons* like *sintillata*, but broader, orange-yellow streaking. *Range* Aru Islands. **LOCALITIES** Brown River and Veimauri Forest Reserves, near Port Moresby, Papua New Guinea.

BROWN LORY *Chalcopsitta duivenbodei* 31cm

Unmistakable; only brown parrot with yellow face and blue rump; black bill. **DISTRIBUTION** northern New Guinea; up to 200m; locally common. **SUBSPECIES** two poorly differentiated subspecies. 1. *C. d. duivenbodei* mid-brown head and back. *Range* northwestern New Guinea, east to Aitape district, Papua New Guinea. 2. *C. d. syringanuchalis* darker brown head and back, sometimes glossed violet. *Range* Aitape district east to Astrolabe Bay, northern Papua New Guinea. **SIMILAR SPECIES** Dusky Lory *Pseudeos fuscata* (plate 9) orange or yellow underparts; buff-white rump; orange bill. **LOCALITY** Puwani River district, south of Vanimo, northwestern Papua New Guinea.

CARDINAL LORY *Chalcopsitta cardinalis* 31cm

Only all-red parrot in range; orange-red bill. **DISTRIBUTION** New Hanover and islands to Buka and Bougainville, easternmost Papua New Guinea, and Solomon Islands south to San Cristobal; up to 1200m; very common. **LOCALITIES** easily seen in most of range.

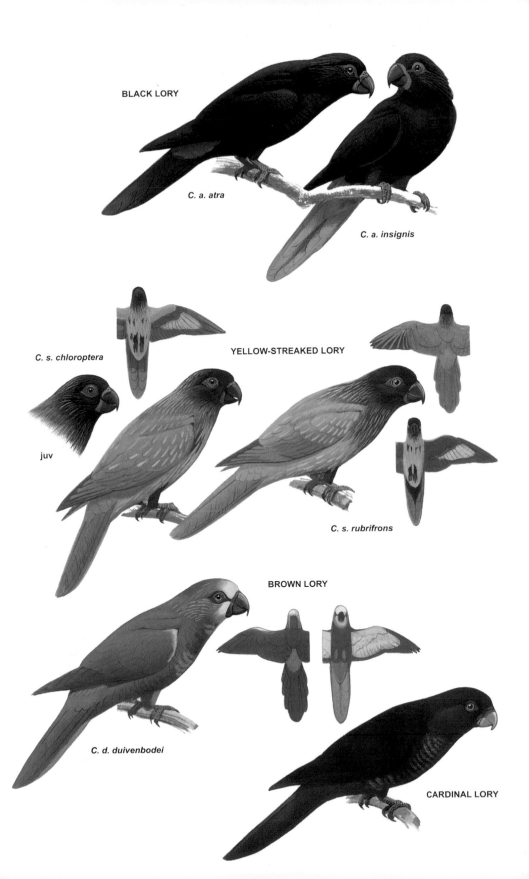

BLACK LORY

C. a. atra

C. a. insignis

C. s. chloroptera

YELLOW-STREAKED LORY

juv

C. s. rubrifrons

BROWN LORY

C. d. duivenbodei

CARDINAL LORY

PLATE 9 *PSEUDEOS* AND *EOS* LORIES (in part)

38

DUSKY LORY *Pseudeos fuscata* 25cm

Midsized lory with short, rounded tail; distinctive brown-and-yellow (yellow phase) or brown-and-orange (orange phase) plumage coloration with buff-white rump; bill and bare skin at base of bill orange; sexes alike, JUV duller; harsh screeching. Forests, secondary growth, plantations, parks or gardens. **DISTRIBUTION** New Guinea, including Yapen Island in Geelvink Bay and western Papuan Islands; up to 2400m; very common. **SIMILAR SPECIES** Brown Lory *Chalcopsitta duivenbodei* (plate 8) uniformly brown underparts; blue rump; black bill. **LOCALITIES** Kau Wildlife Area, Baitabag, Madang Province, and Lake Kutubu, Southern Highlands, Papua New Guinea.

EOS LORIES (in part)

Midsized red lories with black and blue markings; short, rounded tail; sexes alike, JUV duller and often with dusky margins to feathers; shrill screeches and chattering. Forests, secondary growth, plantations, village gardens, mangroves; pairs or small groups where scarce, larger flocks where common.

BLACK-WINGED LORY *Eos cyanogenia* 31cm

Only *Eos* lory with all-black wing-coverts and thighs; violet-blue ear-coverts. **DISTRIBUTION** islands in Geelvink Bay, except Yapen Island, West Papua, Indonesia; up to 460m; vulnerable. **SIMILAR SPECIES** Black-capped Lory *Lorius lory* (plate 18) black crown and green wings; blue underwing-coverts and yellow underwing-band. **LOCALITIES** Biak-Utara and Pulau Supiori Nature Reserves, West Papua.

VIOLET-NECKED LORY *Eos squamata* 27cm

One of two *Eos* lories lacking blue ear-coverts; variable violet-blue neck collar. **DISTRIBUTION** western Papuan Islands and North Moluccas, Indonesia; up to 1220m; common. **SUBSPECIES** three identifiable subspecies.1. *E. s. squamata* neck collar well developed in some birds, almost lacking in others; abdomen deep purple; scapulars dull purple tipped black. *Range* Gebe, Waigeu, Batanta, and Misool, western Papuan Islands. 2. *E. s. riciniata* broad neck collar extending up to hindcrown; scapulars red. *Range* Morotai to Bacan and Damar, North Moluccas, Widi Islands, and Mayu Island in Molucca Sea. 3. *E. s. obiensis* like *riciniata*, but scapulars black. *Range* Obi, North Moluccas. **SIMILAR SPECIES** Chattering Lory *Lorius garrulus* (plate 19) green wings and thighs; green underwing-coverts and orange underwing-band. **LOCALITIES** Kali Batu Putih and Akejailolo, Halmahera, North Moluccas.

BLUE-STREAKED LORY *Eos reticulata* 31cm

Unmistakable; only red lory with prominent blue streaking on back. Most wooded habitats, favoring low, fairly open monsoon forest. **DISTRIBUTION** Yamdena and Larat, Tanimbar Islands, and Babar Island, Indonesia; introduced to Kai Islands, where possibly extirpated, and Damar Island; fairly common on Yamdena; near-threatened. **LOCALITY** easily seen on Yamdena.

DUSKY LORY

orange phase

yellow phase

Juv

Juv

BLACK-WINGED LORY

VIOLET-NECKED LORY

E. s. squamata

E. s. riciniata

BLUE-STREAKED LORY

PLATE 10 *EOS* LORIES (in part)

40

RED AND BLUE LORY *Eos histrio* 31cm

Only *Eos* lory with combination of blue breast-band and blue band from eye through ear-coverts. **DISTRIBUTION** Sangihe, Talaud, and Miangas Islands, Indonesia; endangered; CITES I. **SUBSPECIES** two well-marked and one slightly differentiated subspecies.
1. *E. h. histrio* blue band from eye through ear-coverts and sides of neck to meet blue of mantle and upper back; upper wing-coverts red tipped black. *Range* Sangihe Islands; possibly survives only on Sangihe. 2. *E. h. talautensis* less black on wing-coverts. *Range* Talaud Islands; escapes recorded on Sangihe Island. 3. *E. h. challengeri* blue band from eye and ear-coverts does not meet blue mantle; less extensive blue breast-band intermixed red. *Range* Miangas Island. **LOCALITY** Last recorded communal nighttime roost near Tuabatu village, central Karakelong, Talaud Islands.

RED LORY *Eos bornea* 31cm
(sometimes *Eos rubra*, but nomenclature unresolved)

Less red and blue markings than other *Eos* species. **DISTRIBUTION** South Moluccas to Kai Islands, Indonesia; up to 1250m; common; introduced to Taiwan. **SUBSPECIES** two well-marked subspecies.
1. *E. b. bornea* variation in size and plumage coloration; secondaries and secondary-coverts tipped black; primaries black with red speculum; lower tertials and undertail-coverts blue. *Range* Boano, Seram, Ambon, Haruku, Saparua, Seramlaut Islands, Watubela Islands, Banda Islands, Tayandu Islands, and Kai Islands. 2. *E. b. cyanonothus* much darker red. *Range* Buru, South Moluccas. **SIMILAR SPECIES** Blue-eared Lory *E. semilarvata* (see below) blue upper cheeks to ear-coverts and band down sides of neck; usually at higher altitudes on Seram. Purple-naped Lory *Lorius domicella* (plate 19) black cap and green wings; blue underwing-coverts and yellow underwing-band. **LOCALITIES** Manusela National Park, Seram, and Danau Rana, Buru, Indonesia.

BLUE-EARED LORY *Eos semilarvata* 24cm

Identified by combination of violet-blue cheeks to ear-coverts and sides of neck together with blue lower underparts; smaller. **DISTRIBUTION** Seram, South Moluccas, Indonesia; mostly above 1200m; common. **SIMILAR SPECIES** Red Lory *E. bornea* (see above) no blue on abdomen or sides of face; usually at lower altitudes on Seram. Purple-naped Lory *Lorius domicella* (see above). **LOCALITY** Gunung Binaia, Manusela National Park, Seram, Indonesia.

RED AND BLUE LORY

E. h. histrio

E. h. challengeri

RED LORY

Juv

E. b. bornea

RED LORY

E. b. cyanonothus

BLUE-EARED LORY

Juv

PLATE 11 *TRICHOGLOSSUS* **LORIES (in part)**

42 Midsized lories with graduated tails and orange bills; sexes alike, JUV resembles
adults but with dark bill. All timbered habitats, including urban parks and gardens;
arboreal; noisy and conspicuous; pairs or flocks coming to flowering trees and
shrubs; shrill screeching.

ORNATE LORIKEET
Trichoglossus ornatus 25cm

Distinctive head pattern featuring dark blue crown to ear-coverts
and scarlet lores to cheeks; yellow band down side of hindneck;
breast red barred dusky blue. **DISTRIBUTION** Sulawesi and most
larger offshore islands, including Togian, Peleng, Banggai, and
Tukangbesi Islands, Indonesia; up to 1500m, mostly 300 to 1000m;
common; possibly introduced to Sangihe Island. **SIMILAR SPECIES** Yellow and Green Lorikeet
T. flavoviridis (plate 14) no red in plumage, so appears all-green; breast green scalloped yellow;
smaller. **LOCALITIES** Lore Lindu and Dumoga-Bone National Parks, Sulawesi.

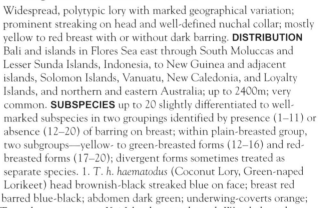

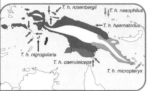

RAINBOW LORIKEET
Trichoglossus haematodus 26cm

Widespread, polytypic lory with marked geographical variation;
prominent streaking on head and well-defined nuchal collar; mostly
yellow to red breast with or without dark barring. **DISTRIBUTION**
Bali and islands in Flores Sea east through South Moluccas and
Lesser Sunda Islands, Indonesia, to New Guinea and adjacent
islands, Solomon Islands, Vanuatu, New Caledonia, and Loyalty
Islands, and northern and eastern Australia; up to 2400m; very
common. **SUBSPECIES** up to 20 slightly differentiated to well-
marked subspecies in two groupings identified by presence (1–11) or
absence (12–20) of barring on breast; within plain-breasted group,
two subgroups—yellow- to green-breasted forms (12–16) and red-
breasted forms (17–20); divergent forms sometimes treated as
separate species. 1. *T. h. haematodus* (Coconut Lory, Green-naped
Lorikeet) head brownish-black streaked blue on face; breast red
barred blue-black; abdomen dark green; underwing-coverts orange;
underwing-band yellow. *Range* Tayandu, westernmost Kai Islands, west through Watubela and
Seramlaut Islands to Seram, Ambon, and Buru, South Moluccas, and western Papuan Islands to
islands in Geelvink Bay, except Biak, and western New Guinea, east in north to Astrolabe Bay,
Papua New Guinea, and in south to upper Fly River, westernmost Papua New Guinea; also Manam
and possibly Schouten Islands, north Papua New Guinea. 2. *T. h. rosenbergii* wide yellow nuchal
collar; abdomen dark purple; underwing-band orange. *Range* Biak Island, West Papua. 3. *T. h.*
micropteryx like *haematodus*, but paler; narrower barring on breast. *Range* eastern Papua New Guinea,
west to Huon Peninsula, central ranges about Lake Kutubu, and Hall Sound; also Misima Island in
Louisiade Archipelago. 4. *T. h. caeruleiceps* crown and sides of head streaked blue; breast finely
barred bluish black; upper abdomen black, lower abdomen orange-red barred bluish black. *Range*
southern New Guinea between lower Fly River, westernmost Papua New Guinea, and Princess
Marianne Straits, southeastern West Papua; also Boigu and Saibai, Torres Strait Islands, Queensland,
Australia. 5. *T. h. nigrogularis* larger than *caeruleiceps*. *Range* Aru and eastern Kai Islands, Indonesia.
6. *T. h. nesophilus* yellow nuchal collar; occiput reddish brown; breast orange-red with little bluish-
black barring; abdomen green. *Range* Hermit and Ninigo Islands, northern Papua New Guinea.

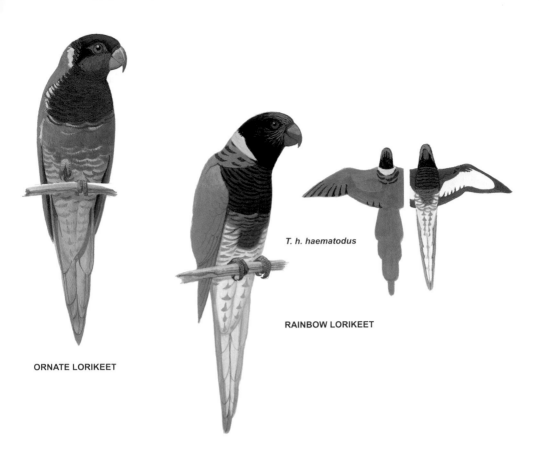

T. h. haematodus

RAINBOW LORIKEET

ORNATE LORIKEET

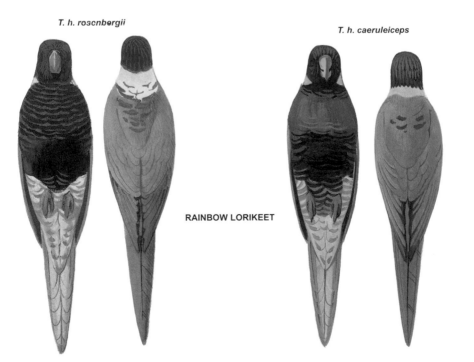

T. h. rosenbergii

T. h. caeruleiceps

RAINBOW LORIKEET

PLATE 12 *TRICHOGLOSSUS* **LORIES (in part)**

44

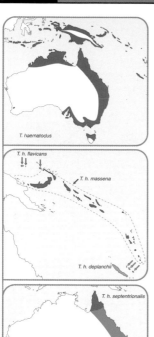

RAINBOW LORIKEET
Trichoglossus haematodus (cont.)

SUBSPECIES (in part; see also plates 11, 13) 7. *T. h. flavicans* plumage variable; upperparts, undertail-coverts and tail bronze-yellow to dull green; yellow nuchal collar; occiput reddish brown; lores and around eyes streaked violet-blue, remainder of head black streaked grayish green; breast bright red with little dark barring. *Range* New Hanover, Admiralty Islands, and apparently Nuguria Islands, eastern Papua New Guinea. 8. *T. h. massena* like *haematodus*, but occiput and nape strongly suffused brown; narrower nuchal collar; breast paler red more narrowly barred bluish black. *Range* Karkar Island and Bismarck Archipelago, eastern Papua New Guinea, east through Bougainville and Solomon Islands to Vanuatu. 9. *T. h. deplanchii* differs from *massena* by more blue streaking on head; less brown on occiput and nape; breast more strongly barred bluish black; bluish-black markings on upper abdomen. *Range* New Caledonia and Loyalty Islands. 10. *T. h. moluccanus* (sometimes treated as separate species) entire head strongly streaked violet-blue; breast yellowish orange with little or no dark barring; abdomen deep purple-blue. *Range* southeastern and eastern Australia from Gulf of Carpentaria and southern Cape York Peninsula, north Queensland, south to southern Victoria and southeastern South Australia, including Kangaroo Island; extralimitally to Tasmania, and introduced to Perth district, southwestern Australia. 11. *T. h. septentrionalis* like *moluccanus*, but brighter, more violet-blue streaking on head; shorter tail. *Range* Torres Strait Islands, except Boigu and Saibai, where replaced by *caeruleiceps*, and Cape York Peninsula, south to Endeavour and Daintree Rivers, north Queensland, where meets *moluccanus*.

T. h. flavicans

T. h. massena

T. h. deplanchii

T. h. moluccanus

PLATE 13 *TRICHOGLOSSUS* LORIES (in part)

46

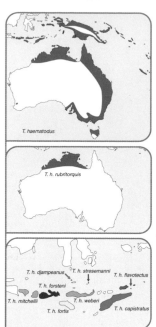

T. haematodus

T. h. rubritorquis

T. h. djampeanus · T. h. stresemanni T. h. flavotectus
T. h. forsteni
T. h. mitchellii · T. h. weberi
T. h. fortis · T. h. capistratus

RAINBOW LORIKEET
Trichoglossus haematodus (cont.)

SUBSPECIES (in part; see also plates 11, 12) 12. *T. h. rubritorquis* (Red-collared Lorikeet, sometimes treated as separate species) head strongly streaked violet-blue except on black throat and foreneck; broad orange-red nuchal collar extending down sides of neck to meet unbarred orange-red breast; hindneck violet-blue marked red; abdomen greenish black. *Range* northern Australia above lat. 18°S from Kimberley division of Western Australia east to Gulf of Carpentaria, north Queensland, but not reaching range of *moluccanus*. 13. *T. h. flavotectus* head green streaked violet-blue on forecrown to cheeks; breast varies from yellow to deep orange; broad yellow nuchal collar; abdomen dark green to greenish black; underwing-coverts yellow variably marked orange. *Range* Wetar, Romang, and Kisar Islands, north-east of Timor; birds from Romang and Kisar variable, some with dark underparts approaching *rubritorquis*. 14. *T. h. capistratus* (Edwards's Lorikeet, sometimes treated as separate species) like *flavotectus*, but nuchal collar more greenish; paler yellow breast; abdomen dark green. *Range* Timor. 15. *T. h. fortis* head blackish brown streaked violet-blue on forecrown to cheeks; lores, throat, line above to behind eye, and occiput green; unbarred breast bright yellow slightly marked orange-red; abdomen dark green, sometimes tinged blue-black; underwing-coverts yellow. *Range* Sumba, Lesser Sunda Islands. 16. *T. h. weberi* (Weber's Lorikeet, sometimes treated as separate species) general plumage green; forehead and lores very slightly streaked greenish blue, remainder of head streaked brighter green; underwing-coverts yellowish green; smaller size. *Range* Flores, Lesser Sunda Islands. 17. *T. h. forsteni* (Forsten's Lorikeet) head greenish black streaked violet-blue on forehead and cheeks; breast uniformly red without dark barring; yellowish-green nuchal collar bordered on hindneck by purple-blue; abdomen purple. *Range* Sumbawa, Lesser Sunda Islands. 18. *T. h. mitchellii* (Mitchell's Lorikeet, sometimes treated as separate species) head blackish brown streaked grayish green on crown to cheeks; occiput suffused rufous; breast uniformly red with little or no barring; small size. *Range* Bali and Lombok, Lesser Sunda Islands. 19. *T. h. djampeanus* doubtfully differentiated from *forsteni* by darker head more strongly streaked brighter violet-blue. *Range* Tanahjampea Island in Flores Sea. 20. *T. h. stresemanni* like *forsteni*, but breast paler orange-red; occiput suffused green; yellowish-orange bases to feathers of mantle. *Range* Kalaotoa Island in Flores Sea. **SIMILAR SPECIES** *Chalcopsitta* lories (plate 8) larger and with broader, rounded tail; no blue on head and different underwing patterns. Scaly-breasted Lorikeet *T. chlorolepidotus* (plate 14) green head and green underbody; orange-red underwings. Olive-headed Lorikeet *T. euteles* (plate 14) olive-yellow head and uniformly green underparts. *Psittaculirostris* fig parrots (plates 33, 34) much shorter, wedge-shaped tail; green underwings. **LOCALITIES** ubiquitous and abundant, so easily seen throughout much of range.

RAINBOW LORIKEET

T. h. rubritorquis

T. h. capistratus

T. h. fortis

T. h. weberi

T. h. forsteni

T. h. mitchellii

PLATE 14 *TRICHOGLOSSUS* LORIES (in part)

48

SCALY-BREASTED LORIKEET
Trichoglossus chlorolepidotus 23cm

Green with yellow scalloped underparts and distinctive orange-red underwings. **DISTRIBUTION** eastern Australia from Cooktown district, north Queensland, south to about lat. 33°S in eastern New South Wales; up to 600m; common; introduced to Melbourne district, southern Victoria. **SIMILAR SPECIES** Rainbow Lorikeet *T. haematodus* (plate 12) blue head and red breast. Musk Lorikeet *Glossopsitta concinna* (plate 16) green underwings; red forecrown and ear-coverts. Swift Parrot *Lathamus discolor* (plate 16) red undertail-coverts and bright red underwings; fine, pointed tail. **LOCALITIES** Easily seen in and around most coastal towns or cities.

POHNPEI LORIKEET
Trichoglossus rubiginosus 24cm

Unmistakable; only all-maroon parrot with olive-yellow wings and tail, and only parrot on Pohnpei. **DISTRIBUTION** Pohnpei, Caroline Islands, Micronesia; up to 600m; easily seen on Pohnpei.

OLIVE-HEADED LORIKEET
Trichoglossus euteles 25cm

Green with olive-yellow head. **DISTRIBUTION** Timor and eastern Lesser Sunda Islands from Lomblen east to Nila and Babar, Indonesia; up to 2400m; common. **SIMILAR SPECIES** Rainbow Lorikeet *T. haematodus* (plate 13) blue head and yellow to orange breast. Iris Lorikeet *Psitteuteles iris* (plate 15) red forecrown and violet band behind eye. **LOCALITY** Gunung Mutis, West Timor, Indonesia.

YELLOW AND GREEN LORIKEET *Trichoglossus flavoviridis* 20cm

Green with yellow scalloped underparts and yellow to brown head markings. **DISTRIBUTION** (see map above) Sulawesi and Sula Islands, Indonesia; mostly 800 to 2000m; common. **SUBSPECIES** two well-marked subspecies. 1. *T. f. flavoviridis* crown olive-yellow; brownish nuchal collar; bare eye-ring pink-orange. *Range* Sula Islands. 2. *T. f. meyeri* crown greenish brown; ear-coverts yellow; bare eye-ring gray. *Range* Sulawesi. **SIMILAR SPECIES** Ornate Lorikeet *T. ornatus* (plate 11) red face and breast; yellow underwing-coverts. *Loriculus* hanging parrots (plate 27) red rump and upper tail-coverts; very short, rounded tail; blue underwings; black bill. **LOCALITIES** Lore Lindu and Dumoga-Bone National Parks, Sulawesi.

MINDANAO LORIKEET
Trichoglossus johnstoniae 20cm

Green with distinctive head pattern and orange bill; sexes alike, JUV duller; *lish...lish* in flight, sharp *chick...chick-it*. **DISTRIBUTION** mountains of Mindanao, southern Philippine Islands. **SUBSPECIES** two poorly differentiated subspecies. 1. *T. j. johnstoniae* forecrown and cheeks rose-red; dark purple band from lores to occiput; underparts yellow scalloped green. *Range* Mounts Apo, Kitanglad, Matutum, and Piapayungan, central Mindanao. 2. *P. j. pistra* face darker, duller red; broader purple band from lores to occiput; more yellowish underparts. *Range* Mount Malindang, western Mindanao. **SIMILAR SPECIES** Guaiabero *Bolbopsittacus lunulatus* (plate 34) dumpy parrot with very short tail; no red on face; usually at lower elevations. Philippine Hanging Parrot *Loriculus philippensis* (plate 28) different pattern of red and blue on face; very short, rounded tail; blue underwings. **LOCALITY** Mount Apo National Park, Mindanao.

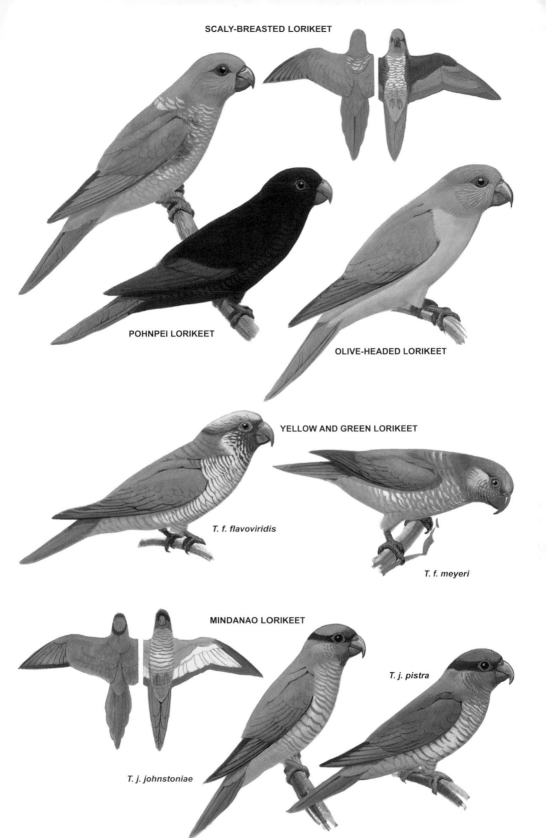

SCALY-BREASTED LORIKEET

POHNPEI LORIKEET

OLIVE-HEADED LORIKEET

YELLOW AND GREEN LORIKEET

T. f. flavoviridis

T. f. meyeri

MINDANAO LORIKEET

T. j. pistra

T. j. johnstoniae

PLATE 15 *PSITTEUTELES* LORIKEETS

50 Small green lorikeets with short, graduated tail, distinctive head pattern and prominently streaked or barred; sexes alike, JUV duller than adults. Forests, secondary growth, open woodlands, plantations; pairs or small flocks; conspicuous in swift flight, but inconspicuous when feeding in treetops; soft screeching or wheezy calls.

VARIED LORIKEET *Psitteuteles versicolor* 19cm

Distinctive red crown and mauve-pink breast; prominent white eye-ring; bill orange-red; no underwing-band; juveniles with dull red forehead, but crown and breast green, no white eye-ring, and brownish bill. **DISTRIBUTION** northern Australia from Kimberley division of Western Australia east to northeastern Queensland; locally common. **SIMILAR SPECIES** Red-collared Lorikeet *Trichoglossus haematodus* (plate 13) blue head and orange-red breast; larger, with longer tail. **LOCALITIES** Kakadu and Nitmulik (Katherine Gorge) National Parks, Northern Territory. Drysdale River National Park, Western Australia.

IRIS LORIKEET *Psitteuteles iris* 20cm

Underparts yellowish green barred darker green; red crown and purple-blue ear-coverts; bill orange-red; no underwing-band; JUV duller with less red on crown, paler blue ear-coverts, and brownish bill. **DISTRIBUTION** Timor and nearby Wetar Island, Lesser Sunda Islands, Indonesia; up to 1500m, mostly above 600m; near-threatened. **SUBSPECIES** two slightly differentiated subspecies. 1. *P. i. iris* crown uniformly red; purple-blue band from behind eye to ear-coverts; cheeks uniformly yellowish green. *Range* Timor. 2. *P. i. wetterensis* crown darker red intermixed grayish blue; darker purple-blue band from eye to ear-coverts; forecheeks suffused dull reddish-brown tinged blue. *Range* Wetar Island. **SIMILAR SPECIES** Olive-headed Lorikeet *Trichoglossus euteles* (plate 14) olive-yellow head without red or purple-blue. Rainbow Lorikeet *Trichoglossus haematodus* (plate 13) dark blue crown and face; yellow breast; larger. **LOCALITY** Gunung Mutis, West Timor, Indonesia.

GOLDIE'S LORIKEET *Psitteuteles goldiei* 19cm

Prominently streaked plumage; crown red; occiput blue; cheeks mauve-pink streaked blue; bill black; yellow underwing-band; JUV with dull red frontal band, but crown and occiput green suffused dull bluish gray. **DISTRIBUTION** highlands of New Guinea east from Weyland Mountains, West Papua; mostly 1000 to 2200m; generally uncommon. **SIMILAR SPECIES** Yellow-streaked Lory *Chalcopsitta sintillata* (plate 8) similar plumage pattern, but much larger with broad, rounded tail. Striated Lorikeet *Charmosyna multistriata* (plate 21) head green; no yellow underwing-band; bicolored bill. Red-fronted Lorikeet *Charmosyna placentis* (plate 22) unstreaked plumage; male with red underwing-coverts, female with green head; red bill. Whiskered Lorikeet *Oreopsittacus arfaki* (plate 25) unstreaked plumage; different head pattern; red underwing-coverts and tail; smaller. **LOCALITY** Goroka township, Eastern Highlands, Papua New Guinea, where regular visitor to flowering eucalypts and grevilleas.

VARIED LORIKEET

Juv

IRIS LORIKEET

♂

♀

♂

P. i. iris

P. i. wetterensis

GOLDIE'S LORIKEET

Juv

PLATE 16 *GLOSSOPSITTA* LORIKEETS AND SWIFT PARROT

52 **Small green lorikeets with short, wedge-shaped tail and fine black or bicolored bill; sexes alike, JUV resembles adults. Lorikeet-like Swift Parrot with long, narrow tail and pale bill; sexes alike, JUV duller than adults. Forests, woodlands, farmlands, orchards, urban parks and gardens; species with different call-notes.**

MUSK LORIKEET *Glossopsitta concinna* 22cm

Distinctive red and blue head pattern and bicolored bill; shrill screeching. **DISTRIBUTION** southeastern Australia, including Tasmania; up to 1600m; common. **SUBSPECIES** two poorly differentiated subspecies. 1. *G. c. concinna* forehead and broad eye-stripe red; crown blue; underwing-coverts green; bill red with black at base. *Range* southeastern Queensland to southeastern South Australia, including Kangaroo Island. 2. *G. c. didimus* less blue on crown. *Range* Tasmania, occasionally King Island. **SIMILAR SPECIES** Other *Glossopsitta* lorikeets much smaller and emit softer, more high-pitched metallic calls. Rainbow Lorikeet *Trichoglossus haematodus* (plate 12) blue head and yellow-orange breast; orange and yellow underwings. Scaly-breasted Lorikeet *T. chlorolepidotus* (plate 14) all-green head and orange underwings. Swift Parrot *Lathamus discolor* (see below) red underwings and narrow, pointed red tail; different call-notes. **LOCALITIES** Grampians National Park, Victoria. Cleland Conservation Park, South Australia.

LITTLE LORIKEET *Glossopsitta pusilla* 15cm

Smaller all-green lorikeet with distinctive red facial mask and black bill; green underwing-coverts; tinkling notes. **DISTRIBUTION** eastern and southeastern mainland Australia; up to 1600m; locally scarce, generally common. **SIMILAR SPECIES** Purple-crowned Lorikeet *G. porphyrocephala* (see below) red underwings; blue breast and purple crown; distinctive *tsit-tsit* call. Musk Lorikeet *G. concinna* (see above) much larger; bicolored bill. Double-eyed Fig Parrot *Cyclopsitta diophthalma* (plate 32) yellow underwing-band; shorter tail and silver-gray bill. **LOCALITIES** Pilliga Nature Reserve, New South Wales. Whipstick National Park, Victoria.

PURPLE-CROWNED LORIKEET
Glossopsitta porphyrocephala 16cm

Blue breast, purple crown and red underwing-coverts; orange-red forehead and orange-yellow ear-coverts; distinctive *tsit-tsit*. **DISTRIBUTION** southeastern and southwestern mainland Australia; up to 600m; common. **SIMILAR SPECIES** Little Lorikeet *G. pusilla* (see above) green underwing-coverts and green breast; red face; different tinkling call. **LOCALITIES** Cleland and Belair National Parks, South Australia. Flinders Chase National Park, Kangaroo Island, South Australia.

SWIFT PARROT *Lathamus discolor* 25cm

Lorikeet-like in habits, so easily misidentified; long, pointed wings, and very narrow, pointed tail conspicuous in flight; red face and red undertail-coverts; dark red bend of wing and underwings; pale bill; *kik-kik-kik* in flight. **DISTRIBUTION** southeastern Australia; migratory, breeding only in Tasmania and wintering in southeastern mainland; up to 1500m; endangered. **SIMILAR SPECIES** *Glossopsitta* lorikeets—see above. Scaly-breasted Lorikeet *Trichoglossus chlorolepidotus* (plate 14) see above. **LOCALITIES** Bruny Island, Tasmania (summer range). Wilson's Promontory National Park and Chiltern State Park Victoria (winter range).

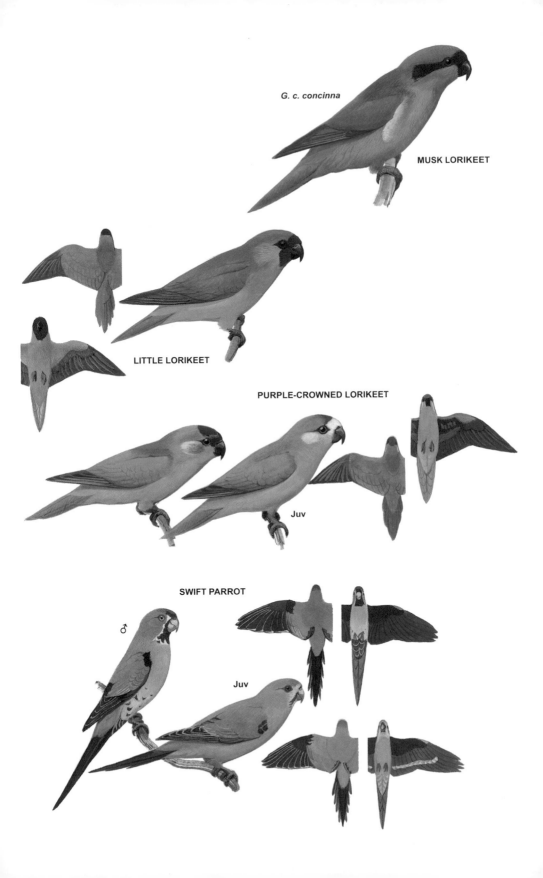

G. c. concinna

MUSK LORIKEET

LITTLE LORIKEET

PURPLE-CROWNED LORIKEET

Juv

SWIFT PARROT

♂

Juv

PLATE 17 *VINI* AND *PHIGYS* LORIKEETS

54 Small stocky green lorikeets with short rounded (*Vini*) or squarish (*Phigys*) tail and streaked, erectile crown feathers (*Vini*) or elongated nuchal feathers (*Phigys*), and orange-red bill; sexes alike, JUV duller and with dark bill. Forests, coconut plantations; pairs or small flocks in fast flight; shrill screeching and soft twittering; attracted to flowering *Erythrina* trees and coconut palms.

BLUE-CROWNED LORIKEET
Vini australis 19cm

Only parrot in range; red throat and abdomen; blue thighs and streaked feathers on crown; JUV less blue on crown, less red on throat and abdomen, and no blue on thighs. **DISTRIBUTION** Samoa and Tonga, including nearby islands in central Polynesia and islands in southern Lau Archipelago, Fiji, to Niue; locally common, generally uncommon. **LOCALITIES** Niuafo'ou Island, Tonga. Apo Cloud Forest Conservation Area, Savaii, Samoa.

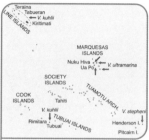

KUHL'S LORIKEET *Vini kuhlii* 19cm

Only parrot in range; red underparts and dark red tail; greenish-yellow lower back to under tail-coverts; streaked green feathers on crown and streaked blue feathers on occiput; JUV underparts dull red with dusky margins to feathers. **DISTRIBUTION** Rimitara and possibly Tubuai, in Tubuai or Austral Islands, central Polynesia; introduced to Teraina, Tabueran, and Kiritimati in Line Islands, Kiribati, and to Atiu, in Cook Islands; endangered. **LOCALITIES** easily seen on Rimitara and Teraina.

STEPHEN'S LORIKEET *Vini stepheni* 19cm

Only parrot in range; red underparts with green and purple pectoral band; streaked green feathers on crown and occiput; tail greenish yellow; JUV underparts green with purple and red markings. **DISTRIBUTION** (see map above) Henderson Island, Pitcairn Group, eastern Polynesia; vulnerable. **LOCALITY** Coastal coconut palms on Henderson Island.

ULTRAMARINE LORIKEET *Vini ultramarina* 18cm

Unmistakable; only small two-tone blue parrot with white facial markings; JUV underparts dark blue, paler on flanks and sides of breast. **DISTRIBUTION** (see map above) Marquesas Islands, French Polynesia; endangered, CITES I. **LOCALITY** Botanic Gardens, Ua Huka, Marquesas Islands.

BLUE LORIKEET *Vini peruviana* 18cm

Unmistakable; only small dark blue parrot with white face and throat; JUV underparts dark grayish blue. **DISTRIBUTION** Society Islands, and western Tuamotu Archipelago, French Polynesia; introduced to Aitutaki, Cook Islands; vulnerable. **LOCALITIES** Amuri village, Aitutaki, Cook Islands. Tatiavoa, Rangiroa, Tuamotu Archipelago.

COLLARED LORY *Phigys solitarius* 20cm

Red and green lorikeet with elongated feathers yellowish-green broadly tipped red covering nape and mantle; dark purple cap and lower underparts; JUV shorter collar of green feathers with little or no red tips. **DISTRIBUTION** larger islands of Fiji Group, including northern Lau Archipelago; common. **SIMILAR SPECIES** Red-throated Lorikeet *Charmosyna amabilis* (plate 21) predominantly green with red throat; narrow, pointed tail; smaller. **LOCALITIES** Colo-i-Suva Forest Park, Viti Levu, and Bouma National Heritage Park, Taveuni, Fiji.

BLUE-CROWNED LORIKEET

KUHL'S LORIKEET

STEPHEN'S LORIKEET

Juv

BLUE LORIKEET

Juv

ULTRAMARINE LORIKEET

Juv

COLLARED LORY

Juv

PLATE 18 *LORIUS* **LORIES (in part)**

56 **Midsized stocky, red and green lories with short, squarish tail; prominent markings on head and underparts; broad underwing-band; red bill; sexes alike, JUV resembles adults. Forests, secondary growth, plantations, village gardens; pairs or small groups; direct flight with rapid, shallow wingbeats.**

PURPLE-BELLIED LORY
Lorius hypoinochrous 26cm

One of two species with black cap and purple on underparts; nasal *whoa-oa*. **DISTRIBUTION** southeastern Papua New Guinea, including eastern Papuan Islands and Bismarck Archipelago; up to 1600m; common. **SUBSPECIES** three slightly differentiated subspecies. 1. *L. h. hypoinochrous* paler red breast contrasting with darker red upper abdomen; lower underparts purple; tail red broadly tipped green; underwing-coverts red tipped black; underwing-stripe yellow; cere white. *Range* Tagula and Misima Islands, Louisiade Archipelago. 2. *L. h. rosselianus* breast and upper abdomen uniformly red. *Range* Rossel Island, Louisiade Archipelago. 3. *L. h. devittatus* underwing-coverts red not tipped black. *Range* Trobriand and Woodlark Islands, D'Entrecasteaux and Bismarck Archipelagos, and southeastern Papua New Guinea, west in north to Huon Gulf and in south to Cape Rodney. **SIMILAR SPECIES** Black-capped Lory *L. lory* (see below) blue band across hindneck; dark gray cere; different call. White-naped Lory *L. albidinucha* (plate 19) uniformly red underparts; yellow line on sides of breast; different call. **LOCALITIES** Goodenough Island, Bismarck Archipelago. Amazon Bay, southeastern Papua New Guinea.

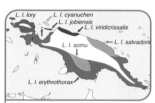

BLACK-CAPPED LORY *Lorius lory* 31cm

Second species with black cap and purple on underparts; raucous whistling and loud, ringing cries. **DISTRIBUTION** New Guinea, including some adjacent islands; up to 1000m, occasionally 1800m; common. **SUBSPECIES** seven subspecies in two groupings— underwing-coverts red (1–3) or blue (4–7). 1. *L. l. lory* blue of underparts extending up sides of breast to meet blue band on hindneck; tail red broadly tipped blue; yellow underwing-band; cere dark gray. *Range* Waigeu, Batanta, Salawati, and Misool, in western Papuan Islands, and Vogelkop Peninsula, West Papua, Indonesia. 2. *L. l. erythrothorax* breast red; blue band on hindneck not meeting blue on abdomen. *Range* southern New Guinea, west in south to southern Geelvink Bay, West Papua, and in north to Huon Peninsula, Papua New Guinea. 3. *L. l. somu* no blue band on hindneck. *Range* western Papua New Guinea on southern side of central range. 4. *L. l. salvadorii* like *erythrothorax*, but underwing-coverts blue. Range northern Papua New Guinea from Astrolabe Bay to Aitape district. 5. *L. l. viridicrissalis* like *salvadorii*, but darker blue-black band on hindneck. *Range* northern West Papua, from Humboldt Bay west to Mamberamo River. 6. *L. l. jobiensis* breast rose-red; no yellow underwing-band. *Range* Yapen and Mios Num Islands, Geelvink Bay, West Papua. 7. *L. l. cyanuchen* no red on nape, blue of hindneck extending up to meet black crown. *Range* Biak Island, Geelvink Bay. **SIMILAR SPECIES** Purple-bellied Lory *L. hypoinochrous* (see above) no blue collar on hindneck; cere white; different call. Larger red *Charmosyna* lorikeets (plates 23, 24) sleek body with long, pointed tail; no broad yellow underwing-band. **LOCALITIES** Brown River Forest Reserve, near Port Moresby, and Kau Wildlife Area, Baitabag, Madang Province, Papua New Guinea.

PURPLE-BELLIED LORY

L. h. devittatus

BLACK-CAPPED LORY

L. l. lory

Juv

L. l. erythrothorax

L. l. somu

BLACK-CAPPED LORY

L. l. salvadorii

L. l. cyanuchen

PLATE 19 *LORIUS* LORIES (in part)

58

PURPLE-NAPED LORY *Lorius domicella* 28cm

Only *Lorius* lory with all-red tail; black cap bordered behind by violet nuchal patch; variable yellow band across breast; blue thighs; underwing-band yellow; melodious call. **DISTRIBUTION** confirmed records only from Seram, North Moluccas, Indonesia; past records from Ambon and Buru, where possibly introduced; mostly 400 to 900m; vulnerable. **SIMILAR SPECIES** Red Lory *Eos bornea* (plate 10) all-red plumage without green wings and black cap. Moluccan King Parrot *Alisterus amboinensis* (♂ , plate 46) long, wedge-shaped tail; no yellow underwing-band; different calls. **LOCALITIES** Manusela National Park and Wae Fufa catchment area, Seram, North Moluccas.

WHITE-NAPED LORY *Lorius albidinucha* 26cm

Only *Lorius* lory with all-red underparts, including thighs; black cap bordered behind by white nuchal patch; faint yellow line on each side of upper breast; tail red broadly tipped green; yellow underwing-band; weak *schweet-schweet*. **DISTRIBUTION** New Ireland, Bismarck Archipelago, eastern Papua New Guinea; mostly 500 to 2000m, replaced in lowlands by *L. hypoinochrous*; near-threatened and little-known. **SIMILAR SPECIES** Purple-bellied Lory *L. hypoinochrous* (plate 18) white cere; lower underparts purple-blue; no yellow line on sides of breast and no white nuchal patch.

YELLOW-BIBBED LORY
Lorius chlorocercus 28cm

Black-capped *Lorius* lory with well-defined yellow pectoral band and crescentic black patch on sides of foreneck; blue thighs; tail red broadly tipped green; underwing-coverts blue and underwing-band orange; shrieking *chuik-lik* or *chu-er-wee*. **DISTRIBUTION** eastern Solomon Islands, from Savo and Guadalcanal to San Cristobal and Rennell; up to 1000m; common. **SIMILAR SPECIES** Cardinal Lory *Chalcopsitta cardinalis* (plate 8) all-red without green wings or black cap. Duchess Lorikeet *Charmosyna margarethae* (plate 23) smaller, slim lorikeet with narrow, finely pointed all-red tail; black only on hindcrown to occiput; different call. **LOCALITY** Komarindi Catchment Conservation Area, Guadalcanal, Solomon Islands.

CHATTERING LORY *Lorius garrulus* 30cm

Only *Lorius* lory with all-red head and green thighs. **DISTRIBUTION** North Moluccas, Indonesia; up to 1300m; endangered. **SUBSPECIES** two well-marked and one slightly differentiated subspecies. 1. *L. g. garrulus* no yellow on red mantle; bend of wing and underwing-coverts green; tail red broadly tipped green; underwing-band rose-red. *Range* Halmahera and Widi Islands, North Moluccas. 2. *L. g. flavopalliatus* well-defined yellow patch on mantle. *Range* Bacan and Obi, North Moluccas. 3. *L. g. morotaianus* like *flavopalliatus*, but on mantle duller yellow patch suffused green. *Range* Morotai and Rau, North Moluccas. **SIMILAR SPECIES** Violet-necked Lory *Eos squamata* (plate 9) red and blue without green wings; smaller with short, rounded tail. Moluccan King Parrot *Alisterus amboinensis* (♂ , plate 46) see above. **LOCALITIES** Kali Batu Putih and Lalobata district, Halmahera, North Moluccas.

PURPLE-NAPED LORY

WHITE-NAPED LORY

YELLOW-BIBBED
LORY

Juv

CHATTERING LORY

L. g. flavopalliatus

L. g. garrulus

PLATE 20 *CHARMOSYNA* **LORIKEETS (in part)**

60 Small to midsized lorikeets with slender, sharply graduated tail and finely-pointed, compressed red or bicolored (in one species) bill; variable underwing-band; some sexual dimorphism, JUV usually resembles ♀. Forests, secondary growth, woodlands, plantations, village gardens; usually small flocks, sometimes large feeding flocks; swift, direct flight; high-pitched calls.

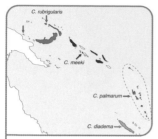

PALM LORIKEET *Charmosyna palmarum* 17cm

One of five small all-green lorikeets with red or blue head markings, and only small lorikeet in range; red chin, lores, and at base of bill; little or no underwing-band and no red at base of tail; iris yellow; sexes alike; *tswit-tswit-tswit*. **DISTRIBUTION** Santa Cruz, Duff, and possibly Reef Islands, easternmost Solomon Islands, and Vanuatu, including Banks Islands but not Torres Islands; mostly above 1000m; vulnerable. **LOCALITIES** Mount Tabwemasana, Espiritu Santo, and proposed Lake Letas Reserve, Gaua, Vanuatu.

RED-CHINNED LORIKEET *Charmosyna rubrigularis* 17cm

Another small all-green lorikeet with red on chin and at base of bill, but not extending to lores; red at base of tail and prominent yellow underwing-band; iris orange; sexes alike; soft *szeep*. **DISTRIBUTION** (see map above) New Britain, New Ireland, and New Hanover, in Bismarck Archipelago, and Karkar Island, eastern Papua New Guinea; mostly 450 to 1300m; common. **SIMILAR SPECIES** Red-flanked Lorikeet *C. placentis* (plate 22) male with red flanks and underwing-coverts; female with yellow-streaked ear-coverts, but no red on face. **LOCALITY** Slopes of Karkar Volcano, Karkar Island, Papua New Guinea.

MEEK'S LORIKEET *Charmosyna meeki* 16cm

Small all-green lorikeet with grayish-blue crown; mantle suffused olive-brown; variable yellowish-white underwing-band; no red at base of tail; sexes alike; high-pitched *tweek-tweek*. **DISTRIBUTION** (see map above) Bougainville Island, eastern Papua New Guinea, and Santa Isabel, Kolombangara, Guadalcanal, and Malaita, in Solomon Islands; mostly 300 to 1200m; near-threatened. **SIMILAR SPECIES** Red-flanked Lorikeet *C. placentis* (plate 22) see above. Duchess Lorikeet *C. margarethae* (plate 23) red head and underparts. **LOCALITY** Slopes of Lake Loloru Crater, south Bougainville, Papua New Guinea.

NEW CALEDONIAN LORIKEET *Charmosyna diadema* 19cm

Known only from ♀; small all-green lorikeet with blue crown; yellow throat and red vent; mantle suffused olive-brown; thighs blue; red at base of tail, but no underwing-band. **DISTRIBUTION** (see map above) New Caledonia, where unconfirmed sightings in 1950s and 1976; critically endangered, possibly extinct. **SIMILAR SPECIES** Rainbow Lorikeet *Trichoglossus haematodus* (plate 12) blue-streaked head; orange-red breast and underwing-coverts; much larger. Red-fronted Parakeet *Cyanoramphus novaezelandiae* (plate 60) and Horned Parakeet *Eunymphicus cornutus* (plate 59) larger and with distinctive head patterns. **LOCALITIES** La Foa to Canala road, southern New Caledonia, and west of Mount Panie, in north, are localities of unconfirmed sightings.

BLUE-FRONTED LORIKEET
Charmosyna toxopei 16cm

Small all-green lorikeet with blue forecrown; red at base of tail; yellow underwing-band; sexes alike; very shrill *ti-ti-ti-ti*. **DISTRIBUTION** Buru, South Moluccas, Indonesia, where only small green lorikeet; mostly 600 to 1000m; critically endangered; rarely recorded. **LOCALITY** Danau Rana, Buru, the provenance of original specimens and locality of unconfirmed sighting.

PALM LORIKEET

RED-CHINNED LORIKEET

MEEK'S LORIKEET

NEW CALEDONIAN LORIKEET

BLUE-FRONTED LORIKEET

PLATE 21 *CHARMOSYNA* LORIKEETS (in part)

62

PYGMY LORIKEET
Charmosyna wilhelminae 13cm

Very small green lorikeet with crown to nape purple-brown, nape streaked blue; breast streaked yellow; mantle suffused olive-brown; lower back and underwings red ($\male$) or green ($\female$); red at base of tail; high-pitched *ts-ts-tsee*. **DISTRIBUTION** highlands of New Guinea, from Arfak Mountains, West Papua, east to Huon Peninsula and Owen Stanley Ranges, eastern Papua New Guinea; mostly 1000 to 2200m; uncommon. **SIMILAR SPECIES** Red-flanked Lorikeet *Charmosyna placentis* and Red-fronted Lorikeet *C. rubronotata* (plate 22) red on crown or face and violet-blue ($\male$) or yellow-streaked ($\female$) ear-coverts; yellow underwing-band; larger. Whiskered Lorikeet *Oreopsittacus arfaki* (plate 25) distinctive facial pattern; red undertail; black bill. *Micropsitta* pygmy parrots (plates 29, 30) short, broad tail, no streaking on breast; different facial pattern; forage on tree trunks. Orange-fronted Hanging Parrot *Loriculus aurantiifrons* (plate 26) very short, rounded tail; red throat; blue underwings; black bill. **LOCALITIES** Goroka district Central Highlands, and Veimauri River Forest Reserve, near Port Moresby, Papua New Guinea.

STRIATED LORIKEET
Charmosyna multistriata 18cm

Small green lorikeet with prominent yellowish streaking on underparts; hindcrown and nape brown, streaked yellow on nape; red vent; red at base of tail; bicolored gray/orange bill; distinctive prolonged whistle. **DISTRIBUTION** southern slopes of central range in western New Guinea, between Snow Mountains, West Papua, and Crater Mountain, Chimbu Province, Papua New Guinea; mostly 180 to 1800m; near-threatened; little known. **SIMILAR SPECIES** Red-flanked Lorikeet *C. placentis* (plate 22) see above. Goldie's Lorikeet *Psitteuteles goldiei* (plate 15) distinctive facial pattern featuring red, mauve-pink, and blue; black bill. **LOCALITY** Tabubil, Ok Tedi River region, westernmost Papua New Guinea.

RED-THROATED LORIKEET
Charmosyna amabilis 18cm

Small green lorikeet with red throat bordered below by narrow yellow band; dark red thighs; red at base of tail, but no underwing-band; high-pitched squeak. **DISTRIBUTION** Viti Levu, Vanua Levu, Taveuni, and possibly Ovalau, Fiji Islands; mainly above 500m; endangered and rarely observed. **SIMILAR SPECIES** Collared Lory *Phigys solitarius* (plate 17) stocky appearance; short, squarish tail; red underparts; larger. **LOCALITIES** Tomaniivi Nature Reserve, Viti Levu, and Des Voeux Peak, Taveuni, Fiji.

PYGMY LORIKEET

STRIATED LORIKEET

RED-THROATED LORIKEET

PLATE 22 *CHARMOSYNA* LORIKEETS (in part)

64

RED-FRONTED LORIKEET
Charmosyna rubronotata 17cm

One of two similar small green lorikeets with blue-streaked (♂) or yellow-streaked (♀) ear-coverts, and ♂ with red on forecrown or throat and on underwing-coverts to sides of breast; JUV like ♀, but ♂ with red underwing-coverts; soft *queet-queet*. **DISTRIBUTION** Salawati and Biak Islands, West Papua, Indonesia, and northwestern New Guinea; up to 900m; uncommon. **SUBSPECIES** two poorly differentiated subspecies. 1. C. r. *rubronotata* ♂ forecrown red; red on upper tail-coverts, but rump green; ♀ forecrown and underwing-coverts to sides of breast green. *Range* Salawati, western Papuan Islands, and northwestern New Guinea, east to Ramu River, Madang Province, Papua New Guinea. 2. C. r. *kordoana* ♂ paler red more extensive on crown; ear-coverts more blue, less violet. *Range* Biak Island, in Geelvink Bay, West Papua. **SIMILAR SPECIES** Red-flanked Lorikeet C. *placentis* (see below) ♂ green forecrown, but red throat; ♀ probably indistinguishable in field. Pygmy Lorikeet C. *wilhelminae* (plate 21) no blue or yellow ear-coverts; yellow streaking on breast; smaller. Fairy Lorikeet *Charmosyna pulchella* (plate 23) red head and underparts. **LOCALITY** Biak Island, Geelvink Bay, West Papua, Indonesia.

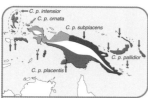

RED-FLANKED LORIKEET
Charmosyna placentis 17cm

Like C. *rubronotata*, but ♂ with green forecrown, red throat and red extending from sides of breast to flanks; ♀ & JUV with green upper tail-coverts; high-pitched *tsss* or *seeet*. **DISTRIBUTION** Moluccas, Kai and Aru Islands, Indonesia, through New Guinea and some adjacent islands to Bougainville and Nuguria Islands, easternmost Papua New Guinea; up to 1600m; most common *Charmosyna* lorikeet. **SUBSPECIES** five subspecies in two groupings—with (1–3) or without (4 & 5) blue on rump. 1. C. *p. placentis* blue patch on rump in both sexes. *Range* South Moluccas through Kai and Aru Islands, Indonesia, to southern New Guinea. 2. C. *p. intensior* smaller blue patch on rump. *Range* North Moluccas, and Gebe in western Papuan Islands, West Papua. 3. C. *p. ornata* larger blue patch on rump. *Range* western Papuan Islands, except Gebe, and adjacent northwestern New Guinea. 4. C. *p. subplacens* no blue on rump. *Range* eastern New Guinea. 5. C. *p. pallidior* like *subplacens*, but generally paler, especially blue ear-coverts in ♂. *Range* Woodlark Island and Bismarck Archipelago to Nuguria and Nissan Islands, Buka and Bougainville, and also Lou and Pak in eastern Admiralty Islands, eastern Papua New Guinea. **SIMILAR SPECIES** Red-fronted Lorikeet C. *rubronotata* (see above) ♂ with red forecrown and upper tail-coverts, but green throat; ♀ probably indistinguishable in field, but with red upper tail-coverts. Red-chinned Lorikeet C. *rubrigularis* (plate 20) no blue or yellow ear-coverts; red only at base of bill. Meek's Lorikeet C. *meeki* (plate 20) no blue or yellow ear-coverts; no red on head. Pygmy Lorikeet C. *wilhelminae* (plate 21) see above. Fairy Lorikeet C. *pulchella* (plate 23) see above. Whiskered Lorikeet *Oreopsittacus arfaki* (plate 25) distinctive facial pattern; red undertail; black bill. **LOCALITIES** Brown River Forest Reserve, near Port Moresby, and Lae Botanic Gardens, Papua New Guinea. Manusela National Park, Seram. South Moluccas, Indonesia.

RED-FRONTED LORIKEET

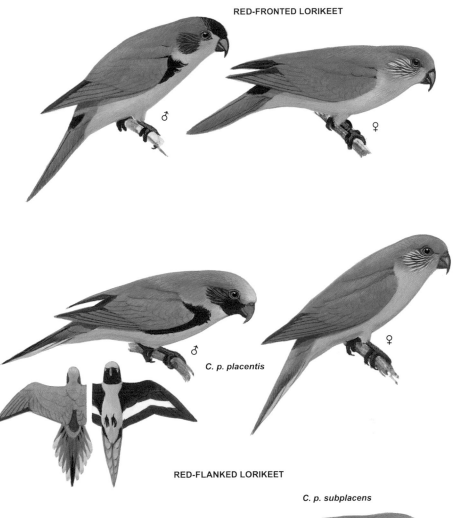

♂

♀

C. p. placentis

♂

♀

RED-FLANKED LORIKEET

C. p. subplacens

♂

PLATE 23 *CHARMOSYNA* **LORIKEETS (in part)**

66

FAIRY LORIKEET *Charmosyna pulchella* 18cm

Smallest of red and green, sexually dimorphic *Charmosyna* lorikeets with black on hindcrown; JUV differs from adults; nasal *ks* or weak *ss*. **DISTRIBUTION** mountains of mainland New Guinea; mostly 750 to 2300m; uncommon. **SUBSPECIES** two well-marked subspecies. 1. *C. p. pulchella* black on hindcrown not reaching eyes; breast streaked yellow; thighs black; sides of lower back to flanks red (♂) or yellow (♀); no underwing-band; tail green tipped red and yellow; JUV crown and nape dull green, breast suffused green with little or no yellow streaking, pale yellow underwing-band, and brownish bill. *Range* mountains of New Guinea, except Cyclops Mountains. 2. *C. p. rothschildi* black on hindcrown extends to eyes; breast green streaked yellow. *Range* Cyclops Mountains and northern slopes of mountains above Idenburg River, West Papua. **SIMILAR SPECIES** Other small *Charmosyna* lorikeets (plates 20–22) predominantly green. Josephine's Lorikeet *C. josefinae* (see below) no yellow streaking on breast; lower underparts black; red tail; larger. **LOCALITIES** Efogi district, Owen Stanley Range, and Wau Ecology Institute, Morobe Province, Papua New Guinea.

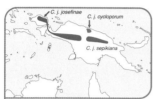

JOSEPHINE'S LORIKEET
Charmosyna josefinae 24cm

Midsized red and green *Charmosyna* lorikeet without breast markings and with red tail; shrill *kris* and nasal *engg*. **DISTRIBUTION** mountains of western and central New Guinea; mostly 750 to 2200m; uncommon and little known. **SUBSPECIES** three slightly differentiated subspecies. 1. *C. j. josefinae* pale lilac-blue streaking on occiput; lower underparts black; blue patch on rump; lower back red (♂) or yellow (♀); JUV green suffusion on black thighs and abdomen, greenish streaking on occiput, and brownish bill. *Range* mountains of Vogelkop east to Snow Mountains, West Papua. 2. *C. j. cycloporum* little or no black on abdomen, and blue streaking on occiput lacking or faintly indicated. *Range* Cyclops Mountains, West Papua. 3. *C. j. sepikiana* more extensive black on abdomen; occiput streaked pale gray. *Range* mountains of western Papua New Guinea, in Sepik River region and in Western Highlands east to about Jimi River valley and Mount Bosavi. **SIMILAR SPECIES** Fairy Lorikeet *C. pulchella* (see above) breast streaked yellow; green and red tail; smaller. Papuan Lorikeet *C. papou* (plate 24) green tail with long, streamer-like central feathers; larger. **LOCALITIES** Ok Tedi River region, Western Province, and Ambua Lodge, Tari Gap, Southern Highlands, Papua New Guinea.

DUCHESS LORIKEET
Charmosyna margarethae 20cm

Smaller red and green *Charmosyna* lorikeet with yellow pectoral band continuing as narrow collar on mantle; sides of rump red (♂) or yellow (♀); JUV dusky margins to feathers of head and underparts, greenish thighs, and brownish bill; *keek-keek-keek*. **DISTRIBUTION** Bougainville, eastern Papua New Guinea, and Gizo, Kolombangara, Guadalcanal, Malaita, and San Cristobal, Solomon Islands; mostly 100 to 1350m; near-threatened. **SIMILAR SPECIES** Meek's Lorikeet *C. meeki* (plate 20) and Red-flanked Lorikeet *C. placentis* (plate 22) green without yellow on breast. Yellow-bibbed Lory *Lorius chlorocercus* (plate 19) stocky appearance with short, squarish tail; black cap; larger. **LOCALITIES** Aku and Buin districts, southern Bougainville, Papua New Guinea. Komarindi Catchment Conservation Area, Guadalcanal, Solomon Islands.

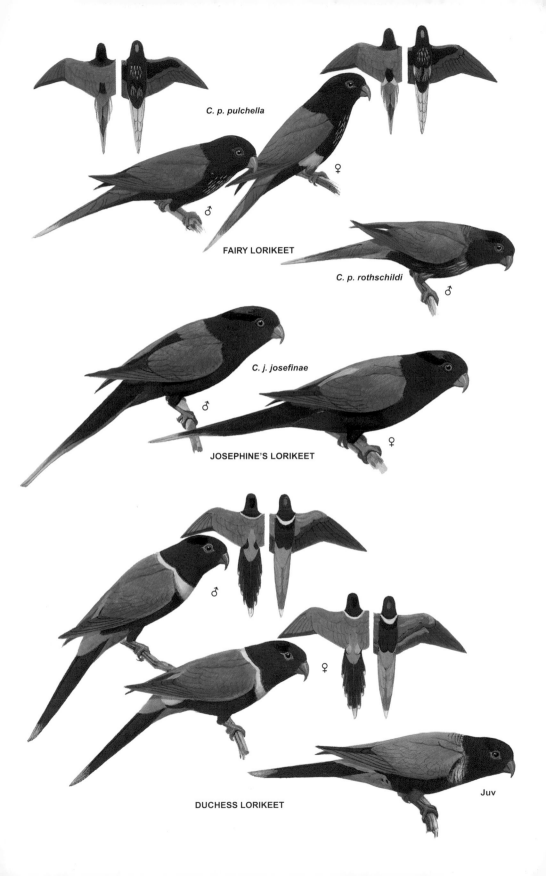

C. p. pulchella

♀

♂

FAIRY LORIKEET

C. p. rothschildi

♂

C. j. josefinae

♂

♀

JOSEPHINE'S LORIKEET

♂

♀

Juv

DUCHESS LORIKEET

PLATE 24 *CHARMOSYNA* LORIKEETS (in part)

68

PAPUAN LORIKEET *Charmosyna papou* 25cm
(without central tail-feathers)

Unmistakable; largest red and green *Charmosyna* lorikeet with elongated, streamer-like central tail feathers; two color phases in most populations; sexes differ, JUV differs from adults; loud *queea* and nasal *taaaa-aaan*. Pairs or small groups, not flocks; jerky actions, often flicking long tail-feathers. **DISTRIBUTION** mountains of mainland New Guinea; mostly 1500 to 3500m; common. **SUBSPECIES** three well-marked and one poorly differentiated subspecies. 1. C. *p. papou* sexes alike and no melanistic phase; black patch on occiput anteriorly streaked blue, and narrow black line across hindneck; yellow patches on flanks and sides of breast; thighs and abdomen black; rump blue, and tail green tipped yellow; JUVdusky margins to feathers of head and underparts, and shorter tail. *Range* mountains of Vogelkop Peninsula, West Papua, Indonesia. 2. C. *p. stellae* (Stella's Lorikeet) in melanistic phase red replaced by black; black from occiput to hindneck; no yellow on flanks or sides of breast; lower back and sides of rump red (♂) or yellow (♀); central tail-feathers green broadly tipped yellow-orange. *Range* mountains of eastern Papua New Guinea, west to Angabunga River and Herzog Mountains. 3. C. *p. goliathina* (Stella's Lorikeet) with melanistic phase; like *stellae*, but central tail-feathers broadly tipped paler yellow. *Range* mountains of western and central New Guinea, from head of Geelvink Bay, West Papua, east to central Papua New Guinea, where meets *stellae*. 4. C. *p. wahnesi* with melanistic phase; like *stellae*, but with yellow band across upper abdomen. *Range* mountains of Huon Peninsula, northern Papua New Guinea. **SIMILAR SPECIES** Josephine's Lorikeet C. *josefinae* (plate 23) shorter, red tail without elongated central feathers; smaller. *Lorius* lories (plate 18) stocky appearance with short, squarish tail; black cap and prominent yellow underwing-band. *Alisterus* king parrots (plate 46) different flight silhouette with long, broad tail; different calls. **LOCALITIES** Ambua Lodge, Tari Gap, Southern Highlands, and Mount Tomba, southern slopes of Hagen Range, Western Highlands, Papua New Guinea.

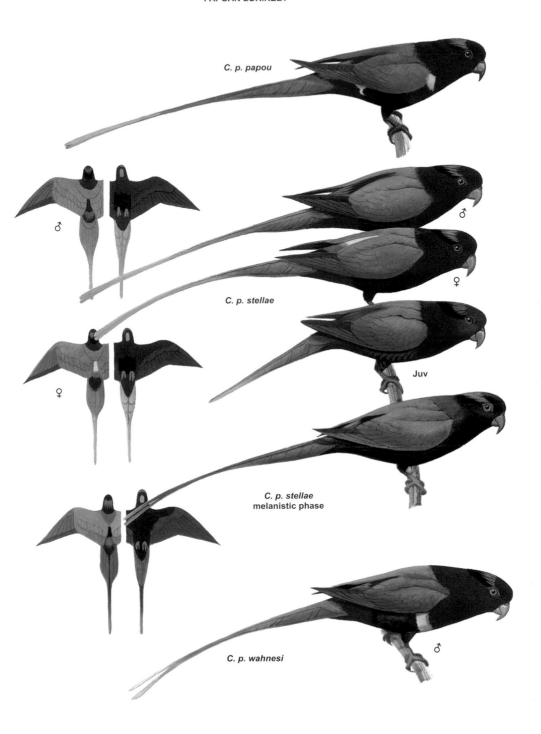

C. p. papou

♂

C. p. stellae

♀

♂

♀

Juv

C. p. stellae
melanistic phase

C. p. wahnesi

♂

PLATE 25 NEW GUINEA HIGHLANDS LORIKEETS

70

WHISKERED LORIKEET
Oreopsittacus arfaki 15cm

Small slim green lorikeet with long, graduated tail, distinctive facial pattern, and red underwing-coverts; sexes differ, JUV duller; soft twittering. **DISTRIBUTION** mountains of mainland New Guinea; mostly 2000 to 3750m; common. **SUBSPECIES** two well-marked and one poorly differentiated subspecies. 1. *O. a. arfaki* forecrown red (♂) or green (♀); lores and cheeks purple with two lines of white streaking; abdomen and lower flanks red; black bill; JUV narrow red (♂) or orange (♀) frontal band, crown green, and cheeks dusty mauve. *Range* mountains of Vogelkop Peninsula, West Papua, Indonesia. 2. *O. a. major* larger than *arfaki*. *Range* Snow Mountains, West Papua. 3. *O. a. grandis* like *major*, but abdomen and lower flanks green. *Range* mountains of Papua New Guinea, west to Huon Peninsula and Victor Emmanuel Range. **SIMILAR SPECIES** Small green *Charmosyna* lorikeets (plate 22) different facial patterns and orange-red bill. Goldie's Lorikeet *Psitteuteles goldiei* (plate 15) neck and underparts heavily streaked; underwing-coverts green; larger. **LOCALITIES** Ambua Lodge, Tari Gap, Southern Highlands, and Mount Tomba, southern slopes of Hagen Range, Western Highlands, Papua New Guinea.

MUSSCHENBROEK'S LORIKEET
Neopsittacus musschenbroekii 23cm

One of two very similar small green lorikeets with red underparts, yellow streaking on head, and red underwings; sexes alike, JUV duller; musical *shree-daloo* and sharp *ks*. **DISTRIBUTION** mountains of mainland New Guinea; mostly 1250 to 2800m. **SUBSPECIES** three poorly differentiated subspecies. 1. *N. m. musschenbroekii* crown and nape olive-brown streaked yellow; cheeks olive-brown streaked pale green; tail green above, orange-yellow below; bill pale yellow. *Range* mountains of Vogelkop Peninsula, West Papua, Indonesia. 2. *N. m. medius* cheeks streaked more yellowish; larger. *Range* Snow Mountains, West Papua, east to mountains of central New Guinea. 3. *N. m. major* like *medius*, but generally paler; cheeks streaked greenish yellow. *Range* mountains of Papua New Guinea from Owen Stanley Range west to Huon Peninsula and Sepik River region. **SIMILAR SPECIES** Emerald Lorikeet *N. pullicauda* (see below) crown green and nape only slightly suffused olive-brown; olive-green undertail; orange bill. Small green *Charmosyna* lorikeets (plate 22) see above. **LOCALITIES** Ambua Lodge, Tari Gap, and Mount Tomba, Papua New Guinea.

EMERALD LORIKEET
Neopsittacus pullicauda 18cm

Differentiated from *N. musschenbroekii* by olive-green undertail and darker, orange bill; calls more high-pitched than *N. musschenbroekii*. **SUBSPECIES** three poorly differentiated subspecies. 1. *N. p. pullicauda* crown green lightly streaked paler; nape slightly suffused olive-brown; cheeks streaked yellow. *Range* mountains of eastern Papua New Guinea west to Sepik River region. 2. *N. p. alpinus* paler orange-red breast contrasting with darker red abdomen; darker green upperparts. *Range* Snow Mountains, West Papua, east to upper Fly River region, western Papua New Guinea. 3. *N. p. socialis* upperparts and sides of head darker green; little olive-brown on nape. *Range* Herzog Mountains and mountains of Huon Peninsula, northern Papua New Guinea. **SIMILAR SPECIES** Musschenbroek's Lorikeet *N. musschenbroekii* and small green *Charmosyna* lorikeets (see above). **LOCALITIES** Ambua Lodge, Tari Gap, and Mount Tomba, Papua New Guinea.

WHISKERED LORIKEET

O. a. arfaki

♂

O. a. grandis

♂

♀

Juv ♂

MUSSCHENBROEK'S LORIKEET

N. m. musschenbroekii

Juv

N. m. major

Juv

EMERALD LORIKEET

N. p. pullicauda

PLATE 26 HANGING PARROTS (in part)

72

Small green parrots with very short, rounded tail, very fine, sharply-pointed bill, and blue underwings; red rump and upper tail-coverts in all but one species; sexes differ, JUV like ♀. Forests, secondary growth, village gardens, plantations; arboreal; pairs or small parties in swift flight; lorikeet-like behavior, coming to flowering trees and shrubs to feed on nectar. Two species in Sulawesi, but elsewhere species replace each other geographically. Other hanging parrots occur in the Afro-Asian Distribution (plate 76).

ORANGE-FRONTED HANGING PARROT
Loriculus aurantiifrons 10cm

Very small black-billed hanging parrot with red throat; shrill *tseo-tseo-tseo*. **DISTRIBUTION** western Papuan Islands, West Papua, Indonesia, and New Guinea; up to 1200m; uncommon. **SUBSPECIES** three poorly differentiated subspecies. 1. *L. a. aurantiifrons* ♂ only hanging parrot with yellow forecrown; ♀ with green forecrown, and cheeks suffused blue; JUV lacks red throat. *Range* Misool, western Papuan Islands. 2. *L. a. batavorum* ♂ with less yellow on forecrown. *Range* Waigeu, western Papuan Islands, and northwestern New Guinea, east in north to Sepik River and in south to Setekwa River. 3. *L. a. meeki* larger than *batavorum*. *Range* Papua New Guinea, west in north to Sepik River region and in south to Fly River region, and on Fergusson, Goodenough, and Karkar Islands. **SIMILAR SPECIES** Small green *Charmosyna* and *Oreopsittacus* lorikeets (plates 22, 25) no yellow on forecrown and no blue underwings; longer, graduated tail. *Cyclopsitta* fig parrots (plates 31, 32) different facial patterns; no blue underwings. *Micropsitta* pygmy parrots (plates 29, 30) different facial patterns; pale bill; normally foraging on tree trunks. **LOCALITIES** Brown River and Veimauri River Forest Reserve, near Port Moresby, and Karkar Island, Papua New Guinea.

GREEN-RUMPED HANGING PARROT **Loriculus tener** 10cm

Only hanging parrot with green rump; red throat; sexes alike. **DISTRIBUTION** (see map above) Bismarck Archipelago, eastern Papua New Guinea; up to 1200m; near-threatened. **SIMILAR SPECIES** Small green *Charmosyna* lorikeets and *Micropsitta* pygmy parrots (see above). **LOCALITIES** not easily seen anywhere.

FLORES HANGING PARROT
Loriculus flosculus 12cm

Small red-billed hanging parrot with brownish suffusion on nape; ♂ with red on throat; sharp *strrt* and *chi-chi-chi*. **DISTRIBUTION** Flores, Lesser Sunda Islands, Indonesia; mostly 400 to 1000m; endangered. **LOCALITIES** Tanjung Kerita Mese Reserve and Gunung Egon, Flores.

YELLOW-THROATED HANGING PARROT **Loriculus pusillus** 12cm

Only hanging parrot with yellow throat in adults; shrill *scree-ee*. **DISTRIBUTION** (see map above) Java and Bali, Indonesia. **SIMILAR SPECIES** in westernmost Java possibly Blue-crowned Hanging Parrot *L. galgulus* (plate 76) blue crown; yellow mantle; black bill. **LOCALITIES** Ujung Kulon and Gunung Pangrango National Parks, Java. Bali Barat National Park, Bali.

GREEN HANGING PARROT
Loriculus exilis 11cm

Another small red-billed hanging parrot; on throat of ♂ red spot surrounded greenish blue; soft *pssst*. **DISTRIBUTION** Sulawesi, Indonesia; up to 1000m; near-threatened. **SIMILAR SPECIES** Maroon-rumped Hanging Parrot *L. stigmatus* (plate 27) ♂ with red forecrown; darker maroon rump; red carpal edge; black bill. **LOCALITIES** Dumoga Bone and Lore Lindu National Parks, and Tangkoko Nature Reserve, Sulawesi.

L. a. aurantiifrons

ORANGE-FRONTED HANGING PARROT

GREEN-RUMPED HANGING PARROT

FLORES HANGING PARROT

Juv

YELLOW-THROATED HANGING PARROT

GREEN HANGING PARROT

PLATE 27 HANGING PARROTS (in part)

74

MAROON-RUMPED HANGING PARROT
Loriculus stigmatus 15cm

Larger black-billed hanging parrot with dark maroon-red rump and upper tail-coverts; mantle suffused orange-brown; red "throat patch" and red on carpal edge of wing; forehead and crown red (♂) or green (♀ & JUV); high-pitched *tsu-tsee-tsee*. **DISTRIBUTION** Sulawesi, including nearby islands, Indonesia; up to 1000m; common.
SIMILAR SPECIES Green Hanging Parrot *Loriculus exilis* (plate 26) no red on crown or carpal edge; red bill; smaller. Yellow and Green Lorikeet *Trichoglossus flavoviridis* (plate 14) underparts green scalloped yellow; greenish-brown crown; longer tail; red bill. **LOCALITIES** Dumoga-Bone and Lore Lindu National Parks, and Bantimurung-Karaenta Nature Reserve, Sulawesi.

MOLUCCAN HANGING PARROT
Loriculus amabilis 11cm

Smaller black-billed hanging parrot differentiated from similar *L. stigmatus* by paler red rump and upper tail-coverts; carpal edge of wing yellow and red; ♂ red forehead and crown; weak *tsee*. **DISTRIBUTION** Morotai, Halmahera, Kasiruta, and Bacan, North Moluccas, Indonesia; up to 450m; fairly common. **SIMILAR SPECIES** Red-flanked Lorikeet *Charmosyna placentis* (plate 22) green crown; violet-blue (♂) or yellow-streaked (♀) ear-coverts; longer tail; red bill. **LOCALITY** Kali Batu Putih, Halmahera.

SULA HANGING PARROT
Loriculus sclateri 15cm

Unmistakable; only black-billed hanging parrot with orange-yellow or scarlet mantle and upper back; sexes alike, JUV undescribed; weak *tsee*. **DISTRIBUTION** Sula, Peleng, and Banggai Islands, Indonesia; up to 450m; common. **SUBSPECIES** two subspecies, but difference may be age-related. 1. *L. s. sclateri* mantle and upper back orange-yellow; red "throat patch" and carpal edge of wing. *Range* Sula Islands. 2. *L. s. ruber* mantle and upper back scarlet. *Range* Peleng, Banggai, and Labobo Islands. **SIMILAR SPECIES** Yellow and Green Lorikeet *Trichoglossus flavoviridis* (plate 14) see above. **LOCALITY** Proposed Taliabu Nature Reserve, Taliabu, Sula Islands.

SANGIHE HANGING PARROT
Loriculus catamene 12cm

Unmistakable; only hanging parrot with red undertail-coverts and red-tipped tail; forecrown red (♂) or green (♀); carpal edge of wing yellowish green; distinctive *tsw...tswee...tsweee-eee*. **DISTRIBUTION** Sangihe Island, Indonesia; up to 900m; endangered. **LOCALITY** Gunung Sahendaruman, Sangihe Island.

♂ MAROON-RUMPED HANGING PARROT

♂ MOLUCCAN HANGING PARROT

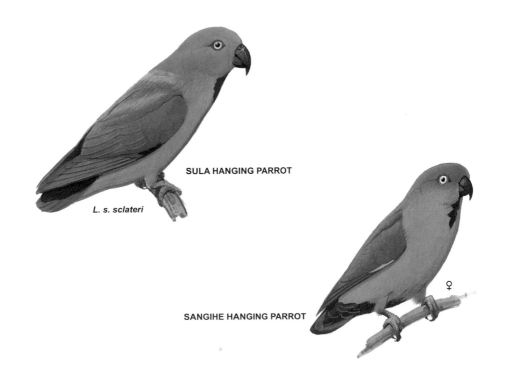

SULA HANGING PARROT

L. s. sclateri

SANGIHE HANGING PARROT

PLATE 28 HANGING PARROTS (in part)

76

PHILIPPINE HANGING PARROT
Loriculus philippensis 14cm

Polytypic hanging parrot with blue on sides of rump and red forehead in both sexes; sharp twittering *twik...twik...twik*, disyllabic *seep-seep*. **DISTRIBUTION** Philippine Islands; mostly up to 1000m, rarely 2500m; possibly extinct on Siquijor and Cebu, but fairly common elsewhere. **SUBSPECIES** 11 slightly differentiated to well-marked subspecies.1. *L. p. philippensis* ♂ forecrown red and occiput dusky yellow; orange-yellow band across nape; throat to center of upper breast red. ♀ throat to center of breast green; forecheeks blue. JUV like ♀, but little or no red on forecrown; dark bill. *Range* Banton, Catanduanes, Luzon, Marinduque, and Polillo, northern Philippines. 2. *L. p. mindorensis* ♂ only forehead red; faint orange band across nape; more extensive blue on sides of rump. ♀ throat to breast yellow. *Range* Mindoro, northern Philippines. 3. *L. p. bournsi* ♂ forehead red, and small yellow patch on forecrown; ♀ faint orange band across nape. *Range* Sibuyan, central Philippines. 4. *L. p. regulus* ♂ red frontal band, and crown golden yellow; ♀ crown green suffused yellow. *Range* Guimaras, Masbate, Negros, Panay, Romblon, Tablas, and Ticao, central Philippines. 5. *L. p. chrysonotus* ♂ hindcrown to upper back orange-yellow; ♀ hindcrown to hindneck dull yellow. *Range* Cebu, central Philippines; may be extinct. 6. *L. p. worcesteri* ♂ crown scarlet, becoming orange on occiput, and mantle slightly suffused orange-yellow; ♀ crown and occiput as in ♂. *Range* Biliran, Bohol, Buad, Calicoan, Leyte, Maripipi, and Samar, central Philippines. 7. *L. p. siquijorensis* ♂ crown to hindneck green; ♀ more extensive blue on cheeks to ear-coverts. *Range* Siquijor, southern Philippines; possibly extinct. 8. *L. p. apicalis* ♂ entire crown scarlet, becoming orange on nape; mantle faintly suffused orange-yellow; rump to upper tail-coverts paler scarlet. ♀ crown and nape as in ♂; mantle faintly suffused orange-yellow. *Range* Bazol, Balut, Dinagat, Mindanao, and Siargao, southern Philippines. 9. *L. p. dohertyi* both sexes like *apicalis*, but mantle strongly suffused orange. *Range* Basilan, southern Philippines. 10. *L. p. camiguinensis* sexes alike; forehead and crown scarlet, but occiput green; lores to above and below eyes, forecheeks, and throat bright blue. *Range* Camiguin Sur, southern Philippines; sometimes treated as separate species. 11. *L. p. bonapartei* ♂ forecrown red, becoming orange on crown and orange-yellow on nape to hindneck; bill black; legs gray. ♀ forecrown to nape as in ♂; lores and forecheeks pale blue; bill black; legs gray. *Range* Sulu Archipelago, southern Philippines; sometimes treated as separate species. **SIMILAR SPECIES** Guaiabero *Bolbopsittacus lunulatus* (plate 34) no red in plumage; greenish-yellow rump; prominent silver/gray bill. Mindanao lorikeet *Trichoglossus johnstoniae* (plate 14) different facial pattern; scalloped green-and-yellow underparts; larger with longer, pointed tail. *Prioniturus* racquet-tailed parrots (plates 40, 41) adults with central tail-racquets, but these lacking in JUV; longer, squarish tail gives different flight silhouette; blue underwings; no red bill; much larger. **LOCALITIES** Quezon National Park, Luzon, Rajah Sukituna National Park, Bohol, and Tawi Tawi, Sulu Archipelago.

PHILIPPINE HANGING PARROT

L. p. mindorensis

♂
♀
Juv

L. p. philippensis

L. p. regulus

♂

L. p. chrysonotus

♂

L. p. worcesteri

♂
♀

L. p. siquijorensis

♂
♀

L. p. apicalis

♂
♀

L. p. camiguinensis

L. p. bonapartei

♂
♀

PLATE 29 PYGMY PARROTS (in part)

78

Diminutive green parrots with stiffened, projecting shafts of tail-feathers and long metatarsals with long claws for foraging "woodpecker-like" on surfaces of tree trunks and limbs; variable sexual dimorphism, JUV duller. Forests, secondary growth, village gardens, plantations; arboreal; easily overlooked because of small size and weak *tseet* or *tsit* calls.

BUFF-FACED PYGMY PARROT
Micropsitta pusio 8.4cm

Buff to pale rufous face; blue crown not extending to eyes. **DISTRIBUTION** northern to southeastern New Guinea, including Bismarck Archipelago; up to 900m; common. **SUBSPECIES** two discernible and two poorly differentiated subspecies. 1. M. *p. pusio* forehead and sides of head buff-brown, paler in ♀; crown and occiput blue, ♀ paler, JUV green. *Range* Bismarck Archipelago, east to Duke of York Island, and southeastern New Guinea, west in north to Astrolabe Bay, and in south to Lake Kutubu, Papua New Guinea. 2. M. *p. beccarii* forehead and face darker brown. *Range* northern New Guinea from Geelvink Bay, West Papua, east to Astrolabe Bay, Papua New Guinea. 3. M. *p. harterti* throat washed blue; duller head markings. *Range* Fergusson Island, D'Entrecasteaux Archipelago, eastern Papua New Guinea. 4. M. *p. stresemanni* like *harterti*, but more yellowish underparts; larger. *Range* Misima and Tagula Islands, Louisiade Archipelago, eastern Papua New Guinea. **SIMILAR SPECIES** Red-breasted Pygmy Parrot M. *bruijnii* ♀ (plate 30) buff-white face contrasting with dark lilac crown; normally at higher elevations. Pygmy Lorikeet *Charmosyna wilhelminae* (plate 21) narrow, sharply pointed tail; streaking on breast; red bill. Orange-fronted Hanging Parrot *Loriculus aurantiifrons* (plate 26) red throat; blue underwings; black bill. **LOCALITIES** Brown River Forest Reserve, near Port Moresby, and Kau Wildlife Area, Baitabag, Madang Province, Papua New Guinea.

YELLOW-CAPPED PYGMY PARROT
Micropsitta keiensis 9.5cm

Like M. *pusio*, but no blue on crown. **DISTRIBUTION** Kai, Aru, and western Papuan Islands, West Papua, and western and southern New Guinea; up to 550m; locally common. **SUBSPECIES** two poorly differentiated and one well-marked subspecies. 1. M. *k. keiensis* crown yellow, brownish on forehead; brown face. *Range* Kai and Aru Islands. 2. M. *p. viridipectus* darker than *keiensis*; crown brownish tinged yellow; face brown. *Range* southern New Guinea between Mimika and Fly Rivers. 3. M. *k. chloroxantha* center of breast and abdomen orange-red (♂), green (♀ & JUV). *Range* western Papuan Islands and Vogelkop and Onin Peninsulas, West Papua. **SIMILAR SPECIES** Red-breasted Pygmy Parrot M. *bruijnii* ♀, Pygmy Lorikeet *Charmosyna wilhelminae*, and Orange-fronted Hanging Parrot *Loriculus aurantiifrons* (see above). **LOCALITIES** Ohalim village, near Elat, Kai Besar, Kai Islands. In coastal mangroves on Salawati, western Papuan Islands, and Sorong district, West Papua. Mount Bosavi district, Western Province, Papua New Guinea.

GEELVINK PYGMY PARROT
Micropsitta geelvinkiana 9cm

Dark gray-brown head or crown. **DISTRIBUTION** islands in Geelvink Bay, West Papua; up to 300m; near-threatened. **SUBSPECIES** two well-marked subspecies. 1. M. *g. geelvinkiana* crown purple-blue extending to eyes; ♂ with yellow patch on occiput; center of underparts orange-yellow, ♀ & JUV paler. *Range* Numfor Island. 2. M. *g. misoriensis* no blue on crown. *Range* Biak Island. **LOCALITY** Biak Utara Protected Area, Biak.

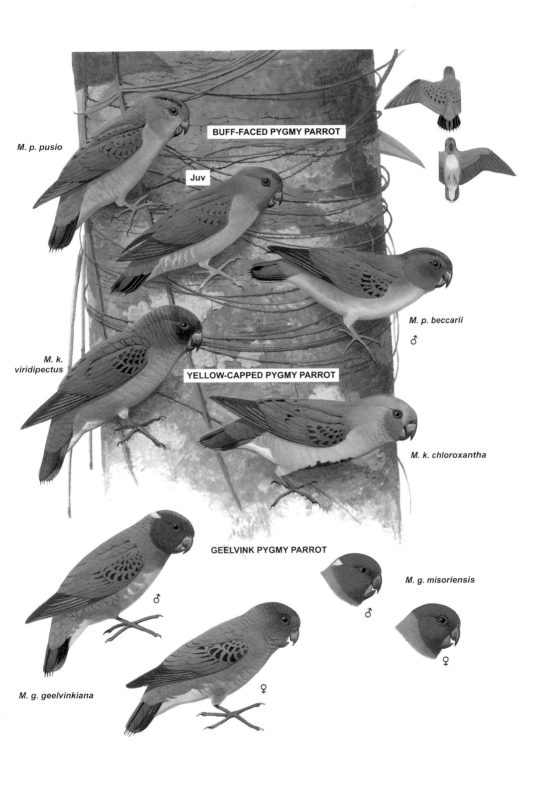

M. p. pusio

BUFF-FACED PYGMY PARROT

Juv

M. p. beccarii
♂

M. k. viridipectus

YELLOW-CAPPED PYGMY PARROT

M. k. chloroxantha

GEELVINK PYGMY PARROT

M. g. misoriensis
♂

M. g. geelvinkiana

♂

♀

♀

PLATE 30 **PYGMY PARROTS (in part)**

80

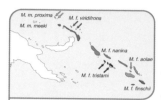

MEEK'S PYGMY PARROT
Micropsitta meeki 10cm

Mottled gray-brown head, and yellow neck to underparts.
DISTRIBUTION Admiralty and St. Matthias Islands, Papua New
Guinea; common. **SUBSPECIES** two discernible subspecies.
1. *M. m. meeki* head gray-brown mottled yellow. *Range* Admiralty
Islands. 2. *M. m. proxima* yellow frontal band meeting yellow
superciliary line. *Range* St. Matthias Islands. **LOCALITY** Manus, Admiralty Islands.

FINSCH'S PYGMY PARROT *Micropsitta finschii* 9.5cm

Only pygmy parrot with green head and neck. **DISTRIBUTION** (see map above) Bismarck
Archipelago, Papua New Guinea, to Solomon Islands; up to 900m; common. **SUBSPECIES** five
discernible subspecies. 1. *M. f. finschii* at base of bill blue ♂, pink ♀; abdomen orange-red ♂, green
♀. *Range* Ugi, San Cristobal, and Rennell, southern Solomon Islands. 2. *M. f. aolae* blue on crown;
♂ green abdomen. *Range* Guadalcanal, Malaita, Florida, and Russell, central Solomon Islands.
3. *M. f. tristami* like *finschii*, but ♂green abdomen. *Range* Vella Lavella, Gizo, Kolombangara, New
Georgia, Rubiana, and Rendova, western Solomon Islands. 4. *M. f. nanina* like *aolae*, but less blue on
crown. *Range* Santa Isabel, Bugotu, and Choiseul, northern Solomon Islands, and Bougainville,
eastern Papua New Guinea. 5. *M. f. viridifrons* like *aolae*, but cheeks blue. *Range* Lihir and Tabar
Islands, New Hanover, and New Ireland, Bismarck Archipelago, eastern Papua New Guinea.
SIMILAR SPECIES Red-breasted Pygmy Parrot M. *bruijnii* ♀ (see below) pale face contrasting with
dark blue crown; normally at higher altitudes. Green-rumped Hanging Parrot *Loriculus tener* (plate
26) red throat-patch; blue underwings; black bill. **LOCALITIES** Slopes of Lake Loloru volcano,
southern Bougainville, Papua New Guinea. Slopes of Mount Kubonitu, Santa Isabel, and Komarindi
Catchment Conservation Area, Guadalcanal, Solomon Islands.

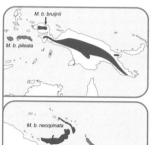

RED-BREASTED PYGMY PARROT
Micropsitta bruijnii 9cm

♂ unmistakable; ♀ & JUV with pale face and blue crown.
DISTRIBUTION mountains of South Moluccas, Indonesia, and New
Guinea to central Solomon Islands; 1000 to 2300m; locally
common. **SUBSPECIES** four subspecies differentiated by plumage
of ♂. 1. *M. b. bruijnii* ♂ buff-red crown, blue nape to sides of breast,
and pale red underparts; ♀ blue crown, whitish face, and green
underparts. *Range* New Guinea from Vogelkop, West Papua, east
to Owen Stanley Range, Papua New Guinea; in Ok Tedi River
region, ♂ possibly with yellow crown and undescribed subspecies.
2. *M. b. pileata* ♂ nape brownish red and narrow blue nuchal collar.
Range Seram and Buru, South Moluccas. 3. *M. b. necopinata* ♂ crown
yellowish brown, darker on nape; blue across foreneck; underparts
orange-red, undertail-coverts yellow. *Range* New Britain and New Ireland, Bismarck Archipelago,
eastern Papua New Guinea. 4. *M. b. rosea* ♂ crown reddish pink. *Range* Bougainville, eastern Papua
New Guinea, and Kolombangara and Guadalcanal, Solomon Islands. **SIMILAR SPECIES** ♀ like
lowland pygmy parrots, but whitish face and dark blue crown. Orange-fronted Hanging Parrot
Loriculus aurantiifrons and Green-rumped Hanging Parrot *L. tener* (plate 26) red throat-patch; blue
underwings; black bill. **LOCALITIES** Gunung Kelapat Muda, Buru, South Moluccas, and near
Nabire, West Papua, Indonesia. Tabubil, Western Province, and Hans Meyer Range, New Ireland,
Bismarck Archipelago, Papua New Guinea.

MEEK'S PYGMY PARROT

M. m. meeki

M. m. proxima

FINSCH'S PYGMY PARROT

♂

M. f. nanina

♂

♀

M. f. finschii

♂

M. f. viridifrons

♂

♀

M. b. bruijnii

RED-BREASTED PYGMY PARROT

♂

M. b. necopinata

♂

M. b. rosea

PLATE 31 FIG PARROTS (in part)

82 Small stocky green parrots with very short, wedge-shaped tail, broad, bulbous bill, and distinctive facial patterns; sexes differ, JUV like ♀. Forests and forest edges; arboreal; *Ficus* seeds principal food, so closely associated with fig trees; swift, direct flight; shrill *tseet*.

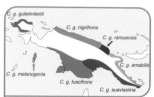

ORANGE-BREASTED FIG PARROT
Cyclopsitta gulielmitertii 13cm

Small fig parrot with black and yellow facial markings; yellow edges to innermost wing-coverts. **DISTRIBUTION** western Papuan and Aru Islands, Indonesia, and New Guinea; up to 1100m, mostly below 500m; common in south, scarce in north. **SUBSPECIES** five well-marked and two doubtful subspecies. 1. *C. g. gulielmitertii* ♂ forecrown dark blue; throat and sides of head pale yellow; breast to upper abdomen orange; no underwing band. ♀ cheeks anteriorly pale yellow and posteriorly black; ear-coverts to sides of neck orange; all green underparts. *Range* Salawati, western Papuan Islands, and western Vogelkop Peninsula, West Papua, Indonesia. 2. *C. g. nigrifrons* like *gulielmitertii*, but forecrown black. *Range* northern New Guinea, between Mamberamo and Sepik Rivers. 3. *C. g. ramuensis* like *amabilis*, but forecrown suffused blue and black. *Range* Ramu River district, northern Papua New Guinea; probably intermediate between *nigrifrons* and *amabilis*. 4. *C. g. amabilis* ♂ forecrown black; upper abdomen to breast and sides of head yellowish white. ♀ breast to upper abdomen orange; lower cheeks black. *Range* Huon Peninsula to Milne Bay, northeastern Papua New Guinea. 5. *C. g. sauvissima* ♂ forecrown dark blue; cheeks black; pale yellow behind and underneath cheeks; breast and upper abdomen orange. ♀ like ♂, but orange behind and underneath cheeks; breast and upper abdomen green tinged russet. *Range* southeastern Papua New Guinea, west to Gulf of Papua. 6. *C. g. fuscifrons* like *sauvissima*, but forecrown and cheeks brownish black. *Range* southern New Guinea, between Fly and Mimika Rivers. 7. *C. g. melanogenia* like *fuscifrons*, but ♀ paler green breast. *Range* Aru Islands, Indonesia; probably not separable from *fuscifrons*. **SIMILAR SPECIES** Double-eyed Fig Parrot *C. diophthalma* (plate 32) red and blue facial pattern; no orange on breast; where sympatric, *C. diophthalma* and *C. gulielmitertii* tend to replace each other locally. *Psittaculirostris* fig parrots (plates 33, 34) different facial patterns with elongated ear-coverts; larger. Small green *Charmosyna* lorikeets (plates 21, 22) slim body shape with longer, sharply pointed tail; no orange on breast; red bill. Orange-fronted Hanging Parrot *Loriculus aurantiifrons* (plate 26) red throat and rump; blue underwings; fine black bill. **LOCALITIES** Brown River and Veimauri Forest Reserves, near Port Moresby, Papua New Guinea.

ORANGE-BREASTED FIG PARROT

C. g. gulielmitertii

♂

♀

♂
Juv

C. g. nigrifrons

C. g. amabilis

♂

♀

♂

♀

C. g. sauvissima

♂

♀

Juv

♂

C. g. fuscifrons

PLATE 32 FIG PARROTS (in part)

84

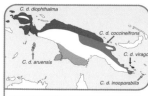

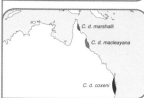

DOUBLE-EYED FIG PARROT
Cyclopsitta diophthalma 15cm

Small fig parrot with red and blue facial markings and orange-red edges to innermost wing-coverts. **DISTRIBUTION** New Guinea, including western and eastern Papuan Islands, Aru Islands, Indonesia, and northeastern Australia; up to 1600m; locally common in northern New Guinea, scarce elsewhere. **SUBSPECIES** six well-marked and two poorly differentiated subspecies. 1. *C. d. diophthalma* ♂ forecrown to cheeks red; orange-yellow band on hindcrown; blue above and in front of eye; mauve-blue band across lower cheeks. ♀ cheeks buff-brown. *Range* western Papuan Islands, West Papua, and northwestern New Guinea. 2. *C. d. coccineifrons* darker red on face; more pronounced yellow band on hindcrown. *Range* northeastern New Guinea, west to about Astrolabe Bay and long. 139°E in central highlands, Papua New Guinea. 3. *C. d. aruensis* ♂ paler greenish-blue above and in front of eye; mauve-blue band from lower cheeks to chin. ♀ red on face replaced by blue; no yellow band on hindcrown. *Range* Aru Islands and southern New Guinea between Fly and Balim Rivers. 4. *C. d. virago* ♂ paler red face; no blue above and in front of eye; mauve-blue spot on lower cheeks. ♀ red spot in center of blue forehead; crown and cheeks green. *Range* Goodenough and Fergusson Islands, D'Entrecasteaux Archipelago, eastern Papua New Guinea. 5. *C. d. inseparabilis* sexes alike; red spot in center of blue forehead; crown and cheeks green. *Range* Tagula Island, Louisiade Archipelago, eastern Papua New Guinea. 6. *C. d. marshalli* (Marshall's Fig Parrot) ♂ like *aruensis*, but paler red face; mauve-blue band across lower cheeks only. ♀ like *aruensis*, but forecrown and lower cheeks darker blue. *Range* eastern Cape York Peninsula, northern Queensland, Australia. 7. *C. d. macleayana* (Red-browed Fig Parrot) ♂ center of forehead and lower cheeks to ear-coverts red; lores to around eyes blue. ♀ lower cheeks buff-brown. *Range* Cooktown to Townsville districts, northeastern Queensland. 8. *C. d. coxeni* (Coxen's Fig Parrot) sexes alike; center of forehead blue; no blue above or in front of eye; larger. *Range* Blackall and Conondale Ranges, or possibly Rockhampton district, southeastern Queensland, south to Hastings River valley, northeastern New South Wales; critically endangered, CITES I. **SIMILAR SPECIES** Orange-breasted Fig Parrot *C. gulielmitertii* (plate 31) yellow and black facial pattern; orange breast; where sympatric *C. diophthalma* and *C. gulielmitertii* tend to replace each other locally. Small green *Charmosyna* lorikeets (plates 21, 22) slim body with longer, sharply pointed tail; red bill. *Psittaculirostris* fig parrots (plates 33, 34) different facial patterns; larger. Orange-fronted Hanging Parrot *Loriculus aurantiifrons* (plate 26) red throat and rump; blue underwings; fine black bill. Little Lorikeet *Glossopsitta pusilla* in Australia (plate 16) entirely red facial mask without blue; fine black bill. **LOCALITIES** Crater Mountain Research Station, Chimbu Province, Papua New Guinea. Iron Range and Daintree National Parks, north Queensland, Australia.

DOUBLE-EYED FIG PARROT

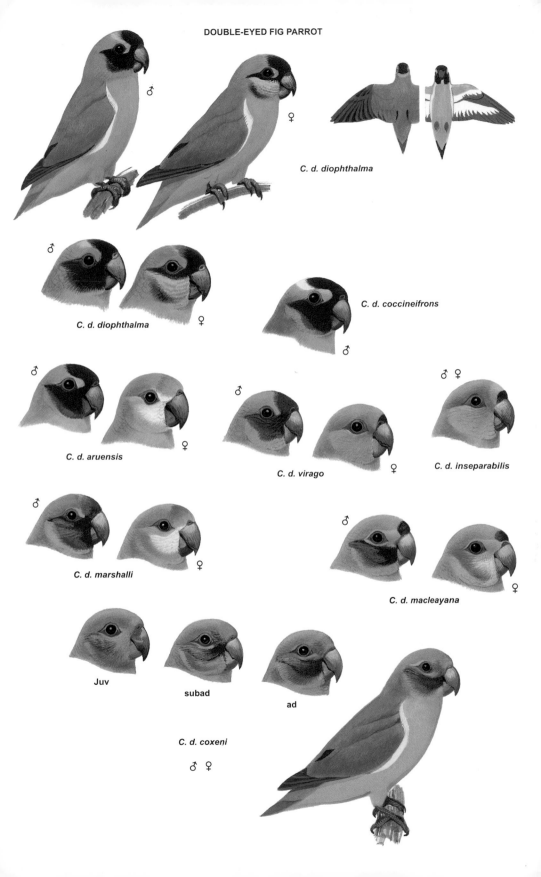

C. d. diophthalma

C. d. diophthalma

C. d. coccineifrons

C. d. aruensis

C. d. virago

C. d. inseparabilis

C. d. marshalli

C. d. macleayana

Juv

subad

ad

C. d. coxeni

♂ ♀

PLATE 33 FIG PARROTS (in part)

86

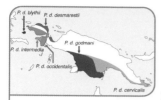

DESMAREST'S FIG PARROT
Psittaculirostris desmarestii 18cm

Only New Guinea parrot with orange or yellow crown; larger fig parrot with red, orange or yellow facial markings and elongated ear-coverts; little or no sexual dimorphism, JUV duller; chattering and clinking notes and repeated *chet-chet*. **DISTRIBUTION** western Papuan Islands, West Papua, Indonesia, and western and southern New Guinea; up to 1300m; locally common. *Psittaculirostris* fig parrots replace each other geographically. **SUBSPECIES** six well-marked subspecies. 1. *P. d. desmarestii* lores and forehead orange-red; crown and nape orange-yellow; blue nuchal collar; hindneck green; blue spot underneath eye; cheeks and ear-coverts green. *Range* northern and eastern Vogelkop Peninsula, West Papua. 2. *P. d. intermedia* crown and nape deeper orange; cheeks and ear-coverts green tipped yellow; little or no blue nuchal collar. *Range* known only from Onin Peninsula, West Papua. 3. *P. d. occidentalis* throat, cheeks, and ear-coverts bright yellow; paler blue underneath eye; no blue nuchal collar. *Range* Salawati and Batanta, western Papuan Islands, and western and southern Vogelkop Peninsula, West Papua. 4. *P. d. blythii* like *occidentalis*, but throat, cheeks and ear-coverts deep orange-yellow; no blue underneath eye in adults, but present in JUV. *Range* Misool, western Papuan Islands. 5. *P. d. godmani* no blue nuchal collar or spot underneath eye; ♂ yellow band on hindneck, ♀ & JUV green hindneck. *Range* southern New Guinea from Mimika River, West Papua, east to Fly River, western Papua New Guinea. 6. *P. d. cervicalis* like *godmani*, but nape and hindneck dark blue; underparts tinged orange-buff; JUV crown and nape green. *Range* southern Papua New Guinea west to Fly and Noord Rivers. **SIMILAR SPECIES** *Cyclopsitta* fig parrots (plates 31, 32) different facial patterns; smaller. Similar sized lorikeets have slim body shape with longer, more pointed tails. *Geoffroyus* parrots (plates 35, 36) different head markings and bill colors; blue underwings; different calls. **LOCALITIES** Sorong and Nabire districts and Timuka golf course, West Papua, Indonesia. Mount Bosavi, Southern Highlands Province, and Bomai and Karimui districts, Chimbu Province, Papua New Guinea.

P. d. desmarestii

Juv

P. d. occidentalis

P. d. blythii

P. d. cervicalis

Juv

♂

♀

Juv

P. d. godmani

PLATE 34 FIG PARROTS (in part) AND GUAIABERO

88

EDWARDS'S FIG PARROT
Psittaculirostris edwardsii **18cm**

Larger fig parrot with yellow-green crown, black nuchal band, red cheeks, and yellow ear-coverts; underparts red (♂), or green (♀ & JUV); *screeett-screeett* in flight. **DISTRIBUTION** northeastern New Guinea from Humboldt Bay, West Papua, east to Huon Gulf, Papua New Guinea; up to 800m; common. **SIMILAR SPECIES** *Cyclopsitta* fig parrots (plates 31, 32) different facial patterns; smaller. Similar sized lorikeets have slim body shape with longer, more pointed tails. *Geoffroyus* parrots (plates 35, 36) different head markings and bill colors; blue underwings; different calls. **LOCALITIES** Maprik and Finschhafen districts, northern Papua New Guinea.

SALVADORI'S FIG PARROT
Psittaculirostris salvadorii **19cm**

Only larger fig parrot with orange (♂) or blue (♀) pectoral band. **DISTRIBUTION** northwestern New Guinea from Humboldt Bay and Cyclops Mountains west to Geelvink Bay, West Papua, Indonesia. **SIMILAR SPECIES** *Cyclopsitta* fig parrots, similar sized lorikeets, and *Geoffroyus* parrots (see above). **LOCALITIES** Nimbokrang district, near Jayapura, and Nabire district, West Papua.

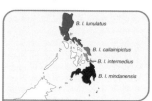

GUAIABERO *Bolbopsittacus lunulatus* **15cm**

Small dumpy green parrot with very short, wedge-shaped tail and proportionately large, broad bill; sexually dimorphic, JUV like ♀. Forests, secondary growth, gardens; arboreal; singly, pairs, small flocks; difficult to detect feeding amidst foliage, but conspicuous in very fast, direct flight at or below canopy; attracted to figs and guavas; high-pitched *zeet-zeet*. **DISTRIBUTION** Philippine Islands; up to 1000m; locally common. **SUBSPECIES** three discernible and one doubtfully distinct subspecies. 1. *B. l. lunulatus* ♂ face, chin, and collar encircling hindneck pale blue; rump and upper tail-coverts greenish-yellow. ♀ chin and lower cheeks pale blue; rump to upper tail-coverts and collar encircling hindneck yellow scalloped black. *Range* Luzon, northern Philippines. 2. *B. l. intermedius* ♂ face and collar encircling hindneck darker blue; ♀ blue only on chin. *Range* Leyte and Panaon, central Philppines. 3. *B. l. callainipictus* like *intermedius*, but more yellowish green. *Range* Samar, central Philippines. 4. *B. l. mindanensis* ♂ green upper cheeks separating blue around eyes and blue lower cheeks, and darker blue collar encircling hindneck; ♀ like *lunulatus*. *Range* Mindanao, southern Philippines. **SIMILAR SPECIES** Mindanao Lorikeet *Trichoglossus johnstoniae* (plate 14) red face and red bill; scalloped green-and-yellow underparts; longer tail; sympatric only on Mindanao, where usually at higher altitudes. Philippine Hanging Parrot *Loriculus philippensis* (plate 28) red forecrown and rump; red bill and feet; blue underwings. *Prioniturus* racquet-tailed parrots (plates 40, 41) adults with central tail-racquets, but these lacking in JUV; longer, squarish tail gives different flight silhouette; blue underwings; much larger. **LOCALITIES** Quezon National Forest Park, Luzon, northern Philippines, and Mount Apo National Park, Mindanao, southern Philippines.

EDWARDS'S FIG PARROT

♂

♀

SALVADORI'S FIG PARROT

♂

♀

GUAIABERO

♂

♀

B. l. lunulatus

Juv

♂

B. l. intermedius

♂

♀

B. l. mindanensis

Juv

PLATE 35 *GEOFFROYUS* **PARROTS (in part)**

90 Midsized stocky green parrots with short, squarish tail and blue underwings; iris yellow-white adults, brown JUV; sexes differ, JUV duller. Forests, woodlands, secondary growth; gardens; arboreal; pairs or small parties; shy; noisy in swift, direct flight above canopy.

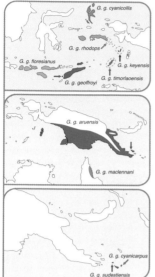

RED-CHEEKED PARROT
Geoffroyus geoffroyi 21cm

Widespread, polytypic *Geoffroyus* parrot with red and blue (♂), brown (♀) or green (JUV) head; distinctive *aank…aank…aank*. **DISTRIBUTION** Moluccas and Lesser Sunda Islands, Indonesia, to New Guinea and Cape York Peninsula, northern Queensland, Australia; up to 800m. Common. **SUBSPECIES** (in part, continued plate 36) 15 subspecies separated in two groupings by absence (1–10) or presence (11–15) of red-brown on lower back to rump. 1. G. g. *geoffroyi* ♂ forehead to cheeks rose-red; crown and occiput blue, not extending to green nape; red-brown blaze on inner wing-coverts; upper mandible red, lower mandible brown; ♀ forehead to occiput and cheeks brown, and bill brown. *Range* Timor, Semau, and Wetar, Lesser Sunda Islands. 2. G. g. *floresianus* darker than *geoffroyi*; mauve-blue (♂) or brown (♀) extending from crown to nape; underwing-coverts darker blue. *Range* Lombok, Sumbawa, Flores, and Sumba, Lesser Sunda Islands. 3. G. g. *cyanicollis* blue collar encircling neck; mantle and upper back suffused bronze-brown. *Range* Morotai, Halmahera, and Bacan, North Moluccas. 4. G. g. *rhodops* darker than *floresianus*; ♂ red face sharply demarcated from blue crown; ♀ head darker brown. *Range* Buru, Seram, Boano, Ambon, Haruku, Saparua, Manawoka, Gorong, and Seramlaut Islands, South Moluccas. 5. G. g. *keyensis* paler yellowish green, especially tail; ♂ hindcheeks rose-red suffused blue; ♀ head paler brown. *Range* Kai Kecil and Kai Besar, Kai Islands, Indonesia. 6. G. g. *timorlaoensis* smaller than *keyensis*. *Range* Tanimbar Islands, Indonesia. 7. G. g. *aruensis* like *geoffroyi*, but darker mauve-blue (♂) or brown (♀) extending from crown to nape; darker blue underwing-coverts. *Range* Aru Islands, Indonesia, and southern New Guinea, in south east of Mimika River, West Papua, and in north west to Huon Peninsula, Papua New Guinea; also Fergusson and Goodenough Islands, eastern Papua New Guinea. 8. G. g. *maclennani* darker than *aruensis*, but underwing-coverts paler blue. *Range* Pascoe River south to Rocky River, eastern Cape York Peninsula, north Queensland, Australia; near-threatened. 9. G. g. *sudestiensis* like *aruensis*, but no red-brown blaze on inner wing-coverts; ♀ crown suffused dark green. *Range* Misima and Tagula Islands in Louisiade Archipelago, eastern Papua New Guinea. 10. G. g. *cyanicarpus* blue edge from bend of wing to outermost primary. *Range* Rossel Island in Louisiade Archipelago.

RED-CHEEKED PARROT

G. g. geoffroyi

♂

♀

Juv

G. g. cyanicollis

♂

♀

G. g. keyensis

♂

♀

G. g. rhodops

♂

♀

G. g. sudestiensis

♂

♀

G. g. maclennani

♂

G. g. aruensis

♂

♀

G. g. cyanicarpus

♂

♀

PLATE 36 *GEOFFROYUS* PARROTS (in part)

92

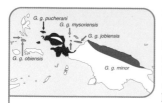

RED-CHEEKED PARROT
Geoffroyus geoffroyi (cont.)

SUBSPECIES (in part, continued from plate 35) 11. G. g. *minor* darker than *aruensis*; lower back to rump red-brown; mantle suffused bronze-brown. *Range* northern New Guinea from Astrolabe Bay area, Papua New Guinea, west to about Mamberamo River, West Papua. 12. G. g. *jobiensis* like *minor*, but underwing-coverts paler blue; lower back to rump brighter red; ♂ red of forehead extending back to forecrown. *Range* Yapen and Mios Num Islands, Geelvink Bay, West Papua. 13. G. g. *mysoriensis* like *minor*, but lower back to rump dark maroon; no bronze-brown suffusion on mantle; underwing-coverts darker blue; ♂ violet-blue extending from crown and nape to hindneck, and red from cheeks to throat; ♀ brown extending from crown and nape to hindneck, and from cheeks to throat. *Range* Biak and Numfoor Islands, Geelvink Bay, West Papua. 14. G. g. *pucherani* like *minor*, but lower back to rump paler brown; underwing-coverts darker blue; little or no bronze-brown blaze on inner wing-coverts. *Range* western Papuan Islands, Indonesia, and western New Guinea east to Etna Bay, West Papua. 15. G. g. *obiensis* like *pucherani*, but blue nape and hindneck extending as collar encircling neck. *Range* Obi and nearby Bisa Island, North Moluccas, Indonesia. **SIMILAR SPECIES** Blue-collared Parrot G. *simplex* (see below) green-headed JUV probably indistinguishable; adults with green head and black bill; different calls; usually at higher elevations. **LOCALITIES** Manusela National Park, Seram, South Moluccas, Indonesia. Brown River Forest Reserve, near Port Moresby, Papua New Guinea. Iron Range National Park, Cape York Peninsula, north Queensland, Australia.

BLUE-COLLARED PARROT
Geoffroyus simplex 22cm

Only *Geoffroyus* parrot with all-green head and black bill; distinctive *kree-kro…kree-kro*. **DISTRIBUTION** mountains of New Guinea; mostly 600 to 2000m; uncommon and shy. **SUBSPECIES** two discernible subspecies. 1. G. s. *simplex* ♂ grayish-blue collar encircling neck, absent in ♀ & JUV. *Range* Arfak and Tamrau Mountains, Vogelkop Peninsula, West Papua. 2. G. s. *buergersi* ♂ wider blue collar. *Range* New Guinea, except Vogelkop Peninsula. **SIMILAR SPECIES** Red-cheeked Parrot G. *geoffroyi* (see above) green-headed JUV probably indistinguishable; usually at lower elevations. **LOCALITIES** nowhere predictably common. Ambua Lodge, Tari Gap, Southern Highlands, and Karimui district, Eastern Highlands, Papua New Guinea.

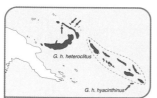

SINGING PARROT *Geoffroyus heteroclitus* 25cm

Only yellow (♂) or gray (♀) -headed parrot in range; distinctive *kreel-kreel* call and *wu-wu…wo-wo…wee-wee…wi-wi* song. **DISTRIBUTION** Bismarck Archipelago, Bougainville and Buka Islands, eastern Papua New Guinea, and Solomon Islands; up to 600m; locally common. **SUBSPECIES** two well-marked subspecies. 1. G. h. *heteroclitus* ♂ head yellow with gray-mauve collar encircling neck; upper mandible yellow, lower mandible gray. ♀ head gray without mauve collar; bill gray. *Range* throughout range except Rennell Island. 2. G. h. *hyacinthinus* ♂ lilac-blue collar extending to mantle and lower breast; ♀ dark gray of head extending to mantle and lower breast. *Range* Rennell Island, southeastern Solomon Islands. **LOCALITIES** Slopes of Lake Loloru, south Bougainville, Papua New Guinea. Komarindi Catchment Conservation Area, Guadalcanal, Solomon Islands.

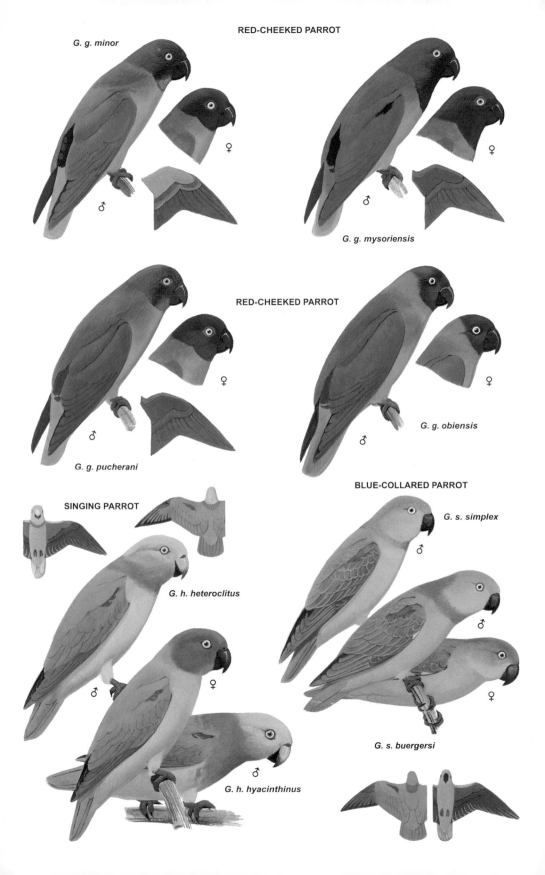

RED-CHEEKED PARROT

G. g. minor

♀

♂

G. g. mysoriensis

♀

♂

RED-CHEEKED PARROT

♀

♂

G. g. obiensis

G. g. pucherani

♀

♂

SINGING PARROT

G. h. heteroclitus

♂   ♀

G. h. hyacinthinus   ♂

BLUE-COLLARED PARROT

G. s. simplex   ♂

♂

♀

G. s. buergersi

PLATE 37 *PSITTACELLA* **TIGER PARROTS (in part)**

94 Small to midsized stocky green parrots with short, rounded tail and distinctive
coloration featuring barred upperparts and red undertail-coverts; sexes differ,
JUV like ♀. Highland forests, scrubby undergrowth and adjacent clearings; arboreal;
quiet and inactive, so easily overlooked; singly, pairs, or small parties, often in low
undergrowth; labored, undulating flight.

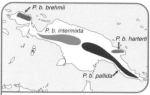

BREHM'S TIGER PARROT
Psittacella brehmii 24cm

Largest tiger parrot ; deep *pak…pak…pak* or plaintive *ee-yur*.
DISTRIBUTION mountains of New Guinea; mainly 1500 to 3000m;
locally common. **SUBSPECIES** four slightly differentiated, isolated
subspecies. 1. *P. b. brehmii* ♂ head olive-brown with yellow line
down side of neck, and all-green breast; ♀ no yellow line down side
of head, and breast yellow barred black. *Range* Vogelkop Peninsula, West Papua, Indonesia.
2. *P. b. intermixta* like *brehmii*, but more yellowish underparts; throat and sides of head paler brown.
Range Snow and Weyland Mountains and Mount Goliath, West Papua. 3. *P. b. pallida* ♂ like *brehmii*;
♀ flanks and sides of abdomen barred black. *Range* east from Sepik River region, Papua New Guinea,
except Huon Peninsula. 4. *P. b. harterti* like *intermixta*, but less yellowish underparts, and paler, more
olive head; ♀ flanks and sides of abdomen faintly barred black. *Range* Huon Peninsula, northern
Papua New Guinea. **SIMILAR SPECIES** Painted Tiger Parrot *P. picta* (see below) rump red or
greenish yellow; ♀ with prominent cheek-patches; smaller. Other *Psittacella* tiger parrots (plate 38)
much smaller. Red-cheeked Parrot *Geoffroyus geoffroyi* (brown-headed ♀; plates 35, 36) no black
and yellow barring; green undertail-coverts; blue underwing-coverts; different calls; usually flying
high above canopy. **LOCALITIES** Ambua Lodge, Tari Gap, Southern Highlands, and Efogi district,
Kokoda Trail, Owen Stanley Ranges, Papua New Guinea.

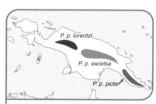

PAINTED TIGER PARROT
Psittacella picta 19cm

Smaller tiger parrot with red or greenish-yellow rump; ♀ with
prominent cheek-patches; nasal *nhrr-a-rehn* and harsh *chee-zeedd*
from eastern birds, and *err-ee* from *lorentzi*. **DISTRIBUTION**
mountains of eastern and central New Guinea; mainly 2500 to
4000m; fairly common. **SUBSPECIES** three well-differentiated,
isolated subspecies. 1. *P. p. picta* ♂ head russet-brown with yellow band across hindneck, throat and
upper breast suffused blue, and rump and upper tail-coverts red; ♀ no yellow band across hindneck,
greenish-blue cheek-patches, and breast yellow barred black. *Range* Wharton and Owen Stanley
Ranges, southeastern Papua New Guinea. 2. *P. p. excelsa* ♂ crown and occiput bright olive-brown;
♀ head bright olive-brown, and cheeks strongly suffused blue. *Range* central highlands of Papua New
Guinea. 3. *P. p. lorentzi* sexes alike; cheeks bluish green; rump and upper tail-coverts greenish-yellow
barred black. *Range* Snow Mountains, West Papua, Indonesia. **SIMILAR SPECIES** Brehm's Tiger
Parrot *P. brehmii* (see above) no red rump or bluish-green cheeks, larger. Other *Psittacella* tiger
parrots (plate 38) no red rump or bluish-green cheeks; much smaller. Red-cheeked Parrot *Geoffroyus
geoffroyi* (brown-headed ♀; plates 35, 36) see above. **LOCALITY** Ambua Lodge, Tari Gap, Southern
Highlands, Papua New Guinea.

BREHM'S TIGER PARROT

♂

♀

♀

P. b. pallida

P. b. brehmii

PAINTED TIGER PARROT

♂ *P. p. picta*

♀

♂

P. p. excelsa

♂

P. p. lorentzi

PLATE 38 *PSITTACELLA* **TIGER PARROTS (in part)**

96

MODEST TIGER PARROT
Psittacella modesta 14cm

One of two small tiger parrots with very similar males, but different females; soft *peep*. **DISTRIBUTION** mountains of western and central New Guinea; mostly 1700 to 2800m; locally common. **SUBSPECIES** three slightly differentiated subspecies in two isolated populations.

1. *P. m. modesta* ♂ head dark brown, on nape to hindneck feathers yellow edged brown, and throat and breast brownish green; ♀ no yellow on nape to hindneck, and breast orange barred brown and dark green. *Range* Vogelkop Peninsula, West Papua, Indonesia. 2. *P. m. collaris* ♂ irregular yellow collar on hindneck; ♀ indistinct yellow markings on hindneck. *Range* southern slopes of Snow Mountains, West Papua. 3. *P. m. subcollaris* ♂ like *collaris*, but narrower, brighter yellow collar on hindneck; ♀ like *modesta*, but head darker brown, sometimes with indistinct yellow collar on hindneck. *Range* northern slopes of Snow Mountains, east to about long. 145°E in central Papua New Guinea. **SIMILAR SPECIES** Madarasz's Tiger Parrot *P. madaraszi* (see below) ♂ probably indistinguishable, but always lacks yellow collar on hindneck; green-headed ♀ without barred underparts. Other *Psittacella* tiger parrots (plate 37) much larger. **LOCALITY** Ambua Lodge, Tari Gap, Southern Highlands, Papua New Guinea.

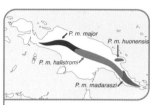

MADARASZ'S TIGER PARROT
Psittacella madaraszi 14cm

Differentiated from *P. modesta* by green-headed ♀ without barred underparts; high-pitched *huwee-hee…whreen* and *ee-o-ee*. **DISTRIBUTION** mountains of central and eastern New Guinea; mostly 1100 to 2500m; uncommon. **SUBSPECIES** four slightly differentiated subspecies in two isolated populations.

1. *P. m. madaraszi* ♂ head brown, feathers on crown to hindneck linearly mottled yellow, and upper breast suffused olive-brown; ♀ head green, suffused blue on forehead, and occiput to hindneck barred orange and black. *Range* southeastern Papua New Guinea, west in north to Mount Misim and in south to Angabunga River. 2. *P. m. huonensis* ♂ head more yellowish brown; ♀ no orange on occiput to hindneck. *Range* Huon Peninsula, northern Papua New Guinea. 3. *P. m. hallstromi* ♂ plumage darker than *madaraszi*; ♀ nape to hindneck strongly suffused reddish orange and more broadly barred black. *Range* central highlands of Papua New Guinea, west to Hindenburg Range, West Papua, Indonesia. 4. *P. m. major* larger than *madaraszi*. *Range* Weyland Mountains and northern slopes of Snow Mountains, West Papua. **SIMILAR SPECIES** Modest Tiger Parrot *P. modesta* (see above) ♂ probably indistinguishable, but mostly with yellow collar on hindneck; brown-headed ♀ with barred underparts. Other *Psittacella* tiger parrots (plate 37) much larger. **LOCALITIES** Satop village area, Huon Peninsula, and Efogi district, Kokoda Trail, Owen Stanley Ranges, Papua New Guinea.

MODEST TIGER PARROT

♂

♀

P. m. modesta

♂

P. m. subcollaris

MADARASZ'S TIGER PARROT

♂

♀

Juv

P. m. madaraszi

PLATE 39 **RACQUET-TAILED PARROTS (in part)**

98

Midsized stocky green parrots with short, rounded tail, central feathers with elongated bare shafts terminating in spatules; blue underwings and undertail; slight to pronounced sexual dimorphism, JUV without tail-racquets. Forests, secondary growth, village gardens; arboreal; pairs, small groups, or sometimes flocks; quiet while feeding, but noisy and conspicuous in swift, direct flight.

YELLOW-BREASTED RACQUET-TAILED PARROT
Prioniturus flavicans 37cm

Crown blue with (♂) or without (♀ & JUV) central red patch; mantle and neck to breast olive-yellow; screeching and high-pitched bugling. **DISTRIBUTION** northern and north-central Sulawesi, south to about 1°30'N, and nearby Banka, Lembeh, Togian, and possibly Banggai Islands, Indonesia; up to 1000m; near-threatened. **SIMILAR SPECIES** Golden-mantled Racquet-tailed Parrot *P. platurus* (see below) smaller; ♂ gray mantle and upper wing-coverts; ♀ no blue on crown; more active in larger flocks; different calls. Blue-backed Parrot *Tanygnathus sumatranus* (plate 43) larger and with longer, round-tipped tail, but no tail-racquets; blue rump; massive bill red (♂) or white (♀). **LOCALITIES** Bogani Nani Wartabone National Park, Tangkoko Batuangus Nature Reserve, and Manembonembo Nature Reserve, north Sulawesi.

GOLDEN-MANTLED RACQUET-TAILED PARROT
Prioniturus platurus 28cm

Distinctive ♂ coloration features gray and orange-yellow on upperparts; nasal *kaaa* and *krrrik* or *krrri* and repeated *queelie*. DISTRIBUTION Sulawesi, including nearby islands, and Talaud and Sula Islands, Indonesia; mostly 1000 to 2300m; common.

SUBSPECIES three well-defined subspecies. 1. *P. p. platurus* ♂ pink-red spot on hindcrown, bordered behind by dull gray-blue to nape, orange-yellow band across upper mantle, and lower mantle and upper wing-coverts dull gray; ♀ predominantly green. *Range* Sulawesi, including Siau, Lembeh, Dodepo, Muna, and Butung Islands, and Togian and Banggai Islands. 2. *P. p. talautensis* ♂ less gray on mantle and upper wing-coverts, and more pronounced red spot on crown bordered behind by more bluish patch to nape; ♀ like *platurus*. *Range* Karakelong and Salebabu, in Talaud Islands. 3. *P. p. sinerubris* ♂ no pink-red spot on crown, upper mantle and median wing-coverts green tinged gray, and bend of wing and lesser wing-coverts washed violet; ♀ undescribed. *Range* recorded from Taliabu and Mangole, in Sula Islands. **SIMILAR SPECIES** Yellow-breasted Racquet-tailed Parrot *P. flavicans* (see above) larger; mantle and neck to breast olive-yellow; ♀ with blue crown; less active and not in large flocks; different calls. Blue-backed Parrot *Tanygnathus sumatranus* (plate 43) see above. **LOCALITIES** Lore Lindu and Dumoga Bone National Parks and Tangkoko Batuangus Nature Reserve, Sulawesi, and proposed Taliabu Nature Reserve, Sula Islands.

BURU RACQUET-TAILED PARROT
Prioniturus mada 32cm

Only racquet-tailed parrot in range; nape to mantle and wing-coverts bluish purple (♂) or green (♀); underparts yellowish green; whistling *si-quie*, repeated *kwii…kwii* and low *squrr-squrr*. **DISTRIBUTION** Buru, South Moluccas, Indonesia; up to 1600m; common, but little known. **SIMILAR SPECIES** Red-cheeked Parrot *Geoffroyus geoffroyi* (green-headed JUV, plate 35) no tail-racquets; different calls. Black-lored Parrot *Tanygnathus gramineus* (plate 43) larger with longer, round-tipped tail, but no tail-racquets; massive bill red (♂) or white (♀). **LOCALITIES** Danau Rana and Gunung Kelapat Muda, Buru.

YELLOW-BREASTED RACQUET-TAILED PARROT

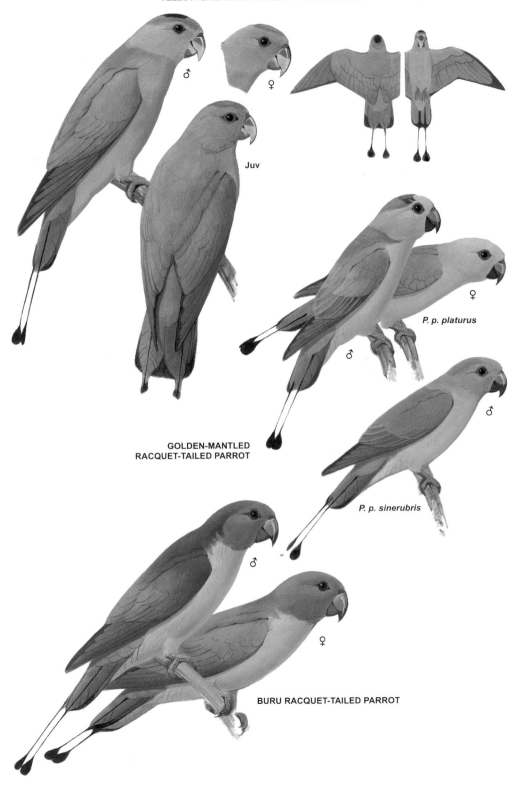

♂

♀

Juv

P. p. platurus

♀

♂

GOLDEN-MANTLED
RACQUET-TAILED PARROT

♂

P. p. sinerubris

♂

♀

BURU RACQUET-TAILED PARROT

PLATE 40 **RACQUET-TAILED PARROTS (in part)**

100

GREEN RACQUET-TAILED PARROT
Prioniturus luconensis 29cm

Only all-green species; little sexual dimorphism, ♀ slightly darker green; harsh *aaaak*, whinnying *we-li-li*, ringing *lin-nng*, disyllabic *your-witt*. **DISTRIBUTION** Luzon and Marinduque, northern Philippine Islands; up to 1000m; vulnerable. **SIMILAR SPECIES** Blue-crowned Racquet-tailed Parrot *P. discurus* (see below) darker green upperparts; blue crown contrasting with bright green face. Mountain Racquet-tailed Parrot *P. montanus* (green-headed ♀, plate 41) darker green head and upperparts; usually at higher elevations. *Tanygnathus* parrots (plates 42, 43) much larger, with longer, round-tipped tail and no tail-racquets; large bill red, or white (♀ *T. sumatranus*). Guaiabero *Bolbopsittacus lunulatus* (plate 34) much smaller with very short, wedge-shaped tail and no tail-racquets. **LOCALITIES** Sierra Madre Mountains and Subic Bay Naval Forest Reserve, Luzon.

BLUE-CROWNED RACQUET-TAILED PARROT
Prioniturus discurus 27cm

A blue-crowned racquet-tailed parrot with blue of crown to nape contrasting with bright green face in both sexes; JUV with little or no blue on crown; harsh screeches and disyllabic squeaky notes. **DISTRIBUTION** northern, eastern, and southern Philippine Islands; up to 1750m, but lower on Mindanao where replaced at higher elevations by *P. waterstradti*; uncommon. **SUBSPECIES** three slightly differentiated subspecies. 1. *P. d. discurus* crown to nape blue sharply demarcated from green face. *Range* Jolo in Sulu Archipelago, and Balut, Mindanao, Olutanga, and Basilan, southern Philippines. 2. *P. d. whiteheadi* less blue on crown and not sharply demarcated from green face. *Range* northern to central Philippines, from southern Luzon to Negros, Cebu, and Bohol, but excluding Mindoro. 3. *P. d. mindorensis* forecrown green; hindcrown and nape darker mauve-blue. *Range* Mindoro, northern Philippines. **SIMILAR SPECIES** Green Racquet-tailed Parrot *P. luconensis* (see above) yellowish green without blue crown. Mountain Racquet-tailed Parrot *P. montanus* (♀, plate 41) face suffused blue, but no blue on crown. Mindanao Racquet-tailed Parrot *P. waterstradti* (♀, plate 41) face suffused blue, but no blue on crown; back suffused olive-brown. *Tanygnathus* parrots (plates 42, 43) see above. Guaiabero *Bolbopsittacus lunulatus* (plate 34) see above. **LOCALITIES** Quezon National Park, Luzon, and Rajah Sikituna National Park, Bohol.

BLUE-HEADED RACQUET-TAILED PARROT
Prioniturus platenae 28cm

Unmistakable; only racquet-tailed parrot with blue head, less extensive in ♀; JUV undescribed. **DISTRIBUTION** Calauit, Busuanga, Culion, Dumaran, Palawan, and Balabac, western Philippine Islands; vulnerable. **SIMILAR SPECIES** only racquet-tailed parrot in range. Blue-naped Parrot *Tanygnathus lucionensis* (plate 42) larger with longer, round-tipped tail; mottled black-and-yellow "wing-patch"; massive red bill. **LOCALITY** St. Paul's Subterranean National Park, Palawan.

GREEN RACQUET-TAILED PARROT

♂

♀

BLUE-CROWNED RACQUET-TAILED PARROT

P. d. mindorensis

P. d. discurus

♂

♀

BLUE-HEADED RACQUET-TAILED PARROT

PLATE 41 · RACQUET-TAILED PARROTS (in part)

102

BLUE-WINGED RACQUET-TAILED PARROT
Prioniturus verticalis 35cm

One of two allopatric racquet-tailed parrots with similar sexual dimorphism. ♂ central red spot on blue crown, ♀ & JUV no red spot or blue crown; rasping *aaaack*, squeaky *lee-aaack*.
DISTRIBUTION Bongao, Manuk Manka, Sibutu, Tawi Tawi, and Tumindao, Sulu Archipelago, southern Philippine Islands; endangered. **SIMILAR SPECIES** only racquet-tailed parrot in range. *Tanygnathus* parrots (plates 42, 43) much larger with longer, round-tipped tail without central tail-racquets; massive red or white (*T. sumatranus* ♀) bill. **LOCALITY** recent records mostly from Tawi Tawi.

MOUNTAIN RACQUET-TAILED PARROT
Prioniturus montanus 30cm

Another racquet-tailed parrot with central red spot on blue crown of ♂, with blue extending to lores and around eyes; back slightly suffused olive-brown, and lateral tail-feathers blue tipped black; ♀ head green; harsh *kak-kak-kak…ak…ak…ak*. **DISTRIBUTION** mountains of Luzon, northern Philippine Islands; mostly 1000 to 2500m; near-threatened. **SIMILAR SPECIES** ♂ readily identified by red crown, but green-headed ♀ like other females. Green Racquet-tailed Parrot *P. luconensis* (plate 40) paler yellowish green without blue on head; usually at lower elevations. Blue-crowned Racquet-tailed Parrot *P. discurus* (♀, plate 40) extensive blue on crown. *Tanygnathus* parrots (plate 42, 43) see above. Guaiabero *Bolbopsittacus lunulatus* (plate 34) much smaller with very short, wedge-shaped tail and no central tail-racquets. **LOCALITY** Mount Puguis area, northern Luzon.

MINDANAO RACQUET-TAILED PARROT
Prioniturus waterstradti 30cm

Both sexes like *P. montanus* ♀, but with brownish suffusion on upperparts and lateral tail-feathers green not blue; calls like *P. montanus*. **DISTRIBUTION** mountains of Mindanao, southern Philippine Islands; mostly 850 to 2500m; near-threatened. **SUBSPECIES** two poorly differentiated subspecies.
1. *P. w. waterstradti* forecrown and face pale blue; upperparts suffused olive-brown; lateral tail-feathers green tipped black. *Range* Mounts Apo, Matutum, and Mayo, southern Mindanao.
2. *P. w. malindangensis* paler blue on forecrown and face; less brownish on mantle. *Range* Mounts Kitanglad and Malindang, and mountains of Misamis Oriental, west-central Mindanao. **SIMILAR SPECIES** Blue-crowned Racquet-tailed Parrot *P. discurus* (plate 40) blue crown and nape sharply demarcated from green face; no olive-brown on upperparts; usually at lower elevations. *Tanygnathus* parrots (plates 42, 43) see above. Guaiabero *Bolbopsittacus lunulatus* (plate 34) see above.
LOCALITY Mount Apo National Park, Mindanao.

BLUE-WINGED RACQUET-TAILED PARROT

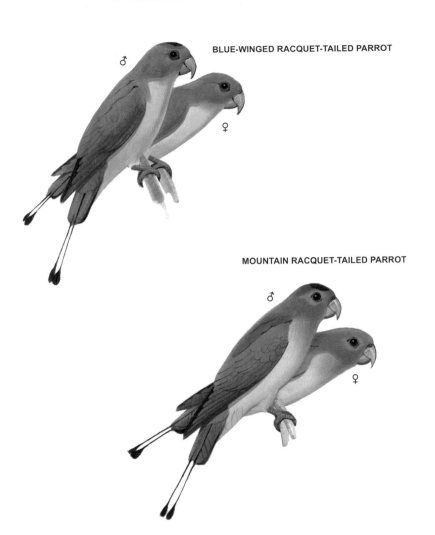

♂

♀

MOUNTAIN RACQUET-TAILED PARROT

♂

♀

MINDANAO RACQUET-TAILED PARROT

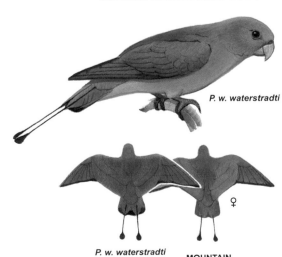

P. w. waterstradti

P. w. waterstradti

♀

**MOUNTAIN
RACQUET-TAILED PARROT**

PLATE 42 *TANYGNATHUS* **PARROTS (in part)**

Midsized to large parrots with large bill and short, rounded tail producing "top-heavy" appearance; little sexual dimorphism, JUV duller. Forests, tall secondary growth, mangroves, plantations; arboreal; singly, pairs, or small groups; quiet while feeding, noisy and conspicuous in labored flight with wingbeats below body level.

GREAT-BILLED PARROT
Tanygnathus megalorhynchos 41cm

Large parrot with mottled black-and-yellow wing-coverts; sexes alike; harsh *kee-rarr*. **DISTRIBUTION** western Papuan, Tanimbar, and Lesser Sunda Islands, and Moluccas to Talaud Islands, Indonesia; probably introduced to southernmost Philippines; up to 1000m; uncommon and favors small islands. **SUBSPECIES** five identifiable subspecies with variation in nominate *megalorhynchos*. 1. *T. m. megalorhynchos* back and rump pale blue; scapulars and lesser wing-coverts black margined blue and green; median wing-coverts black margined yellow; bill red (♂), or white (♀ & JUV). *Range* Talaud and Sangihe Islands to islands off Sulawesi and in Flores Sea, North Moluccas, and western Papuan Islands; probably introduced to Balut and Sarangani, southernmost Philippines. 2. *T. m. affinis* scapulars green; bend of wing blue. *Range* South Moluccas. 3. *T. m. sumbensis* like *megalorhynchos*, but darker blue rump. *Range* Sumba, Lesser Sunda Islands. 4. *T. m. hellmayri* like *affinis*, but bend of wing green. *Range* Roti and Semau Islands and southwestern Timor, Lesser Sunda Islands. 5. *T. m. subaffinis* like *affinis*, but rump green tinged blue. *Range* Babar Island and Tanimbar Islands. **SIMILAR SPECIES** Blue-backed Parrot *T. sumatranus* and Black-lored Parrot *T. gramineus* (plate 43) no black-and-yellow mottled wing-coverts. *Prioniturus* racquet-tailed parrots (plate 39) much smaller with elongated central tail-feathers and blue underwings. Eclectus Parrot *Eclectus roratus* (green ♂; plates 44, 45) no black-and-yellow mottled wing-coverts; red underwing-coverts; different call. **LOCALITIES** Manusela National Park, Seram, and Hitu Peninsula, Ambon, South Moluccas, and Yamdena, Tanimbar Islands.

BLUE-NAPED PARROT
Tanygnathus lucionensis 31cm

Smaller *Tanygnathus* parrot, and only species with blue hindcrown to nape; no sexual dimorphism; repeated *aaa-kkk* and *kek-a-kee*, drawn-out *eeeee*. **DISTRIBUTION** Philippine Islands, Talaud and Sangihe Islands, Indonesia, and introduced to islands off Sabah, north Borneo; up to 300m; near-threatened. **SUBSPECIES** much variation, but blue-backed (1 & 2) and green-backed (3) subspecies. 1. *T. l. lucionensis* hindcrown to nape blue; median wing-coverts black margined yellow-orange; back and rump to upper tail-coverts blue. *Range* Luzon and Mindoro, northern Philippine Islands. 2. *T. l. hybridus* like *lucionensis*, but hindcrown green and nape paler blue tinged violet. *Range* Polillo, northern Philippine Islands. 3. *T. l. talautensis* similar to *lucionensis*, but back and rump to upper tail-coverts green. *Range* central to southern Philippine Islands, Talaud and Sangihe Islands, Indonesia, and introduced to the Maratua, Mantanani, and Siamil Islands, off northern Sabah, north Borneo, with feral population reported at Kota Kinabalu, Sabah. **SIMILAR SPECIES** Great-billed Parrot *T. megalorhynchos* (see above) no blue on hindcrown to nape; larger with massive bill. Blue-backed Parrot *T. sumatranus* (plate 43) no blue on hindcrown to nape; no mottled black-and-yellow wing-coverts. **LOCALITIES** St. Paul's Subterranean National Park, Palawan, southern Philippine Islands. Karakelong, Talaud Islands, where said to be fairly common.

GREAT-BILLED PARROT

T. m. megalorhynchos

Juv

T. m. affinis

BLUE-NAPED PARROT

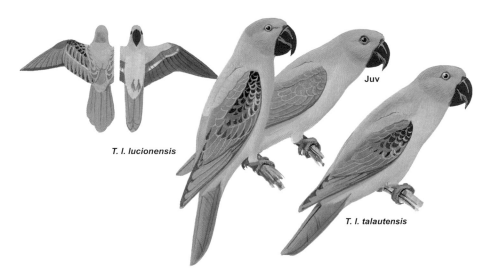

T. l. lucionensis

Juv

T. l. talautensis

PLATE 43 *TANYGNATHUS* **PARROTS (in part)**

106

BLUE-BACKED PARROT
Tanygnathus sumatranus 32cm

Smaller plainly-colored *Tanygnathus* parrot; loud *nyak…nyak*.
DISTRIBUTION Sulawesi and nearby islands, Talaud and Sangihe
Islands, Indonesia, and Philippine Islands; mostly below 500m;
common in Sulawesi, uncommon to rare elsewhere. **SUBSPECIES**
two identifiable and four poorly differentiated subspecies.
1. *T. s. sumatranus* rump and lower back blue; lesser- and primary-
coverts green margined blue; bill red (♂) or white (♀ & JUV).
Range Sulawesi and nearby islands, north to Talaud and east to Sula
Islands. 2. *T. s. sangirensis* like *sumatranus*, but more blue on bend of
wing and lesser wing-coverts. *Range* Talaud and Sangihe Islands, Indonesia. 3. *T. s. burbidgii* rump
and lower back dark blue; head not darker green; mantle not margined dark blue. *Range* Sulu
Archipelago, southern Philippines. 4. *T. s. everetti* like *burbidgii*, but head darker green and mantle
margined dark blue; rump and lower back darker blue. *Range* central to southern Philippines.
5. *T. s. duponti* like *everetti*, but more yellowish neck forming distinct collar. *Range* Luzon, northern
Philippines. 6. *T. s. freeri* like *everetti*, but more yellowish and paler blue margins on mantle. *Range*
Polillo, northern Philippines. **SIMILAR SPECIES** Great-billed Parrot *T. megalorhynchos* (plate 42)
mottled black-and-yellow wing-coverts; red bill in both sexes; larger. Blue-naped Parrot *T. lucionensis*
(plate 42) blue hindcrown to nape; rump and lower back green; mottled wing-coverts. *Prioniturus*
racquet-tailed parrots (plates 39, 40) much smaller with elongated central tail-feathers and blue
underwings. **LOCALITIES** Dumoga-Bone and Lore Lindu National Parks, Sulawesi. Tawi Tawi, Sulu
Archipelago, southern Philippine Islands.

BLACK-LORED PARROT
Tanygnathus gramineus 40cm

Only *Tanygnathus* parrot with blue forehead and crown; black line
from lores to eye; bill red (♂) or white (♀); JUV undescribed.
DISTRIBUTION Buru, South Moluccas, Indonesia; mostly 600 to
1700m; vulnerable and little known. **SIMILAR SPECIES** Great-
billed Parrot *T. megalorhynchos* (plate 42) mottled wing-coverts;
rump and lower back blue; larger with massive bill. Buru Racquet-tailed Parrot *Prioniturus mada*
(plate 39) much smaller with elongated central tail-feathers and blue underwings. Red-cheeked
Parrot *Geoffroyus geoffroyi* (green-headed JUV, plate 35) much smaller with shorter, squarish tail;
blue underwing-coverts and different call. **LOCALITY** Gunung Kelapat Muda, Buru.

BLUE-BACKED PARROT

♂　　　　*T. s. sumatranus*

♀

♂

T. s. everetti

BLACK-LORED PARROT

♂

♀

PLATE 44 **ECLECTUS PARROT (in part)**

108

ECLECTUS PARROT *Eclectus roratus* 35cm

Large stocky parrot with short, squarish tail; distinctive coloration with extraordinary sexual dimorphism—green ♂, red ♀, JUV like adults; hairlike feathers produce sleek, glossy appearance; screeching *kraach-kraak* and flute-like *chu-wee*. Forests, tall secondary growth, plantations; arboreal; singly, pairs, small groups; noisy and conspicuous; distinctive flight with wingbeats below body level.

DISTRIBUTION Moluccas and Lesser Sunda Islands, Indonesia, east through New Guinea to Solomon Islands and Cape York Peninsula, northern Australia; up to 1300m; common; introduced to Gorong Islands, Indonesia, and Palau Archipelago, Micronesia.

SUBSPECIES (in part, see plate 45) nine subspecies separated into three groupings by females having underparts and mantle dull purple (1 & 2), entirely red (3 & 4), or deep blue (5–9); subspecific identification of males difficult. 1. *E. r. roratus* (Grand Eclectus Parrot) ♂ green with red underwing-coverts to sides of underbody; central tail-feathers green above narrowly tipped yellowish white; upper mandible orange, lower mandible black. ♀ red with underwing-coverts to lower breast, abdomen and broad band across mantle dull purple; undertail-coverts tipped yellowish; black bill. *Range* Buru, Seram, Ambon, Haruku, and Saparua, South Moluccas. 2. *E. r. vosmaeri* ♂ more yellowish green, and central tail-feathers blue edged green on outer webs; ♀ like *roratus*, but undertail-coverts yellow, and tail broadly tipped yellow. *Range* North Moluccas. 3. *E. r. cornelia* ♂ like *roratus*, but upperparts paler green, and larger; ♀ entirely red with mauve-blue underwing-coverts, and larger. *Range* Sumba, Lesser Sunda Islands. 4. *E. r. riedeli* ♂ like *roratus* but foreneck and lower cheeks more bluish green, tail broadly tipped yellow, and smaller; ♀ like *cornelia* but undertail-coverts yellow, and smaller. *Range* Tanimbar Islands, Indonesia.

E. r. roratus

♂   ♀

E. r. vosmaeri

♂   ♀

E. r. cornelia

♀

E. r. riedeli

♂   ♀

PLATE 45 **ECLECTUS PARROT (in part) AND PESQUET'S PARROT**

110

ECLECTUS PARROT *Eclectus roratus* (cont.)

SUBSPECIES (in part, see plate 44) 5. *E. r. polychloros* (Red-sided Eclectus Parrot) ♂ like *roratus*, but head and neck darker green; ♀ head and breast brighter red, narrow feathered eye-ring blue, underwing-coverts to lower breast, abdomen, and broad band across mantle deep blue, and no yellow on undertail-coverts. *Range* Kai and western Papuan Islands, Indonesia, east through New Guinea, including offshore islands and Boigu, Dauan, and Saibai, northernmost Torres Strait Islands, Australia, to Trobriand Islands and D'Entrecasteaux and Louisiade Archipelagos, eastern Papua New Guinea; introduced to Gorong, Banda, and Tayandu Islands, Indonesia, and Palau Archipelago, Micronesia. 6. *E. r. biaki* smaller than *polychloros*; ♀ with brighter red upperparts. *Range* Biak Island, West Papua. 7. *E. r. aruensis* like *polychloros*, but ♂ tail more broadly tipped yellow, and ♀ tail brighter red. *Range* Aru Islands, Indonesia. 8. *E. r. macgillivrayi* larger than *polychloros* and with longer tail; ♂ less yellowish green. *Range* Massy Creek and McIllwraith Range north to Pascoe River, eastern Cape York Peninsula, north Queensland, Australia; near-threatened. 9. *E. r. solomonensis* smaller than *polychloros*; ♂ more yellowish green, and outer webs of primaries more brilliant violet-blue; ♀ head and breast brighter red, with breast, lower abdomen, and broad band across mantle brighter blue. *Range* Admiralty Islands and Bismarck Archipelago to Buka and Bougainville Islands, eastern Papua New Guinea, and Solomon Islands. **SIMILAR SPECIES** Palm Cockatoo *Probosciger aterrimus* (plate 1) and Pesquet's Parrot *Psittrichas fulgidus* (see below) similar in distant flight, but predominantly black plumage. Great-billed Parrot *Tanygnathus megalorhynchos* (plate 42) slimmer body with longer, round-tipped tail; mottled black-and-yellow wing-coverts; massive red bill. Superficially similar *Lorius* lories coexist with ♀ Eclectus Parrot, but are much smaller with green wings, and give different calls. **LOCALITIES** Langgaliru-Manipeu, Sumba, Lesser Sunda Islands, and Manusela National Park, Seram, South Moluccas, Indonesia. Brown River Forest Reserve, near Port Moresby, Papua New Guinea. Iron Range National Park, Cape York Peninsula, north Queensland, Australia.

PESQUET'S PARROT *Psittrichas fulgidus* 46cm

Unmistakable; large black-and-red parrot with bare facial skin and projecting, hook-tipped bill giving "vulture-like" appearance; behind eye red spot present (♂ & JUV) or absent (♀); screeching *aaa-aar...aaa-aar*. Lower montane forests and tall secondary growth; arboreal; singly, pairs, small groups; noisy and conspicuous.
DISTRIBUTION mountains of mainland New Guinea; mostly 600 to 1500m; scarce and declining. **SIMILAR SPECIES** Palm Cockatoo *Probosciger aterrimus* (plate 1) similar in distant flight, but no red on wings or underparts. Eclectus Parrot *Eclectus roratus* (see above) similar in distant flight, but green or red plumage. **LOCALITIES** Crater Mountain Research Station, Chimbu Province, and Lake Kopiago, Southern Highlands, Papua New Guinea.

ECLECTUS PARROT

♂

♀

E. r. polychloros

♂

♀

PESQUET'S PARROT

♂

E. r. solomonensis

PLATE 46 — *ALISTERUS* KING PARROTS

112 **Midsized red-and-green parrots with long, broad tail; slight to pronounced sexual dimorphism, JUV like ♀. Forests, woodlands, farmlands, plantations; largely arboreal; pairs, small groups, sometimes flocks; quiet while feeding, but noisy and conspicuous in flight.**

MOLUCCAN KING PARROT
Alisterus amboinensis 35cm

Only king parrot without pale blaze across wing; sexes alike, juveniles differ; high-pitched *kree* (♂), screeching *chack-chack* (♂♀). **DISTRIBUTION** Peleng, Sula, and Moluccan Islands to western Papuan Islands, Indonesia, and western New Guinea; up to 1600m; uncommon. **SUBSPECIES** four well-marked and two slightly differentiated subspecies. 1. *A. a. amboinensis* head and underparts red; mantle to upper tail-coverts deep blue; wings green; tail blue-black, outer feathers edged pink; upper mandible orange, lower black. JUV green mantle and brownish bill. *Range* South Moluccas, except Buru. 2. *A. a. sulaensis* green across upper mantle; no pink in tail. *Range* Sula Islands. 3. *A. a. versicolor* like *sulaensis*, but mantle uniformly blue. *Range* Peleng Island. 4. *A. a. buruensis* mantle mostly green; bill black in both sexes. *Range* Buru, South Moluccas. 5. *A. a. hypophonius* wings blue; no pink in tail. *Range* Halmahera, North Moluccas. 6. *A. a. dorsalis* like *amboinensis*, but no pink in tail. *Range* western Papuan Islands and northwestern New Guinea. **SIMILAR SPECIES** *Tanygnathus* parrots (plates 42, 43) no red in plumage; different calls. *Lorius* lories (plates 18, 19) smaller with short, squarish tail; different calls. **LOCALITIES** Manusela National Park, Seram, South Moluccas, and Kali Batu Putih, Halmahera, North Moluccas, Indonesia.

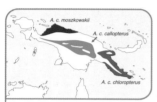

PAPUAN KING PARROT
Alisterus chloropterus 36cm

Sexually dimorphic, ♂ with yellow-green blaze across wing; shrill *keek*, high-pitched *eree…eree*. **DISTRIBUTION** New Guinea, east of Weyland Mountains, West Papua; up to 2600m; locally common. **SUBSPECIES** three well-marked subspecies. 1. *A. c. chloropterus* ♂ head and underparts red, upper mantle to nape blue, and mantle to back black; ♀ & JUV head and upperparts green. *Range* eastern New Guinea. 2. *A. c. callopterus* ♂ blue on mantle not extending up to nape. *Range* central New Guinea. 3. *A. c. moszkowskii* ♀ head and underparts red. *Range* northwestern New Guinea. **SIMILAR SPECIES** Papuan Lorikeet *Charmosyna papou* (plate 24) long, narrow tail with streamer-like central feathers; different calls. *Lorius* lories (plate 18) see above. **LOCALITIES** Brown River Forest Reserve, near Port Moresby, and Wau Ecology Institute, Morobe Province, Papua New Guinea.

AUSTRALIAN KING PARROT
Alisterus scapularis 43cm

Largest king parrot, with silver-green blaze across wing; flute-like *pwee-eet* (♂), shrill *crassak-crassak* (♂♀). **DISTRIBUTION** eastern Australia; up to 1600m; common. **SUBSPECIES** two subspecies differentiated by size. 1. *A. s. scapularis* ♂ head and underparts red, and back and wings green; ♀ & JUV head and breast green. *Range* coastal southeastern and central-eastern Australia. 2. *A. s. minor* smaller size. *Range* northeastern Australia. **SIMILAR SPECIES** Red-winged Parrot *Aprosmictus erythropterus* (plate 47) pale green head and underparts; red "wing-patch." Crimson Rosella *Platycercus elegans* (green JUV, plate 51) violet-blue cheeks; much blue in wings; different call. **LOCALITIES** Healesville Wildlife Sanctuary, near Melbourne, Victoria, and Lamington National Park, southeastern Queensland.

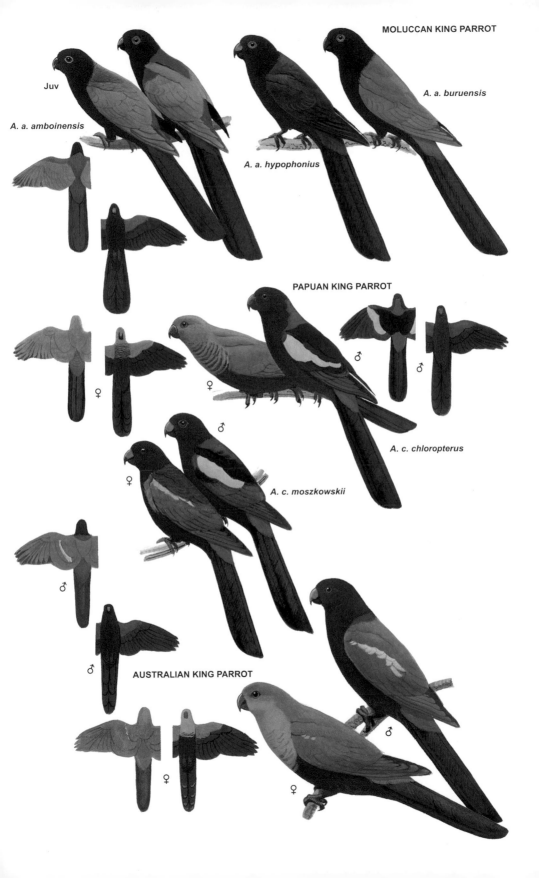

MOLUCCAN KING PARROT

Juv

A. a. amboinensis

A. a. hypophonius

A. a. buruensis

PAPUAN KING PARROT

♂

♂

♀

♀

A. c. chloropterus

♂

A. c. moszkowskii

♀

♂

♂

AUSTRALIAN KING PARROT

♂

♀

♀

PLATE 47 *APROSMICTUS* **PARROTS**

114 Midsized stocky, pale green parrots with short, broad tail and prominent red "wing-patch"; sexes differ, JUV like ♀. Open forest, secondary growth, woodlands; largely arboreal, but will feed on ground; pairs, small flocks; noisy and conspicuous in erratic flight with flapping wingbeats.

OLIVE-SHOULDERED PARROT
Aprosmictus jonquillaceus 35cm

One of two very similar, but geographically isolated parrots; identified by green, not black back of ♂; harsh squawks. **DISTRIBUTION** Timor, Roti, and Wetar, Lesser Sunda Islands, Indonesia; up to 2800m; near-threatened. **SUBSPECIES** two poorly differentiated subspecies. 1. *A. j. jonquillaceus* ♂ mantle and upper back green margined blue, with greenish-yellow inner wing-coverts, and red outer wing-coverts; ♀ & JUV mantle and upper back dull green not margined blue, and inner wing-coverts duller yellowish green. *Range* Timor and nearby Roti Island. 2. *A. j. wetterensis* ♂ inner wing-coverts duller, less yellowish-green. *Range* Wetar Island. **SIMILAR SPECIES** Great-billed Parrot *Tanygnathus megalorhynchos* (plate 42) wing-coverts mottled black and yellow, not red, and massive red bill. Red-cheeked Parrot *Geoffroyus geoffroyi* (green-headed JUV, plate 35) no red on upper wing-coverts; blue underwing-coverts; smaller with shorter, squarish tail; different calls. Olive-headed Lorikeet *Trichoglossus euteles* (plate 14) no red on upper wing-coverts; olive-yellow head; smaller with narrow, pointed tail; different calls. **LOCALITY** Bipolo district and Camplong Reserve, West Timor, Indonesia.

RED-WINGED PARROT
Aprosmictus erythropterus 35cm

Identified by black back of ♂; sharp *crillik-crillik* and *chik-chik-chik*. **DISTRIBUTION** northern and northeastern Australia and southern New Guinea; up to 600m; common. **SUBSPECIES** two poorly differentiated subspecies. 1. *A. e. erythropterus* ♂ mantle, upper back, and scapulars black, upper wing-coverts crimson-red, and hindcrown faintly tinged blue; ♀ & JUV mantle, upper back, scapulars, and inner wing-coverts green, and outer wing-coverts dull red. *Range* northeastern Australia south to inland northern New South Wales and northeastern South Australia. 2. *A. e. coccineopterus* ♂ hindcrown more strongly tinged blue; upper wing-coverts paler red; smaller. *Range* northern Australia, including larger offshore islands, from Kimberley division of Western Australia east to northwestern Queensland, where intergrades with *erythropterus*; also southern New Guinea from Digul River, southeastern West Papua, to Fly River, southwestern Papua New Guinea. **SIMILAR SPECIES** Australian King Parrot *Alisterus scapularis* (plate 46) no red "wing-patch," but red head (♂) and underparts. Papuan King Parrot *Alisterus chloropterus* (plate 46) greenish-yellow "wing-patch"; red head (♂) and underparts. **LOCALITIES** Kakadu National Park, Northern Territory, and Carnarvon National Park, Queensland, Australia. Bensbach Lodge, Trans-Fly region, southwestern Papua New Guinea.

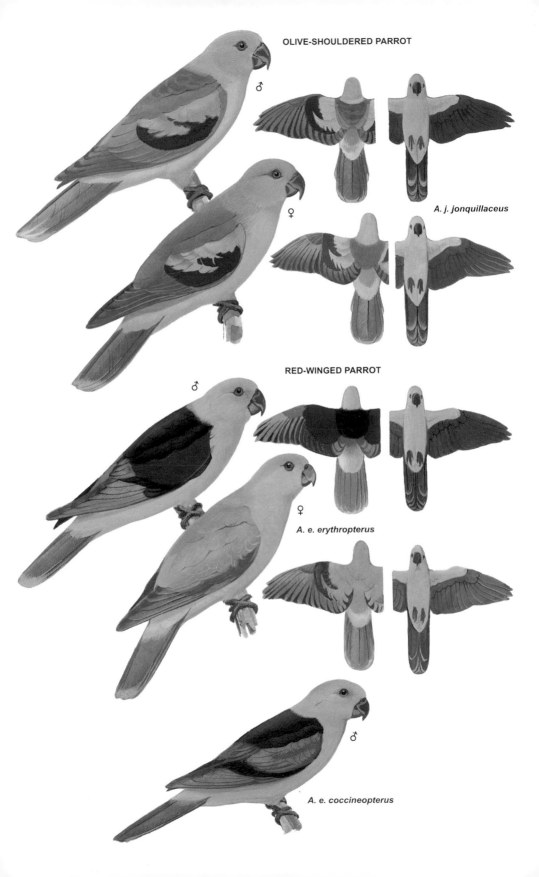

OLIVE-SHOULDERED PARROT

A. j. jonquillaceus

RED-WINGED PARROT

A. e. erythropterus

A. e. coccineopterus

PLATE 48 *POLYTELIS* PARROTS

116 **Sleek, midsized parrots with very long, strongly graduated tail and long, narrow wings giving "streamlined" flight silhouette; slight to pronounced sexual dimorphism, JUV duller. Open forests, riverine woodlands, arid scrublands; pairs, small groups, sometimes flocks at food source; feeding in trees or on ground; seasonal or nomadic movements; swift, direct flight; species replace each other geographically.**

SUPERB PARROT *Polytelis swainsonii* 40cm

Only green *Polytelis* parrot; sexes differ; ♂ yellow face and throat bordered by scarlet band across foreneck; ♀ & JUV face and throat grayish green, no scarlet band on foreneck, and thighs orange; warbling *quee-eel...quee-eel*. **DISTRIBUTION** inland southeastern Australia, from north-central New South Wales south to mid-Murray River and tributaries in northernmost Victoria, mostly below lat. 33°S and breeds only in south of range; up to 650m; vulnerable. **LOCALITIES** Barmah State Park, Murray River, northern Victoria. Cuba State Forest, Murrumbidgee River, southern New South Wales.

REGENT PARROT *Polytelis anthopeplus* 40cm

Unmistakable; only long-tailed yellow parrot with red-and-black wing-coverts and orange bill; rolling *carrack...carrack*. **DISTRIBUTION** southwestern and inland southeastern Australia; locally common in southwest, endangered in southeast. **SUBSPECIES** two poorly differentiated subspecies. 1. *P. a. anthopeplus* ♂ head and underparts dull olive-yellow, mantle and upper back dull olive-green, dull yellow "wing-patch," inner secondary-coverts dull red tipped yellow, and tail black; ♀ head and underparts dull greenish olive, smaller yellow "wing-patch," little red on secondary-coverts, and tail dull olive edged pink on lateral feathers. *Range* southwestern Australia, north to Lake Moore district and east to Balladonia district. 2. *P. a. monarchoides* ♂ head and underparts bright jonquil yellow, brighter yellow "wing-patch," and dark red on secondary-coverts; ♀ head and underparts olive-yellow. *Range* inland southeastern Australia centered on mid Murray and lower Darling Rivers and tributaries. **SIMILAR SPECIES** Yellow Rosella *Platycercus elegans flaveolus* (plate 52) blue cheeks; blue, but no red in wings; white bill; slower, undulating flight; different calls. **LOCALITIES** Hattah-Kulkyne and Wyperfeld National Parks, northwestern Victoria, and Danggali Conservation Park, South Australia.

PRINCESS PARROT *Polytelis alexandrae* 40cm

Unmistakable; distinctive pastel coloration featuring olive upperparts, rose-pink throat, and prominent yellow-green "wing-patch"; ♀ & JUV duller with shorter tail; prolonged harsh clattering. Common cagebird with color mutations. **DISTRIBUTION** interior of western and central Australia, but limits to range not determined; near-threatened, and seldom encountered in vast, remote range. **SIMILAR SPECIES** Cockatiel *Nymphicus hollandicus* (plate 7) similar in distant flight, but gray coloration with white "wing-patch"; smaller. **LOCALITIES** Presence in any locality not predictable, but most recent sightings are along Canning Stock Route, Western Australia.

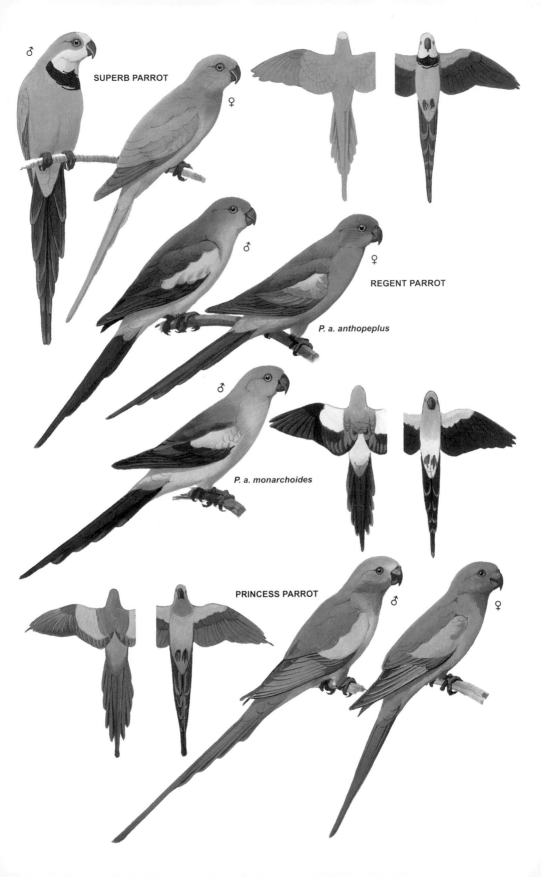

SUPERB PARROT

♂

♀

REGENT PARROT

♂

♀

P. a. anthopeplus

♂

P. a. monarchoides

PRINCESS PARROT

♂

♀

PLATE 49 *BARNARDIUS* PARROTS

118 Midsized stocky green parrots with narrow yellow collar encircling hindneck, and belonging to Australasian "broad-tailed" group characterized by long, but broad, strongly graduated tail; little sexual dimorphism, duller JUV with white underwing-stripe; two species often considered conspecific as Australian Ringneck Parrot. Open forest, woodlands, mallee, arid scrublands; pairs, small groups; mostly ground-feeders; undulating flight.

MALLEE RINGNECK PARROT
Barnardius barnardi 35cm

Eastern *Barnardius* parrot with green head; piping *kwink…kwink* or *pit-tink…pit-tink*. **DISTRIBUTION** interior of eastern Australia; common, but wary. **SUBSPECIES** two well-marked subspecies. 1. *B. b. barnardi* red frontal band; mantle and upper back blue-black (♂) or green tinged olive-brown (♀ & JUV); orange-yellow abdominal band. *Range* interior of southeastern and central-eastern Australia north to inland southern Queensland, where merges with *macgillivrayi*. 2. *B. b. macgillivrayi* (Cloncurry Parrot) paler green, including mantle and upper back; no red frontal band; abdomen lemon yellow. *Range* northwestern Queensland and easternmost Northern Territory, south from lat. 18°30'S to Diamantina River, where intergrades with *barnardi*. **SIMILAR SPECIES** conspicuously larger than green *Psephotus* parrots in range. **LOCALITIES** Corella Dam, Mary Kathleen, northwestern Queensland (*macgillivrayi*). Kinchega National Park, western New South Wales, and Hattah-Kulkyne National Park, northwestern Victoria.

PORT LINCOLN PARROT
Barnardius zonarius 37–42cm

Unmistakable; only black-headed, green parrot in range; lower-pitched *pit-tink…pit-tink*, or *pit-tink-tink…pit-tink-tink* (*semitorquatus*). **DISTRIBUTION** southern, central, and western Australia, west of Flinders Ranges, South Australia, where some hybridization with *B. barnardi*; common and not wary, occurring in urban parklands. **SUBSPECIES** three discernible subspecies, but much intergradation. 1. *B. z. zonarius* no red frontal band; throat and breast dark green sharply demarcated from yellow abdomen; smaller size (37cm). *Range* widespread in central, south-central, and northwestern Australia; broad zone of intergradation with *semitorquatus* in southwest. 2. *B. z. occidentalis* paler coloration, particularly grayish-black head and lemon yellow abdomen. *Range* north-central Western Australia, south and west of Great Sandy Desert. 3. *B. z. semitorquatus* (Twenty-eight Parrot) red frontal band; abdomen to undertail-coverts green; larger (42cm). *Range* southwestern corner of Western Australia. **LOCALITIES** Kulliparu and Hincks Conservation Parks, Eyre Peninsula, South Australia, and West Macdonnell National Park, Northern Territory (*zonarius*). Kings Park, Perth city, Western Australia (*semitorquatus*).

MALLEE RINGNECK PARROT

♂

Juv

B. b. barnardi

B. b. macgillivrayi

♂

PORT LINCOLN PARROT

B. z. semitorquatus

♂

B. z. zonarius

♂

B. z. zonarius × B. z. semitorquatus

♂

PLATE 50 **RED-CAPPED PARROT AND ROSELLAS (in part)**

120

RED-CAPPED PARROT
Purpureicephalus spurius 37cm

Unmistakable; midsized distinctively colored parrot with long graduated tail and peculiarly narrow, projecting bill; variable sexual dimorphism, JUV differs. Forests, woodlands, farmlands, orchards, urban parklands; closely associated with marri *Corymbia calophylla*, the main food tree; adults in pairs, juveniles often in small flocks; feeds in trees and on ground; slightly undulating flight with greenish-yellow rump prominent; harsh *krurr-rak...krurr-rak* (♂♀) and *chek-a-chek* (♂). ♂ forehead to nape deep crimson; cheeks and rump greenish yellow; foreneck to abdomen deep purple-blue; thighs to undertail-coverts red. ♀ variable; some birds like ♂ with green in crown and duller blue breast, others much duller. JUV rust-colored frontal band; crown to nape green; breast and abdomen dull cinnabar-brown. **DISTRIBUTION** southwestern Western Australia north to about lat. 31°30'S and in far southeast to about long. 122°E. **LOCALITIES** Yanchep and Yalgorup National Parks, near Perth, Stirling Range National Park, and orchards in Byford and Denmark districts, Western Australia.

ROSELLAS (in part)

Small to midsized parrots in *Platycercus* known collectively as rosellas, and identified by well-defined cheek-patches and "mottled" backs; sexes alike in two species groups—blue-cheeked group with distinctive JUV plumage and white-cheeked group with JUV like adults, but lone yellow-cheeked species shows pronounced sexual dimorphism and distinctive JUV plumage. Forests, woodlands, farmlands, urban parks or gardens; adults in pairs, juveniles often in small flocks; feeding in trees and on ground; noisy in undulating flight.

GREEN ROSELLA *Platycercus caledonicus* 37cm

Largest rosella, and only predominantly yellow rosella with dark green back; broad red frontal band; violet-blue cheek-patches; mantle and back greenish black, feathers narrowly margined dark green; rump yellow-olive; JUV predominantly dull green with narrow red frontal band and violet-blue cheek-patches; *cussik...cussik* in flight, *kwik-kweek...kwik-kweek* while perched. **DISTRIBUTION** endemic to Tasmania and islands in Bass Strait; up to 1500m; common. **LOCALITIES** Cradle Mountain–Lake St. Clair and Southwest National Parks, and often seen in outskirts of Hobart and Launceston, Tasmania.

RED-CAPPED PARROT

♂

Juv

GREEN ROSELLA

♂

Juv

PLATE 51 ROSELLAS (in part)

122

P. elegans

P. e. nigrescens

P. e. elegans

P. e. melanopterus →

CRIMSON AND YELLOW ROSELLAS
Platycercus elegans 36cm

Midsized polytypic rosellas with violet-blue cheek-patches and red, yellow, or intermixed red-and-yellow coloration; sexes alike, but distinctive JUV; *cussik-cussik…cussik-cussik* in flight, also piping *kwik-kweek-kwik* or *kwink-kweek*. **DISTRIBUTION** eastern and southeastern mainland Australia; up to 1900m; common to abundant; introduced to North Island, New Zealand, and Norfolk Island. **SUBSPECIES** (in part, continued plate 52) five well-marked and one slightly differentiated subspecies; extensive intergradation between three subspecies with hybrid population in Adelaide district, South Australia; more divergent subspecies often treated as separate species, especially in aviculture. 1. *P. e. elegans* (Crimson Rosella) rich crimson-red; violet-blue cheek-patches; mantle and back black, feathers broadly edged crimson-red; flight feathers and tail blue; JUV predominantly dull green, forehead and crown to throat red, violet-blue cheek-patches, and thighs to undertail-coverts red. *Range* eastern and southeastern Australia, in central-eastern Queensland where there are isolated populations, and from southeastern Queensland to southern Victoria and southeastern South Australia; introduced to North Island, New Zealand, and Norfolk Island. 2. *P. e. nigrescens* darker crimson-red, and smaller; JUV like adults, but mantle and upper back black, feathers edged dark red and washed dull green. *Range* northeastern Queensland, from Windsor and Atherton Tableland south to Seaview–Paluma Ranges. 3. *P. e. melanopterus* like *elegans*, but feathers of mantle and upper back more narrowly edged crimson-red; JUV like *elegans*. *Range* Kangaroo Island, South Australia.

P. e. elegans

CRIMSON AND YELLOW ROSELLAS

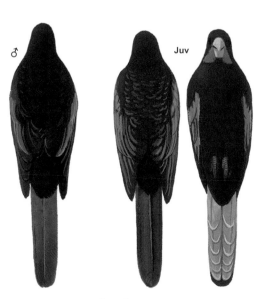

P. e. nigrescens

P. e. melanopterus

PLATE 52 ROSELLAS (in part)

124

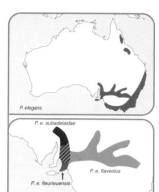

P. elegans

P. e. subadelaidae

P. e. flaveolus

P. e. fleurieuensis

CRIMSON AND YELLOW ROSELLAS
Platycercus elegans (cont.)

SUBSPECIES (in part, continued from plate 51) 4. *P. e. fleurieuensis* deep orange-red, mantle and upper back grayish black with feathers broadly edged dull orange-red, flight feathers paler blue, and tail greenish blue; JUV like *elegans*, but duller, less olive-green, and all red markings replaced by dull orange-red. *Range* confined to Fleurieu Peninsula, South Australia, north to Bungala and Inman Rivers where intergrades with *subadelaidae* throughout Mount Lofty Ranges and Adelaide Plains, and east through lower Murray River region where merges with *flaveolus*; extensive *fleurieuensis* × *subadelaidae* × *flaveolus* population with variable plumage coloration. 5. *P. e. subadelaidae* forehead, crown and face dull orange-red, hindcrown to hindneck dull yellow, underparts dull yellow variably suffused orange-red, mantle and upper back grayish black with feathers broadly edged dull yellow, rump olive-yellow, and tail greenish blue; JUV like *fleurieuensis*, but paler yellowish green, and throat, upper breast and thighs to undertail-coverts dull orange-yellow variably suffused dull orange-red. *Range* restricted to southern Flinders Ranges, South Australia, north to about lat. 32°S; intergrades with *fleurieuensis* throughout Mount Lofty Ranges, Adelaide Plains, and lower Murray River region. 6. *P. e. flaveolus* (Yellow Rosella) bright pale yellow, lores and frontal band orange-red, mantle and upper back black with feathers broadly edged yellow, and rump olive-yellow; JUV pale orange-red frontal band, upperparts pale olive-green, and underparts dull olive-yellow. *Range* riparian distribution centered on Murray–Murrumbidgee–Lachlan Rivers system, from lower Murray River, southeastern South Australia, where intergrades with *fleurieuensis* and *subadelaidae*, east to Riverina region, southern New South Wales, where meets *elegans*, and north on Darling River to Kinchega National Park, western New South Wales. 7. *P. e. fleurieuensis* × *P. e. subadelaidae* × *P. e. flaveolus* (Adelaide Rosella) plumage coloration variable; lores and crown orange-red, nape and sides of head dull orange-yellow, mantle and upper back black with feathers broadly edged olive-yellow and tinged dull orange-red, and underparts yellow variably suffused orange-red; JUV dull olive-green, forehead to crown, upper breast and thighs to undertail-coverts vary from orange-yellow to dull orange-red, and rump yellowish olive. *Range* Mount Lofty Ranges and Adelaide Plains, southeastern South Australia, south to Bungala and Inman Rivers where intergrades with *fleurieuensis*, north to about lat. 32°S where intergrades with *subadelaidae*, and east to lower Murray River region where intergrades with *flaveolus*. **SIMILAR SPECIES** Fully colored adults unmistakable, but green juveniles can be confused with Australian King Parrot *Alisterus scapularis* ♀ & JUV (plate 46) which have no violet-blue cheek-patches and no blue in wings; broad, all-green tail, and different calls. **LOCALITIES** Lane Cove and Royal National Parks, near Sydney, New South Wales, and Dandenong Ranges National Park, near Melbourne, Victoria (Crimson Rosella). Belair National Park and Cleland Conservation Park, near Adelaide, South Australia (Adelaide Rosella). Hattah-Kulkyne National Park, northwestern Victoria (Yellow Rosella).

P. e. fleurieuensis

♂

P. e. subadelaidae

♂

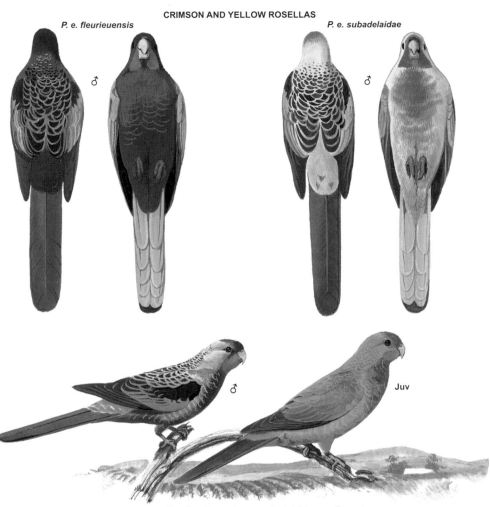

♂

Juv

P. e. fleurieuensis × *e. subadelaidae* × *e. flaveolus*

P. e. flaveolus

♂

Juv

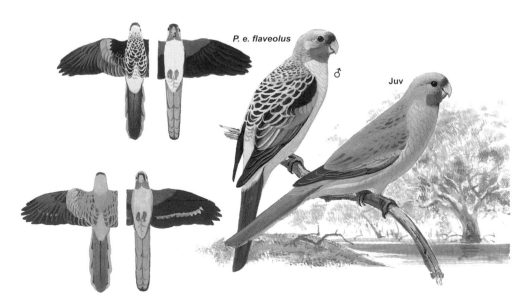

PLATE 53 ROSELLAS (in part)

126

EASTERN ROSELLA *Platycercus eximius* 30cm

Unmistakable; midsized white-cheeked rosella with red head; little sexual dimorphism, JUV duller; piping *kwink…kwink*, sharp *chit-chut…chit-chut* in flight. **DISTRIBUTION** southeastern Australia, including Tasmania; up to 1250m; scarce in Tasmania, abundant elsewhere; introduced to New Zealand. **SUBSPECIES** three slightly differentiated subspecies. 1. *P. e. eximius* mantle and upper back black mottled yellowish green; rump pale green. *Range* northeastern New South Wales south to southeastern South Australia; introduced to New Zealand. 2. *P. e. elecica* (Golden-mantled Rosella) rump greenish blue; back black mottled golden yellow (♂) or greenish yellow (♀). *Range* northeastern New South Wales north to southeastern Queensland. 3. *P. e. diemenensis* larger white cheek-patches; head and breast darker red. *Range* eastern Tasmania; near-threatened. **LOCALITIES** easily seen in most of mainland range.

PALE-HEADED ROSELLA
Platycercus adscitus 30cm

Unmistakable; midsized rosella with variable pale coloration; sexes alike; JUV duller; calls like *P. eximius*. **DISTRIBUTION** northeastern Australia; up to 700m; uncommon in north, common in south. **SUBSPECIES** two subspecies differentiated mainly by rump color, but broad zone of intergradation. 1. *P. a. adscitus* upper cheeks white, lower cheeks blue; abdomen blue; back black mottled yellow; rump yellow. *Range* restricted to Cape York Peninsula, north Queensland. 2. *P. a. palliceps* much variation; rump greenish blue; back black mottled golden yellow, sometimes suffused blue; breast yellowish white or entire underparts blue extending to lower cheeks. *Range* northeastern Queensland south to northeastern New South Wales. **LOCALITIES** Pilliga Nature Reserve, northern New South Wales. Carnarvon Gorge and Lakefield National Parks, Queensland.

WESTERN ROSELLA *Platycercus icterotis* 26cm

Smallest rosella and with yellow cheek-patches; strong sexual dimorphism, JUV duller; soft *kwink-kwink*, much quieter than other rosellas. **DISTRIBUTION** southwestern Western Australia (see map below); uncommon. **SUBSPECIES** two subspecies differentiated by rump and tail color. 1. *P. i. icterotis* ♂ red with back black mottled green, sometimes suffused red, and rump and tail green; ♀ dull orange-red variably suffused green, and pale yellow cheek-patches; JUV green with little orange-red on head and underparts. *Range* coastal and subcoastal southwestern Australia. 2. *P. i. xanthogenys* ♂ rump gray-olive, back mottled red and gray, paler yellow cheeks, and tail dull blue tinged olive; ♀ & JUV olive-gray variably suffused orange-red, and tail dull blue. *Range* interior of southwestern Australia; near-threatened. **LOCALITIES** Yanchep, Stirling Range, and D'Entrecasteaux National Parks, Western Australia.

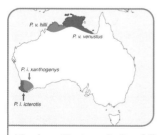

NORTHERN ROSELLA
Platycercus venustus 28cm

Midsized pale yellow rosella with black head, and only rosella in range; sexes alike, JUV duller; calls like *P. eximius*. **DISTRIBUTION** northern Australia; locally common, but generally uncommon. **SUBSPECIES** two subspecies separated by intensity of black scalloping. 1. *P. v. venustus* underparts strongly scalloped black; back black mottled pale yellow; cheek-patches white with little blue. *Range* far northwestern Queensland west to Victoria River, western Northern Territory. 2. *P. v. hilli* underparts finely scalloped black; back black, feathers more broadly edged pale yellow; lower cheeks blue. *Range* Kimberley division of Western Australia east to Victoria River, western Northern Territory. **LOCALITIES** Nitmiluk (Katherine Gorge) and Kakadu National Parks, Northern Territory. Drysdale River National Park, Western Australia.

EASTERN ROSELLA

PALE-HEADED ROSELLA

♂

♂

P. e. eximius

Juv

Juv

P. a. adscitus

♂

P. e. diemenensis

♂

P. e. elecica

♂

P. a. palliceps

WESTERN ROSELLA

♂

P. i. xanthogenys

P. v. venustus

NORTHERN ROSELLA

P. v. hilli

♂

♀

P. i. icterotis

Juv

PLATE 54 — *PSEPHOTUS* PARROTS (in part)

128   **Small parrots with uniformly colored backs and long, graduated tails; no well-marked cheek-patches; two species groups (subgenera): (i) pronounced sexual dimorphism in adults and JUV—*P. haematonotus* and *P. varius*, (ii) pronounced sexual dimorphism, but JUV like female—*P. chrysopterygius* and *P. dissimilis*; third species in second group, Paradise Parrot *P. pulcherrimus* from central-eastern Australia, now extinct (see plate 145). Open woodlands, mallee, arid scrublands, farmlands, urban parklands; ground-feeders; pairs, family parties, sometimes flocks at food source; slightly undulating flight.**

RED-RUMPED PARROT
Psephotus haematonotus 27cm

One of two similar *Psephotus* parrots with bright green ♂ and dull olive ♀; differentiated from *P. varius* by absence of head markings and "wing-patch"; whistling *su-weet…su-weet*. Conspicuous in more open country; often in wintering flocks. **DISTRIBUTION** southeastern mainland Australia, chiefly inland; up to 1000m; very common. **SUBSPECIES** two subspecies separated by intensity of plumage coloration. 1. *P. h. haematonotus* ♂ head, breast and upperparts bright green, abdomen yellow, rump red, and undertail-coverts white; ♀ head, upperparts and breast olive, rump green, and undertail-coverts white washed pale blue; JUV like adults, but much duller. *Range* inland southern Queensland to Victoria and eastern South Australia. 2. *P. h. caeruleus* ♂ paler, more bluish green, and rump paler orange-red; ♀ head, upperparts and breast brownish gray, and abdomen to undertail-coverts white. *Range* Lake Eyre basin and Cooper–Strzelecki Creeks drainages in far southeastern Northern Territory, southwestern Queensland, northwestern New South Wales, and eastern South Australia. **SIMILAR SPECIES** Mulga Parrot *P. varius* (see below) ♂ with yellow frontal band and "wing-patch," and red abdomen; ♀ with red "wing-patch" and green lower underparts. Bluebonnet *Northiella haematogaster* (plate 55) superficially like ♀ *P. haematonotus*, but blue face and underparts yellow marked red; different call. **LOCALITIES** easily seen in or around most urban centers in range.

MULGA PARROT   *Psephotus varius* 28cm

Differentiated from similar *P. haematonotus* by prominent head markings and "wing-patch"; ♂ yellow frontal band and russet-red patch on hindcrown, and abdomen to thighs yellow variably marked red; ♀ russet-red patch on hindcrown, dull red "wing-patch," and abdomen to undertail-coverts pale green; JUV like adults, but much duller; soft *twit-twit*. Quiet and inconspicuous, keeping to covering vegetation; does not form flocks. **DISTRIBUTION** interior of southern mainland Australia, mostly south of lat. 24°S. **SIMILAR SPECIES** Red-rumped Parrot *P. haematonotus* (see above) no head markings or "wing-patch"; more conspicuous in open habitats; different calls. Bluebonnet *Northiella haematogaster* (plate 55) superficially like ♀ *P. varius*, but blue face and underparts yellow marked red. **LOCALITIES** Currawinya National Park, southwestern Queensland. Yathong Nature Reserve, western New South Wales, and Hattah-Kulkyne National Park, northwestern Victoria.

RED-RUMPED PARROT

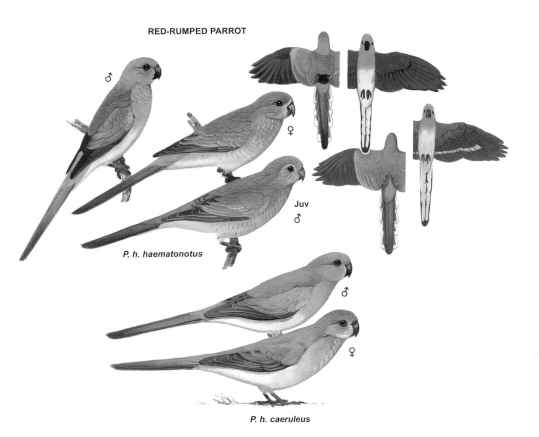

♂

♀

Juv
♂

P. h. haematonotus

♂

♀

P. h. caeruleus

MULGA PARROT

♂

♀

Juv
♂

PLATE 55 *PSEPHOTUS* **PARROTS (in part) AND BLUEBONNET**

130

GOLDEN-SHOULDERED PARROT
Psephotus chrysopterygius 26cm

One of two similar *Psephotus* parrots with black crown, brown upperparts, and yellow "wing-patch" in males; green females very similar. ♂ yellow frontal band, crown black above eyes, median wing-coverts yellow forming small "wing-patch," and abdomen to undertail-coverts orange-red tipped white; ♀ & JUV pale yellow frontal band, crown and occiput tinged brown, and feathers on center of abdomen margined pale red; *few-weep...few-weep* or mellow *fee-oo...fee-oo*. Usually in presence of terrestrial termitaria, where digs nesting tunnel. **DISTRIBUTION** far northeastern Australia, where two or possibly three isolated populations on Cape York Peninsula, north Queensland; endangered, CITES I. **LOCALITY** Artemis Station, Cape York Peninsula (private property).

HOODED PARROT *Psephotus dissimilis* 26cm

Like *P. chrysopterygius*, but ♂ forehead to nape and below eyes black, no yellow frontal band, larger yellow "wing-patch," and undertail-coverts orange-red margined white, but abdomen green; ♀ & JUV forehead to nape and below eyes grayish brown, and undertail-coverts pink margined white, but abdomen pale blue; *chu-weet...chu-weet* or sharp *chissik-chissik*. Usually in presence of terrestrial termitaria where digs nesting tunnels. **DISTRIBUTION** northern Australia, in Northern Territory south to about lat. 15°30'S, and from Daly and Mary Rivers east to western shores of Gulf of Carpentaria; scarce and declining, CITES I. **LOCALITY** Edith Falls, Nitmiluk (Katherine Gorge) National Park, Northern Territory.

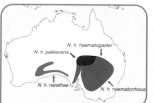

BLUEBONNET *Northiella haematogaster* 30cm

Small parrot with long, graduated tail, and only olive-gray parrot with blue face; sexes alike, JUV like adults; harsh *cluck-cluck*, flute-like *cloote-cloote*, and loud *yak-yak-yak*. Arid and semiarid woodlands or scrublands, mallee, timbered grasslands; feeds on ground; pairs, small parties; wary; erratic, undulating flight. **DISTRIBUTION** interior of southeastern and central-southern Australia; common. **SUBSPECIES** three well-marked subspecies with intergradation, and one distinctive isolate. 1. *N. h. haematogaster* lower underparts yellow with variable red on abdomen; yellowish-olive "wing-patch." *Range* eastern South Australia, east of Lake Eyre basin and North Flinders Ranges, to northwestern Victoria, western New South Wales, and southwestern Queensland. 2. *N. h. haematorrhous* red abdomen to undertail-coverts; brownish-red "wing-patch"; bend of wing pale green (♂) or mauve-blue (♀). *Range* interior of northern New South Wales and southern Queensland. 3. *N. h. pallescens* general plumage coloration much paler than *haematogaster*. *Range* Lake Eyre basin and lower Cooper–Strzelecki Creeks drainages in far southwestern Queensland, northeastern South Australia, and extreme northwestern New South Wales. 4. *N. h. narethae* (Naretha Bluebonnet) two-tone blue face; abdomen and thighs yellow; undertail-coverts red; outer wing-coverts orange-red, duller in ♀; smaller. *Range* southeastern Western Australia west to extreme southwestern South Australia; sometimes considered a separate species. **SIMILAR SPECIES** Red-rumped Parrot *Psephotus haematonotus* and Mulga Parrot *P. varius* females (plate 54) superficially like *N. haematogaster*, but no blue face and no yellow or red on lower underparts; different calls. **LOCALITIES** Kinchega National Park, western New South Wales, and Hattah-Kulkyne National Park, northwestern Victoria. Queen Victoria Spring Nature Reserve, Western Australia (Naretha Bluebonnet).

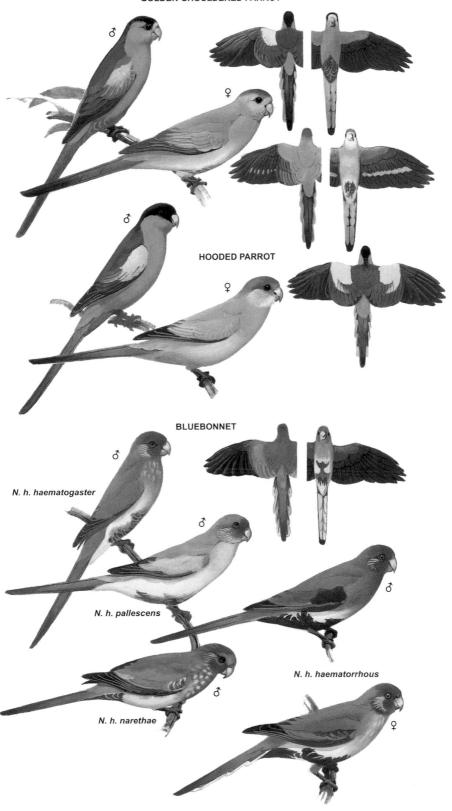

GOLDEN-SHOULDERED PARROT

♂

♀

HOODED PARROT

♂

♀

BLUEBONNET

♂

N. h. haematogaster

♂

N. h. pallescens

♂

N. h. haematorrhous

♂

N. h. narethae

♀

PLATE 56 *NEOPHEMA* **PARROTS (in part)**

132 Small green parrots with long, graduated tail, lateral tail-feathers mostly yellow; two species groups (subgenera) separated by plumage patterns and sexual dimorphism: (i) blue frontal band, no red markings, sexes alike, JUV duller—*N. chrysostoma*, *N. elegans*, *N. petrophila*, and—*N. chrysogaster*; (ii) blue face, red in plumage of males, sexes differ, JUV like ♀—*N. pulchella* and *N. splendida*. Savanna woodland, low shrublands, grasslands; coastal heathlands and saline flats; small parties, flocks; ground-feeders; swift, erratic flight.

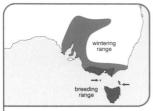

BLUE-WINGED PARROT
Neophema chrysostoma 21cm

One of two very similar olive-green *Neophema* parrots with blue frontal band and blue wing-coverts; appears duller glaucous than *N. elegans*. ♂ crown bronze-yellow, blue frontal band not extending behind eye, upper wing-coverts cobalt-blue, and orange suffusion sometimes on yellow abdomen; ♀ crown dull olive-green, and upper wing-coverts duller blue suffused green; JUV no blue frontal band; metallic tinkling, sharp *tsit-tsit*. **DISTRIBUTION** southeastern Australia; migratory, breeds in Tasmania and southern Victoria, winters northward; up to 1200m; common. **SIMILAR SPECIES** often associates with other similar neophemas, so identification difficult; differences set out below. **LOCALITIES** Cradle Mountain–St. Clair National Park, Tasmania, and Lower Glenelg National Park, southwestern Victoria (breeding range). Coorong National Park, southeastern South Australia (wintering range).

ELEGANT PARROT *Neophema elegans* 22cm

Appears brighter, more yellowish than *N. chrysostoma*; blue frontal band extends behind eye; less, paler blue on wing-coverts; feeble whistle, sharp *tsit-tsit*. **DISTRIBUTION** southwestern and southeastern mainland Australia; mostly below 500m; locally common. Often in mixed flocks with similar species; differences set out above and below. **LOCALITIES** Chinocup Nature Reserve and Lake Dumbleyung, Western Australia. Coorong and Flinders Ranges National Parks, South Australia.

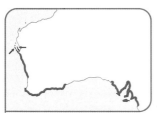

ROCK PARROT *Neophema petrophila* 22cm

Dullest *Neophema* parrot, and only olive-green species with blue face; confined to seaboard and nearshore islands, rarely found more than a few hundred meters from sea. **DISTRIBUTION** coastal southwestern and southern mainland Australia; common. **LOCALITIES** Aldinga Scrub and Nuyts Archipelago, South Australia. Rottnest and Lancelin Islands, and Eyre Bird Observatory, Western Australia.

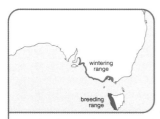

ORANGE-BELLIED PARROT
Neophema chrysogaster 22cm

Only *Neophema* parrot with blue frontal band and bright grass-green upperparts; consistent orange abdominal patch; *tzit* contact call, diagnostic alarm *chitter-chitter* repeated so rapidly to give buzzing sound. Similar neophemas sometimes show orange on abdomen, but not bright green upperparts. **DISTRIBUTION** migratory; breeds western Tasmania, winters on saline flats in coastal southeastern mainland Australia from southeastern South Australia to central Victoria; mostly below 100m; critically endangered, CITES I. **LOCALITIES** Melaleuca Inlet, Southwest National Park, Tasmania (breeding range). Point Wilson, Port Phillip Bay, near Melbourne, Victoria, and Carpenter Rocks, southeastern South Australia (wintering range).

BLUE-WINGED PARROT

♂

♀

Juv

ELEGANT PARROT

♂

Juv

ROCK PARROT

♂

Juv

ORANGE-BELLIED PARROT

♂

Juv

PLATE 57 *NEOPHEMA* PARROTS (in part) AND BUDGERIGAR

134

TURQUOISE PARROT
Neophema pulchella 20cm

One of two allopatric blue-faced *Neophema* parrots with bright grass-green upperparts and red in plumage of ♂; common cagebird with color mutations; ♂ with red "shoulder-patch," absent in ♀ & JUV; soft *tseet-tseet*, high-pitched *zit-zit-zit* alarm call. Favors forest or woodland edges adjoining grasslands; pairs, small parties, wintering flocks; tame. **DISTRIBUTION** southeastern mainland Australia, from southwestern Queensland disjunctly south to eastern and north-central Victoria; up to 700m; near-threatened. **LOCALITIES** Mount Kaputar and Warrumbungles National Parks, northern New South Wales. Chiltern and Warby Ranges State Parks, northern Victoria.

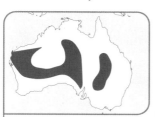

SCARLET-CHESTED PARROT
Neophema splendida 20cm

Another blue-faced *Neophema* parrot with red in plumage of ♂; common cagebird with color mutations; foreneck and center of breast scarlet (♂) or green (♀ & JUV); soft *whick-up…whick-up*. Favors arid *Eucalyptus–Acacia* scrublands, attracted to recently burned areas; quiet and unobtrusive; tame and curious. **DISTRIBUTION** interior of southern mainland Australia, from far southwestern Queensland, western New South Wales, and northwestern Victoria west to southeastern Western Australia; scarce, but periodic irruptions with increasing numbers. **LOCALITIES** Vokes Hill Corner, Unnamed Conservation Park, and Gluepot Reserve, South Australia. Neale Junction and Great Victoria Desert Nature Reserves, eastern Western Australia.

BOURKE'S PARROT
Neopsephotus bourkii 19cm

Unmistakable; small *Neophema*-like parrot with distinctive brown and pink coloration; slight sexual dimorphism, JUV duller; common cagebird with color mutations; soft *chu-wee…chu-wee*. Favors dry mulga *Acacia aneura* woodlands; pairs, small groups; ground-feeder; quiet and inconspicuous, so easily overlooked; active after nightfall and before dawn; tame; fluttering flight. **DISTRIBUTION** interior of central and southern mainland Australia in two apparently isolated populations separated between long. 139°E and 140°E by Flinders Ranges and Simpson Desert. **LOCALITIES** Idalia and Currawinya National Parks, western Queensland. Sturt National Park, western New South Wales. Gibson Desert Nature Reserve, eastern Western Australia.

BUDGERIGAR *Melopsittacus undulatus* 18cm

Unmistakable; only small long-tailed green parrot with boldly barred upperparts; sexes alike, JUV resembles adults; domesticated cagebird with many color mutations; warbling *chedelee…chedelee*. Most open habitats in arid and semiarid regions, favoring eucalypts near water for nesting; noisy, conspicuous flocks in fast twisting flight; ground-feeder. **DISTRIBUTION** interior of mainland Australia, generally south of lat. 17°S; abundant, but highly nomadic; feral populations in Florida, USA. **LOCALITIES** Idalia and Diamantina National Parks, western Queensland. Innamincka and Strzelecki Regional Reserves, northeastern South Australia. Uluru National Park, southern Northern Territory.

TURQUOISE PARROT

♂

♀

SCARLET-CHESTED PARROT

♂

♀

♂

♀

BOURKE'S PARROT

BUDGERIGAR

♂

Juv

PLATE 58   GROUND AND NIGHT PARROTS

136   **Two midsized terrestrial parrots identified by cryptic plumage pattern of green mottled black and yellow; long or short tail comprising narrow, pointed feathers; long tarsi; sexes alike, JUV resembles adults.**

GROUND PARROT *Pezoporus wallicus* 30cm

Unmistakable; sleek parrot with long, sharply graduated tail, short in JUV; calls before dawn and at dusk, *tee…tee…tee…stit* followed by ascending *tee…tee…tee-teeee-ee*. Coastal and contiguous plateau heathlands and sedgelands, where seldom seen unless flushed "quail-like" from cover; shy, elusive; swift, "pheasant-like" flight with uptilted tail. **DISTRIBUTION** coastal southwestern and southeastern Australia, including Tasmania; up to 1300m; stronghold in Tasmania; endangered, CITES I. **SUBSPECIES** two isolated, poorly differentiated subspecies. 1. *P. w. wallicus* bright green with red frontal band, absent in JUV; lower underparts yellow barred black. *Range* disjunctly along coastal southeastern Australia, from about lat. 25°30'S in southeastern Queensland to southern Victoria and extreme southeastern South Australia; also Tasmania and Hunter Island in Bass Strait. 2. *P. w. flaviventris* lower underparts greenish yellow with indistinct, interrupted greenish-brown barring. *Range* disjunctly in three populations along southern coast of Western Australia, from Bow River east to West River and Fitzgerald River National Park, and from Alexander Bay east to Cape Arid. **LOCALITIES** Cooloola National Park, southeastern Queensland. Bundjalung National Park and Nadgee Nature Reserve, eastern New South Wales. Southwest National Park, Tasmania.

NIGHT PARROT *Pezoporus occidentalis* 23cm

Unmistakable; dumpy yellowish-green parrot with short, slightly graduated tail; no red frontal band; reported calls are low whistle, squeaking and croaking notes. Scrublands, spinifex *Triodia* grasslands, and samphire flats in arid zone; most mysterious of Australian birds; nocturnal and secretive; most reports of single birds, but small parties seen coming to drink; flight said to be erratic without undulation. **DISTRIBUTION** reported from scattered localities in arid interior of mainland Australia; recent specimens and reliable reports from western Queensland, between lat. 23°00'S and 20°40'S; critically endangered, CITES I; presumed extinct until rediscovered in 1979 and dessicated carcasses found in 1990 and 2006. **LOCALITIES** no potentially productive sites, but searches continuing in western Queensland, particularly along Boulia–Mount Isa road.

GROUND PARROT

P. w. wallicus

Juv

P. w. flaviventris

NIGHT PARROT

PLATE 59 *PROSOPEIA* AND *EUNYMPHICUS* PARROTS

138 **Collectively termed "shining parrots" because of sheen on back and wings; large parrots with long, broad tail and heavy bill; sexes alike, JUV duller. Forests, tall secondary growth, plantations, gardens; mostly arboreal; singly, pairs, small groups; noisy and conspicuous in flapping flight above canopy, but inconspicuous amidst foliage.**

MASKED SHINING PARROT
Prosopeia personata 47cm

Unmistakable; only green shining parrot, and only large green parrot in range; black face and yellow down center of underparts; grating *raaa* or *krark*, sometimes *kreee*, can be heard from afar and often heard before seen. **DISTRIBUTION** Viti Levu, Fiji Islands; up to 1200m; vulnerable. **LOCALITIES** Colo-i-Suva Forest Park and Tamanivi Nature Reserve, Viti Levu, Fiji.

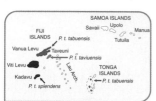

RED SHINING PARROT
Prosopeia tabuensis 45cm

Unmistakable; only large green-and-red parrot in range; loud *nea...nea*, soft *ra-ra-ra-ra*. **DISTRIBUTION** Fiji Islands, introduced to Tonga; up to 1250m; scarce on Viti Levu, locally common elsewhere. **SUBSPECIES** two slightly differentiated and one well-marked subspecies. 1. *P. t. tabuensis* head and underparts deep maroon; variable blue collar across upper mantle, broad and well defined in some birds, but narrow and incomplete in others. *Range* Vanua Levu, Kioa, Koro, and Gau, in Fiji Islands; introduced to Tongatapu, 'Eua, Late and Tofu, in Tonga Group. 2. *P. t. taviuensis* no blue collar across upper mantle; smaller. *Range* Tavieuni, Qamea, and Laucala, in Fiji Islands. 3. *P. t. splendens* (sometimes considered separate species) head and underparts bright scarlet; broad blue band across upper mantle. *Range* Kadavu and nearby Ono, and also Viti Levu, where apparently introduced, Fiji Islands. **LOCALITIES** Vunisea district, Kadavu, and Silktail Lodge, Taveuni, Fiji Islands.

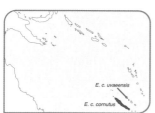

HORNED PARAKEET
Eunymphicus cornutus 32cm

Midsized green parrot with long, broad tail and elongated feathers on crown forming non-erectile "crest"; sexes alike, JUV duller; raucous *ko-kot...ko-kot*. Forests, plantations; arboreal; pairs, family parties, small flocks; shy; noisy in slightly undulating flight above canopy, but quiet and inconspicuous amidst foliage. **DISTRIBUTION** New Caledonia and nearby Ouvéa Island; endangered, CITES I. **SUBSPECIES** two well-marked subspecies. 1. *E. c. cornutus* forehead and forecrown red; ear-coverts and hindneck yellow; face black; coronal "crest" of two elongated feathers black tipped red. *Range* New Caledonia, mainly above 450m. 2. *E. c. uvaeensis* red only on center of forehead; face blackish green; ear-coverts and hindneck green; "crest" of six elongated feathers green. *Range* Ouvéa, in Loyalty Islands. **SIMILAR SPECIES** Red-fronted Parakeet *Cyanoramphus novaezelandiae* (plate 60) no elongated crown feathers; red forecrown and stripe behind eye, but no black face; red patch on side of rump; smaller. **LOCALITIES** Rivière-Bleue Forest Reserve and Reserve Speciale de Faune et de Flore de la Nodela, New Caledonia. Northern region of Ouvéa, Loyalty Islands.

MASKED SHINING PARROT

P. t. tabuensis

RED SHINING PARROT

P. t. splendens

E. c. cornutus

E. c. uvaeensis

HORNED PARAKEET

PLATE 60 *CYANORAMPHUS* **PARROTS (in part)**

140 **Small to midsized green parrots with long, graduated tail and distinctive bicolored silver-gray/dark-gray bill; longer tarsi facilitate walking on ground; ♂ larger than ♀, with larger, broader bill, JUV like adults. Forests, secondary growth, scrublands, tussock grasslands on treeless islands; pairs, small parties, flocks at food source; ground-feeders; fairly swift, slightly undulating flight; distinctive, bleating calls. Two species—Black-fronted Parakeet *C. zealandicus* from Tahiti, Society Islands, and Raiatea Parakeet *C. ulietanus* from Raiatea, Society Islands—now extinct (see plate 145).**

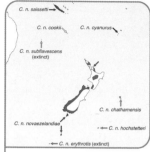

RED-FRONTED PARAKEET
Cyanoramphus novaezelandiae 27cm

Polytypic small to midsized *Cyanoramphus* parrot with red markings on head, and red patch on each side of rump; repeated *kek-kek-kek-kek* or *kek-kik...kek-kik*. **DISTRIBUTION** New Zealand and outlying islands, Norfolk Island, and New Caledonia; up to 1250m; some populations endangered, others locally common; CITES I. **SUBSPECIES** six extant subspecies differentiated by size and slight plumage differences; two subspecies—*C. n. erythrotis* from Macquarie Island and *C. n. subflavescens* from Lord Howe Island, Australia— now extinct. 1. *C. n. novaezelandiae* forecrown, stripe from lores to behind eye, and patch on side of rump red; outer webs of flight feathers violet-blue. *Range* New Zealand, including most offshore islands, and Auckland Islands. 2. *C. n. cyanurus* blue suffusion to upperparts and tail; frontal band red, becoming orange-red on forecrown; underparts bluish green. *Range* Kermadec Islands, north of New Zealand. 3. *C. n. chathamensis* bright emerald-green face; underparts more yellowish green; slightly larger. *Range* Chatham Islands, east of New Zealand. 4. *C. n. hochstetteri* overall more yellowish green; head markings and patch on sides of rump paler orange-red. *Range* Antipodes Islands, south of New Zealand. 5. *C. n. cookii* (sometimes considered separate species) larger than *novaezelandiae*, with broader, heavier bill; forecrown and stripe from lores to behind eye darker red. *Range* Norfolk Island; critically endangered. 6. *C. n. saissetti* (sometimes considered separate species) like *novaezelandiae*, but face and underparts more yellowish; paler, brighter red head markings. *Range* New Caledonia. **SIMILAR SPECIES** Other *Cyanoramphus* parrots (plate 61) different markings on head and sides of rump. Horned Parakeet *Eunymphicus cornutus* (plate 59) New Caledonia; elongated crown feathers forming non-erectile "crest"; black face; no red behind eye or on sides of rump. Crimson Rosella *Platycercus elegans* green JUV (plate 51) on Norfolk Island; violet-blue cheek-patches; red undertail-coverts; pale bill; larger; different calls. **LOCALITIES** Little Barrier and Kapiti Islands, New Zealand. Mount Pitt National Park, Norfolk Island. Rivière-Bleue Forest Reserve, New Caledonia.

C. n. novaezelandiae

C. n. cyanurus

C. n. hochstetteri

C. n. cookii

C. n. saissetti

PLATE 61 *CYANORAMPHUS* **PARROTS (in part)**

142

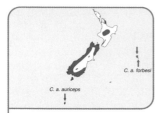

YELLOW-FRONTED PARAKEET
Cyanoramphus auriceps 23cm

Differentiated from similar C. *novaezelandiae* by yellow crown and lack of red stripe behind eye; repeated *ki-ki-ki-ki*, weaker and more high-pitched than C. *novaezelandiae*. **DISTRIBUTION** New Zealand, including offshore and outlying islands; up to 1250m; uncommon to endangered. **SUBSPECIES** two subspecies separated by size and slight plumage differences. 1. C. *a. auriceps* red frontal band to lores and eye, but not behind eye; forecrown yellow; red patch on each side of rump. *Range* New Zealand, including some offshore islands, and outlying Auckland Islands. 2. C. *a. forbesi* (sometimes considered separate species) sides of face bright emerald-green; red frontal band not extending to eye; larger. *Range* outlying Chatham Islands; endangered, CITES I. **SIMILAR SPECIES** easily confused with two sympatric *Cyanoramphus* parrots. Red-fronted Parakeet C. *novaezelandiae* (plate 60) red forecrown and stripe behind eye. Orange-fronted Parakeet C. *malherbi* (see below) probably indistinguishable in field; orange frontal band and orange patch on each side of rump. **LOCALITIES** Little Barrier and Hen and Chicken Islands, North Island, and Nelson Lakes and Arthur's Pass National Parks, South Island, New Zealand.

ORANGE-FRONTED PARAKEET
Cyanoramphus malherbi 20cm

Smallest *Cyanoramphus* parrot, differentiated from similar C. *auriceps* by paler yellow forecrown, orange frontal band, and orange patch on each side of rump; sometimes considered a color morph of C. *auriceps*, and interbreeding recorded; calls not different from C. *auriceps*. **DISTRIBUTION** very few scattered recent records from north of South Island, New Zealand; early reports from North Island and Stewart Island; mostly 600 to 900m; endangered. **SIMILAR SPECIES** Probably indistinguishable from Yellow-fronted Parakeet C. *auriceps* in field (see above). Red-fronted Parakeet C. *novaezelandiae* (plate 60) see above. **LOCALITIES** recent records from Hawdon River valley, in Arthur's Pass National Park, and in Hope River district, North Canterbury, South Island, New Zealand.

ANTIPODES GREEN PARAKEET
Cyanoramphus unicolor 30cm

Largest *Cyanoramphus* parrot, identified by all-green plumage without red or yellow markings; deep, resonant *kok-kok-kok-kok*. **DISTRIBUTION** outlying Antipodes Islands, New Zealand; common. **SIMILAR SPECIES** Red-fronted Parakeet C. *novaezelandiae* (plate 60) see above. **LOCALITY** easily seen on Antipodes Islands, but access only with permit from New Zealand wildlife authorities.

YELLOW-FRONTED PARAKEET

C. a. auriceps

C. a. forbesi

ORANGE-FRONTED PARAKEET

ANTIPODES GREEN PARAKEET

PLATE 62 *NESTOR* and *STRIGOPS* PARROTS

144 Large stocky parrots with narrow, projecting bill and short, squarish tail; shafts of
tail-feathers with projecting spine-like tips; sexes alike, JUV resembles adults.
N. notabilis in montane woodlands, highland valleys, subalpine scrublands, and
alpine grasslands; *N. meridionalis* in lowland forests; adults in pairs or small parties,
juveniles sometimes in flocks; feeding in trees and on ground; strong flight with
flapping wingbeats. Third species—Norfolk Island Kaka *N. productus* from Norfolk
Island—now extinct (see plate 146).

KEA *Nestor notabilis* 48cm

Unmistakable; olive-green parrot with orange-red back to rump
and underwing-coverts; undersides of flight feathers barred orange-
yellow; far-carrying *keee-aah*. Noisy, conspicuous, and highly
inquisitive; tame and extremely bold around ski lodges, causing
damage to buildings, tents, or parked cars; often in circling flight in
blustery winds. **DISTRIBUTION** mountains of South Island, New
Zealand, from Fiordland north to Nelson and Marlborough Provinces; mostly 600 to 3000m;
vulnerable. **SIMILAR SPECIES** Kaka *N. meridionalis* (see below) predominantly brownish with
russet-red on lower underparts; different calls; usually at lower elevations. **LOCALITIES** Mount
Cook and Arthur's Pass National Parks, South Island, New Zealand.

KAKA *Nestor meridionalis* 45cm

Distinctive brown-and-red coloration; massive, projecting bill with
elongated upper mandible; grating *kraa-aa*, ringing *uu-wiiaa, chok ...
chok ... chok*. Strips away bark or digs into decaying wood to extract
insect larvae, so presence indicated by accumulation of debris and
discarded branches. **DISTRIBUTION** New Zealand; mostly 450 to
1000m; vulnerable. **SUBSPECIES** two slightly differentiated
subspecies. 1. *N. m. meridionalis* back, wings, and tail olive-brown; forehead to occiput grayish-white;
neck, abdomen, rump, and underwing-coverts brownish red. *Range* South and Stewart Islands and
larger offshore islands. 2. *N. m. septentrionalis* generally duller; forehead to occiput pale gray. *Range*
North Island and some larger offshore islands. **SIMILAR SPECIES** Kea *N. notabilis* (see above) olive-
green without red on neck or underparts; different calls; occurs only on South Island, normally at
higher elevations and in different habitats. **LOCALITIES** Halfmoon Bay, Stewart Island, Nelson
Lakes National Park, South Island, and Kapiti Island Reserve, North Island, New Zealand.

KAKAPO *Strigops habroptila* 65cm

Unmistakable: very large bulky parrot with distinctive mottled
plumage; upperparts green irregularly mottled black, brown, and
yellow, underparts yellow irregularly mottled paler yellow and buff-
brown; yellowish-brown facial disc gives "owl-like" appearance;
♀ much smaller than ♂; JUV duller, less yellowish, particularly
brownish underparts, and pointed, not rounded tip to outermost
primaries; loud booming by ♂ at display arenas, also squeals and shrieks. Forest substrate and
adjoining scrublands; solitary; terrestrial, nocturnal, and flightless; well-defined tracks and
excavated, bowl-like display arenas telltale signs of presence of birds. **DISTRIBUTION** no known
natural populations; surviving birds translocated to predator-free Maud, Inner Chetwode, Pearl, and
Codfish Islands, New Zealand; critically endangered, CITES I. **LOCALITIES** access to islands only
with permission from New Zealand wildlife authorities.

KEA

KAKA

N. m. meridionalis

N. m. septentrionalis

KAKAPO

Juv

ad

Juv

PLATE 63 *CORACOPSIS* PARROTS

146

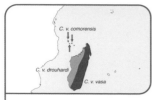

VASA PARROT *Coracopsis vasa* 50cm

Larger of two similar all-brownish parrots with long, rounded tail; sexes alike, JUV resembles adults; prolonged *pee-aw*, raucous *kraaar*, *cho-cho-chi-chi-chi* song. Most wooded habitats, favoring less dense forest and brush than *C. nigra*; noisy and gregarious; normally small flocks, but large flocks at nighttime roost and food source; feeds in trees and on ground; tame; active on moonlit nights; corvid-like flight with flapping wingbeats. **DISTRIBUTION** Madagascar and Comoro Islands; up to 1000m; common. **SUBSPECIES** three slightly differentiated subspecies. 1. *C. v. vasa* dark grayish brown; undertail-coverts gray streaked black; breeding ♀ without feathers on head exposing orange-yellow skin. *Range* eastern Madagascar, westward intergrading with *drouhardi*. 2. *C. v. drouhardi* undertail-coverts and undertail grayish white. *Range* western Madagascar. 3. *C. v. comorensis* underparts more brownish; undertail-coverts grayish brown. *Range* Grand Comoro, Moheli, and Anjouan, in Comoro Islands. **SIMILAR SPECIES** Black Parrot *C. nigra* (see below) smaller; different flight pattern; different calls. **LOCALITIES** Montagne d'Ambre National Park and Ambohitantely Special Reserve, Madagascar.

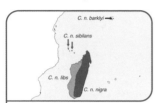

BLACK PARROT *Coracopsis nigra* 35cm

Smaller than similar *C. vasa*; sexes alike, JUV resembles adults; whistling *wee-too-wee* and *too-it...too-it*, sharp *wit-wit-wit*, harsh *caark*. Woodlands, favoring denser forest and brush than *C. vasa*; on Praslin Island closely associated with remnant palm forest; small flocks; feeds more in trees than *C. vasa*, but also on ground; wary; active on moonlit nights; graceful flight with rhythmic wingbeats and long glides. **DISTRIBUTION** Madagascar, Comoro Islands, and Praslin Island in Seychelles; up to 2000m; generally common, though less numerous than *C. vasa*; endangered on Praslin. **SUBSPECIES** four slightly differentiated subspecies. 1. *C. n. nigra* blackish brown; outer webs of primaries gray. *Range* eastern Madagascar, westward intergrading with *libs*. 2. *C. n. libs* paler; underparts more brownish. *Range* western Madagascar. 3. *C. n. sibilans* paler brown; no gray on outer webs of primaries. *Range* Grand Comoro and Anjouan, Comoro Islands. 4. *C. n. barklyi* like *sibilans*, but bluish gray on outer webs of primaries; crown inconspicuously streaked. *Range* Praslin Island, Seychelles; critically endangered. **SIMILAR SPECIES** Vasa Parrot *C. vasa* (see above) larger; different flight pattern; different calls. **LOCALITIES** Montagne d'Ambre National Park and Ranomafana Forest Reserve, Madagascar. Vallée de Mai Reserve, Praslin Island, Seychelles.

VASA PARROT

nesting
♀

C. v. vasa

C. v. drouhardi

C. v. comorensis

BLACK PARROT

C. n. nigra

C. n. barklyi

PLATE 64 GRAY PARROT AND *POICEPHALUS* PARROTS (in part)

148

GRAY PARROT *Psittacus erithacus* 33cm

Unmistakable; large stocky gray parrot with short, squarish red tail; sexes alike, JUV resembles adults; harsh *kraark...kraark, rak...rak...rak, kree-aark...kree-aark*. Forests, secondary growth, mangroves, plantations; noisy, conspicuous flocks; arboreal; swift flight with shallow wingbeats. **DISTRIBUTION** West and central Africa; up to 2200m; locally common, scarce in west. **SUBSPECIES** two well-differentiated subspecies. 1. *P. e. erithacus* pale gray; bare face white; tail red; bill black. *Range* central Africa, from western Uganda to southeastern Ivory Coast. 2. *P. e. timneh* darker gray; tail dark maroon; bill dark red tipped black; smaller. *Range* Guinea-Bissau to southern Ivory Coast. **LOCALITIES** Yapo Forest, southern Ivory Coast. Dzanga-Ndoki National Park, Central African Republic. Rwenzori National Park, western Uganda.

POICEPHALUS PARROTS

Midsized to large stocky parrots with short, squarish tail and stout bill; slight to strong sexual dimorphism, JUV duller. Forests, woodlands, secondary growth, scrublands, plantations; mostly arboreal; pairs, small groups, larger flocks at food source; swift, direct flight.

BROWN-NECKED PARROT
Poicephalus robustus 33cm

Largest *Poicephalus* parrot; green with brownish or silvery-gray head; slight sexual dimorphism; raucous *zzk-eek* or *zwree-enk*. **DISTRIBUTION** southern and west-central Africa; up to 3750m; uncommon, endangered in south. **SUBSPECIES** one well-marked and two slightly differentiated subspecies. 1. *P. r. robustus* (Cape Parrot) head and neck olive-brown; ♀ pink-red forecrown. *Range* eastern Republic of South Africa, from Knysna to southeastern Transvaal and western Swaziland; endangered and dependent on montane *Podocarpus* forest. 2. *P. r. suahelicus* (sometimes considered separate species) head and neck silvery gray. *Range* northern Transvaal, Republic of South Africa, and northern Namibia to central Tanzania and southern Congo River basin. 3. *P. r. fuscicollis* like *suahelicus*, but more bluish green. *Range* Gambia and southern Senegal to northern Ghana and Togo. **SIMILAR SPECIES** Jardine's Parrot *P. gulielmi* (see below) green head and neck. **LOCALITIES** Hlabeni and Ingeli Forests, southern KwaZulu-Natal, and Kruger National Park (north), Transvaaal, Republic of South Africa. Kiang West National Park, Gambia.

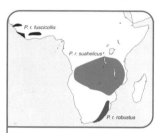

JARDINE'S PARROT *Poicephalus gulielmi* 32cm

Midsized green parrot with red forecrown; sexes alike; high-pitched *scru-ee-at...char-reek...scru-ur-reeeat*, whistling *teee-oo*. **DISTRIBUTION** central and west-central Africa; up to 3500m; common. **SUBSPECIES** three slightly differentiated subspecies. 1. *P. g. gulielmi* forecrown orange-red. *Range* Congo River basin, from southwestern Uganda to northern Angola and southeastern Nigeria. 2. *P. g. fantiensis* forecrown orange. *Range* Liberia to southern Ghana. 3. *P. g. massaicus* only forehead orange-red. *Range* highlands of southern Kenya and northern Tanzania. **SIMILAR SPECIES** Brown-necked Parrot *P. robustus* (see above) head and neck silvery gray. **LOCALITIES** Arusha National Park, Tanzania. Korup National Park, Cameroon. Bia National Park, Ghana.

YELLOW-FACED PARROT
Poicephalus flavifrons 32cm

Midsized green parrot with yellow face; sexes alike; harsh shrieks. **DISTRIBUTION** highlands of Ethiopia; 1000 to 3000m; locally common. **SUBSPECIES** two poorly differentiated subspecies. 1. *P. f. flavifrons* face and crown yellow. *Range* highlands in central Ethiopia. 2. *P. f. aurantiiceps* face yellow suffused orange. *Range* highlands in southwestern Ethiopia. **LOCALITIES** Menagesha Forest, near Addis Ababa, and Bale Mountains National Park, Ethiopia.

GRAY PARROT

P. e. erithacus

P. e. timneh

BROWN-NECKED PARROT

♂

♀

P. r. robustus

P. r. fuscicollis

♂

JARDINE'S PARROT

P. g. massaicus

P. g. gulielmi

P. g. fantiensis

YELLOW-FACED PARROT

Juv

P. f. flavifrons

Juv

P. f. aurantiiceps

PLATE 65 *POICEPHALUS* **PARROTS (in part)**

150

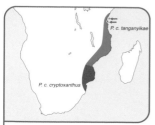

BROWN-HEADED PARROT
Poicephalus cryptoxanthus 22cm

Slightly smaller of two similar green *Poicephalus* parrots with brown head; sexes alike; strident *chree-oo…chree-oo*, sharp *kreek*, shrill *chaa-chaa…chaa-chaa*. **DISTRIBUTION** southeastern Africa; up to 1200m; common, rare on Zanzibar. **SUBSPECIES** two poorly differentiated subspecies. 1. *P. c. cryptoxanthus* head and neck dusky brown, not extending to upper breast; underwing-coverts yellow. *Range* northeastern Republic of South Africa to southern Moçambique and southeastern Zimbabwe. 2. *P. c. tanganyikae* paler green; ear-coverts silvery brown. *Range* Moçambique, north from about Save River, and southern Malawi to eastern Tanzania, including Zanzibar and Pemba Islands, and coastal Kenya north to about lat. 2°S. **SIMILAR SPECIES** Meyer's Parrot *P. meyeri* (see below) yellow on bend of wing, thighs, and mostly on crown; rump greenish blue. **LOCALITIES** Kruger National Park, Transvaal, Republic of South Africa. Lengwe National Park, Malawi. Selous Game Reserve, Tanzania.

NIAM-NIAM PARROT *Poicephalus crassus* 25cm

Slightly larger than similar *P. cryptoxanthus*; head and neck olive-brown, extending to upper breast; underwing-coverts green; favors *Adina-Syzygium* forest; sharp *scree-oot*. **DISTRIBUTION** west-central Africa, between lat. 4°N and 7°N, from southernmost Chad, possibly eastern Cameroon, to northernmost Democratic Republic of Congo and southwestern Sudan; up to 1000m; locally common, but little known. **SIMILAR SPECIES** Meyer's Parrot *P. meyeri* (see above). **LOCALITY** Bozoum district, western Central African Republic.

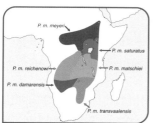

MEYER'S PARROT *Poicephalus meyeri* 21cm

One of three smaller brown *Poicephalus* parrots, but only one with yellow on crown; diagnostic combination of green underparts and yellow on bend of wing; sexes alike; high-pitched *chee-chee-chee*, growling *kraw-er…kraw-er, klink-kleep…chee-wee…chee-wee* from duetting pairs. **DISTRIBUTION** central and eastern Africa; up to 2200m; most common parrot in much of range. **SUBSPECIES** six subspecies in two groupings: paler subspecies (1–3) from drier regions, darker subspecies (4–6) from humid regions. 1. *P. m. meyeri* yellow across crown; lower underparts green; thighs, bend of wing, lesser wing-coverts and underwing-coverts yellow. *Range* northeastern Cameroon and southern Chad to southern Sudan, western Ethiopia and southwestern Eritrea. 2. *P. m. transvaalensis* paler; little or no yellow on crown; lower underparts more bluish. *Range* southern Zambia and northern Moçambique to northeastern Botswana and south to Transvaal, Republic of South Africa. 3. *P. m. damarensis* paler than *transvaalensis*; no yellow on crown. *Range* northwestern Botswana west to central Namibia and southern Angola. 4. *P. m. reichenowi* darker brown; no yellow on crown. *Range* northern and central Angola to southwestern Democratic Republic of Congo. 5. *P. m. matschiei* like *reichenowi*, but with yellow across crown; rump bright blue. *Range* southeastern Democratic Republic of Congo, central and northern Zambia, and northern Malawi to southwestern Tanzania. 6. *P. m. saturatus* like *matschiei*, but head and upperparts darker brown; rump green. *Range* western Tanzania to Uganda and central Kenya. **SIMILAR SPECIES** Rüppell's Parrot *P. rueppellii* (plate 66) no green on underparts; rump brown (♂) or bright blue (♀). Other smaller *Poicephalus* parrots (plate 66) and Brown-headed Parrot *P. cryptoxanthus* (see above) no yellow on crown or "shoulders." **LOCALITIES** Kafue National Park, Zambia. Cangandala National Park, Angola. Lake Baringo, Kenya.

BROWN-HEADED PARROT

P. c. cryptoxanthus

NIAM-NIAM PARROT

Juv

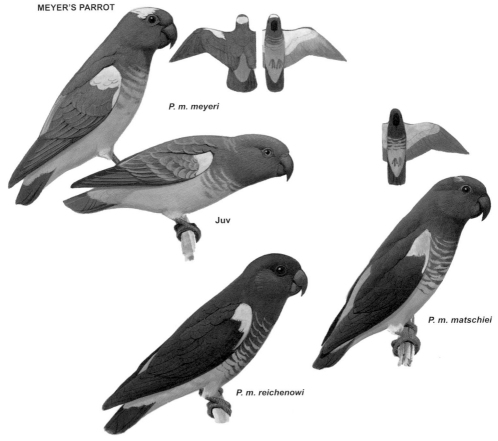

MEYER'S PARROT

P. m. meyeri

Juv

P. m. matschiei

P. m. reichenowi

PLATE 66 *POICEPHALUS* **PARROTS (in part)**

152

RÜPPELL'S PARROT
Poicephalus rueppellii 22cm

Another small brown *Poicephalus* parrot with yellow underwing-coverts, thighs, and "shoulders," but not on crown; only species without green in plumage; strong reverse sexual dimorphism—rump and lower underparts brown (♂) or blue (♀); JUV duller than ♀; sharp *quaw*, shrill alarm screech. **DISTRIBUTION** southwestern Africa from Luanda, coastal northern Angola, south to central Namibia; up to 1250m; common. **SIMILAR SPECIES** Meyer's Parrot *P. meyeri* (plate 65) green underparts; green to greenish-blue rump. **LOCALITIES** Iona National Park and Namibe Regional National Park, Angola. Waterburg Plateau National Park and Namib Desert Park, Namibia.

RED-BELLIED PARROT
Poicephalus rufiventris 22cm

Only small brown *Poicephalus* parrot without yellow thighs or "shoulders"; strong sexual dimorphism—♂ lower breast to abdomen and underwing-coverts orange, ♀ lower breast to abdomen and underwing-coverts green; high-pitched *cree-eeak...cree-eeak*, guttural *cree-krat*. **DISTRIBUTION** Somalia to southern and eastern Ethiopia and extreme northeastern Tanzania; mostly 800 to 2000m; common. **SIMILAR SPECIES** Meyer's Parrot *P. meyeri* (plate 65) yellow thighs, "shoulders," and underwing-coverts. **LOCALITIES** Tarangire National Park, northeastern Tanzania. Tsavo East National Park, southeastern Kenya. Nechisar National Park, southern Ethiopia.

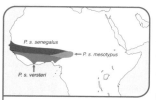

SENEGAL PARROT
Poicephalus senegalus 23cm

Unmistakable; only green *Poicephalus* parrot with gray head and yellow breast; sexes alike; *scree-eeat...scree-eeat*, metallic *cree-lelele...cree-lelele...stee-ow...stee-ow*, whistling *pew-eeo...pew-eeo*. **DISTRIBUTION** central-western Africa; up to 1200m; common. **SUBSPECIES** three poorly differentiated subspecies. 1. *P. s. senegalus* head brownish gray, darker on crown; lower breast and abdomen yellow centrally tinged orange; underwing-coverts and undertail-coverts yellow. *Range* dry zone of West Africa, mostly north of lat. 9°N, from southern Mauritania east to southern Burkina Faso and northern Nigeria. 2. *P. s. versteri* lower breast to abdomen centrally orange-red. *Range* humid zone of western Africa, north of rainforest belt to about lat. 9°N, from Liberia and Ivory Coast east to western Nigeria. 3. *P. s. mesotypus* paler than *senegalus*; green of breast extending farther down toward orange abdomen. *Range* eastern and northeastern Nigeria to northern Cameroon, southwestern Chad, and possibly extreme northeastern Central African Republic. **LOCALITIES** Kiang West National Park, Gambia. Mole National Park, northern Ghana. Parcs Nationaux du W, southern Niger, northern Benin, and northern Burkina Faso. Gashaka-Gumti National Park, eastern Nigeria.

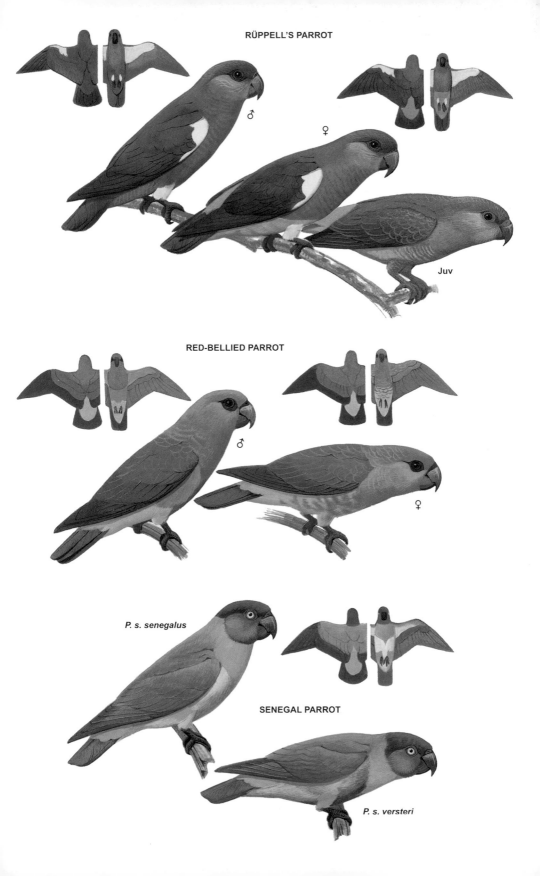

RÜPPELL'S PARROT

♂

♀

Juv

RED-BELLIED PARROT

♂

♀

P. s. senegalus

SENEGAL PARROT

P. s. versteri

PLATE 67 **AFRICAN LOVEBIRDS (in part)**

154 Small stocky green parrots with stout bill and very short, rounded tail; species in
two groupings—(i) with white eye-ring and sexes alike, (ii) without white eye-ring
and sexes differ, plus one intermediate species (*A. roseicollis*) and one aberrant,
little known species (*A. swindernianus*). Forests, open woodlands, savanna
grasslands, cultivation; noisy flocks; two species arboreal, others feed on ground;
swift, direct flight.

GRAY-HEADED LOVEBIRD
Agapornis canus 14cm

Unmistakable; only gray-headed (♂) or all-green (♀) lovebird with
white bill, and only small green parrot in range; shrill *plee...plee*,
subdued chattering. **DISTRIBUTION** Madagascar; up to 1000m;
common; introduced to Comoro Islands, Rodrigues, Reunion, and
Seychelles. **SUBSPECIES** two slightly differentiated subspecies.
1. *A. c. canus* head, neck, and breast pale gray (♂) or green (♀); underwing-coverts black (♂) or
green (♀); JUV like adults, but ♂ head gray suffused green. *Range* Madgascar, except southwest;
introduced to Comoro Islands, Rodrigues, Reunion, and Seychelles. 2. *A. c. ablectaneus* darker bluish
green; ♂ head and breast gray tinged violet. *Range* southwestern Madagascar. **LOCALITIES**
Tsimbazaza Park, in center of Antananarivo, and Toliara-Morombe road, Madagascar. Suburbs of
Victoria, Mahé, Seychelles (feral population).

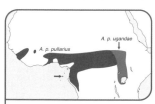

RED-FACED LOVEBIRD
Agapornis pullarius 15cm

Only red-faced, red-billed lovebird without white eye-ring;
twittering *si-si-si-si-si*, trilling *screet-eet...screet-eet*, whistling
tchiri...tchiri. **DISTRIBUTION** central and central-western Africa; up
to 1500m; uncommon. **SUBSPECIES** two poorly differentiated
subspecies. 1. *A. p. pullarius* face and throat orange-red (♂) or orange
(♀); rump cobalt-blue; underwing-coverts black (♂) or green (♀). *Range* western and central
Africa, east to southern Sudan and westernmost Uganda; also on São Tomé Island, in Gulf of
Guinea. 2. *A. p. ugandae* rump paler blue. *Range* east-central Africa. **SIMILAR SPECIES** Black-
winged Lovebird *A. taranta* (see below) red only on forecrown (♂) or no red on head (♀); green
rump; usually at higher elevations. Fischer's Lovebird *A. fischeri* (plate 68) prominent white eye-ring;
yellow neck and breast. Black-collared Lovebird *A. swindernianus* (plate 69) no red on face; black
bill; exclusively arboreal in forests. **LOCALITIES** Gashaka-Gumti National Park, Nigeria. Bamingui-
Bangoran National Park, Central African Republic.

BLACK-WINGED LOVEBIRD
Agapornis taranta 17cm

Largest lovebird; ♂ red forecrown and around eyes, black flight
feathers and underwing-coverts; ♀ only all-green, red-billed
lovebird; JUV like ♀, but black underwing-coverts in young ♂;
high-pitched *kseek...kseek*. Commonly associated with highland
Juniperus-Podocarpus forests; arboreal. **DISTRIBUTION** highlands of
Ethiopia and southern Eritrea; mostly 1600 to 3800m; fairly common. **SIMILAR SPECIES** Red-faced
Lovebird *A. pullarius* (see above) red (♂) or orange (♀) face and throat; green flight feathers; blue
rump; usually at lower elevations. **LOCALITIES** wooded areas in and around Addis Ababa, Ethiopia,
and Asmara, Eritrea. Simien Mountains National Park, Ethiopia.

GRAY-HEADED LOVEBIRD

A. c. canus

♂

♀

Juv
♂

A. c. ablectaneus

♂

RED-FACED LOVEBIRD

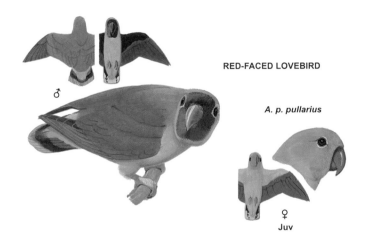

♂

A. p. pullarius

♀
Juv

BLACK-WINGED LOVEBIRD

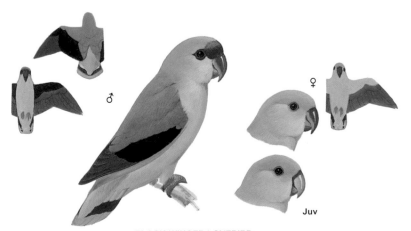

♂

♀

Juv

PLATE 68 | **AFRICAN LOVEBIRDS (in part)**

156

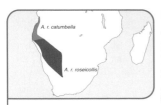

PEACH-FACED LOVEBIRD
Agapornis roseicollis 15cm

Familiar cagebird with many color mutations; only red-faced lovebird with pale bill, and only small green parrot in range; white feathered eye-ring; sexes alike, JUV duller; shrill, metallic *shreek...shreek*. **DISTRIBUTION** southwestern Africa; up to 1500m; feral populations at Simonstown, Republic of South Africa, and in Phoenix and Tucson, Arizona, U.S.A. **SUBSPECIES** two poorly differentiated subspecies. 1. *A. r. roseicollis* forecrown to behind eyes red, becoming rose-pink on face and upper breast; rump blue. *Range* northern Cape Province, Republic of South Africa, and Namibia inland to Lake Ngami, Botswana. 2. *A. r. catumbella* forecrown deeper red and cheeks more reddish. *Range* southwestern Angola; feral population in Quiçama National Park, northwestern Angola. **LOCALITIES** Namibe Regional Nature Park and Quiçama National Park, Angola. Waterburg Plateau National Park, Namib Desert Park, and Hobatere Game Reserve (western side of Etosha National Park), Namibia.

MASKED LOVEBIRD
Agapornis personatus 15cm

Common cagebird with color mutations; only black-faced lovebird with yellow breast and wide yellow collar across hindneck; upper tail-coverts dull blue; bare white eye-ring; high-pitched twittering. **DISTRIBUTION** northern and central Tanzania, from north of Mount Meru south to Mbeya and Njombe districts, with single record from Taveta, southeastern Kenya; mostly 1100 to 1800m; common; feral populations, some mixed with *A. fischeri* and hybrids, at Tanga and Dar es Salaam, coastal Tanzania, and at Nairobi, Mombasa, and Naivasha, Kenya. **SIMILAR SPECIES** Fischer's Lovebird *A. fischeri* (see below) orange-red face and throat; no yellow collar on hindneck. *A. personatus* × *A. fischeri* hybrids (see below) brownish-black face with orange-brown forehead; narrow greenish-yellow band bordering orange-red throat. **LOCALITIES** Lake Manyara, northern Tanzania. Lake Naivasha, Kenya (feral population).

FISCHER'S LOVEBIRD *Agapornis fischeri* 15cm

Common cagebird with color mutations; sexes alike, JUV duller; dull olive crown to hindneck; face and throat to upper breast reddish orange; lower breast yellow; upper tail-coverts blue; bare white eye-ring; sharp *chirrik...chirrik*, more melodious *chirreek*, shrill *tingk...tingk*. **DISTRIBUTION** north-central Tanzania, from Kome and Ukerewe Islands, southern Lake Victoria, and southern shores of lake, and Serengeti and Arusha National Parks south to Nzega and Singida districts, reaching Rwanda and Burundi possibly as irregular vagrants, though populations may be feral; mostly 1100 to 2000m; near threatened; feral populations, some mixed with *A. personatus* and hybrids, at Tanga and Dar es Salaam, coastal Tanzania, and at Mombasa, Nairobi, Athi River, Naivasha, and Isiolo, Kenya. **SIMILAR SPECIES** Masked Lovebird *A. personatus* (see above) black face; broad yellow collar across hindneck. *A. fischeri* × *A. personatus* hybrids (see below) brownish-black face with orange-brown forehead; narrow greenish-yellow band bordering orange-red throat. **LOCALITIES** Arusha, Ndutu, and Serengeti National Parks, northern Tanzania. Lake Naivasha, Kenya (feral population).

FISCHER'S LOVEBIRD × MASKED LOVEBIRD HYBRID

Resembles *A. personatus*, but with orange markings on brownish-black head; foreneck to upper breast orange, and rump to upper tail-coverts dull olive suffused mauve-blue; bare white eye-ring. **DISTRIBUTION** present with Fischer's and Masked Lovebirds in urban feral populations in East Africa.

A. r. roseicollis

Juv

PEACH-FACED LOVEBIRD

MASKED LOVEBIRD

FISCHER'S LOVEBIRD

**FISCHER'S LOVEBIRD ×
MASKED LOVEBIRD HYBRID**

PLATE 69 AFRICAN LOVEBIRDS (in part)

158

NYASA LOVEBIRD *Agapornis lilianae* 14cm

Smaller than similar A. *fischeri*, and with green upper tail-coverts; forehead and throat orange-red, merging into salmon-pink on crown, face, and upper breast; bare white eye-ring; sexes alike, JUV duller with blackish suffusion on green cheeks; chattering likened to rattling of metal chain. Closely associated with mopane (*Colophospermum*) woodland. **DISTRIBUTION** discrete, separated populations in southernmost Tanzania, Zambia–Zimbabwe border district, northwestern Moçambique, southern Malawi, and southeastern Zambia to northern Zimbabwe; also in Lundazi district, northeastern Zambia, where probably introduced, and records of aviary escapees in southern Namibia; up to 1000m; near-threatened. **LOCALITIES** Mana Pools National Park, Zimbabwe. South Luangwa National Park, Zambia.

BLACK-CHEEKED LOVEBIRD
Agapornis nigrigenis 14cm

Smaller than similar A. *personatus*, and with green upper tail-coverts; forehead and forecrown reddish brown; lores, throat, and cheeks brownish black; upper breast dull pink; bare white eye-ring; sexes alike, JUV resembles adults; high-pitched chattering. Closely associated with mopane (*Colophospermum*) woodland. **DISTRIBUTION** southwestern Zambia, sporadically in northwestern Zimbabwe, and rarely reaching extreme northeastern Namibia and northernmost Botswana; mostly 600 to 1000m; vulnerable. **LOCALITY** Kafue National Park (southern sector), Zambia.

BLACK-COLLARED LOVEBIRD
Agapornis swindernianus 13cm

Unmistakable; smallest lovebird, and only species with black nuchal collar and black bill; sexes alike, JUV duller without black collar; twittering notes, shrill *chinga…chinga*. Feeds in upper stages of forest, mainly in *Ficus* trees, where detection difficult. **DISTRIBUTION** West and central Africa; up to 1800m, mostly below 800m; uncommon to scarce and declining. **SUBSPECIES** two well-marked and one poorly differentiated subspecies. 1. A. s. *swindernianus* narrow black nuchal collar; neck below collar dull yellow; lower back to upper tail-coverts deep mauve-blue. *Range* disjunctly in Liberia, Ivory Coast, and Ghana. 2. A. s. *zenkeri* below black collar neck reddish brown, extending as suffusion on breast. *Range* central Africa, from southern Cameroon to Gabon and east to western Democratic Republic of Congo and southwestern Central African Republic. 3. A. s. *emini* like *zenkeri*, but less extensive reddish brown on neck and breast. *Range* central Democratic Republic of Congo to far western Uganda; may be isolated from *zenkeri*. **SIMILAR SPECIES** Red-faced Lovebird A. *pullarius* (plate 67) red or orange-red face and throat; red bill; prefers open country. **LOCALITIES** Bia National Park, Ghana. Yapo Forest and Azagny National Park, Ivory Coast. Makokou-Belinga district, northeastern Gabon.

NYASA LOVEBIRD

Juv

BLACK-CHEEKED LOVEBIRD

BLACK-COLLARED LOVEBIRD

Juv

A. s. swindernianus

A. s. emini

PLATE 70 *PSITTACULA* PARROTS (in part)

160 **Midsized to large green parrots with narrow attenuated feathers in long, strongly graduated tail; long, pointed wings distinctive in flight; slight to pronounced sexual dimorphism, JUV duller than adults. Forests, secondary growth, plantations, cultivation; arboreal, but will raid crops; pairs, small groups, large flocks at food source or nighttime roost. Two species—Newton's Parakeet *P. exsul* from Rodrigues, Mascarene Islands, and Seychelles Parakeet *P. wardi* from Seychelles Islands—now extinct (see plate 145).**

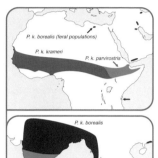

ROSE-RINGED PARAKEET
Psittacula krameri 40cm

Familiar cagebird with many color mutations; large parrot with rose-pink collar encircling hindneck (♂); all-green ♀ & JUV; screeching *kee-ak…kee-ak*, softer *chee…chee*, shrill *ak-ak-ak* when alarmed. **DISTRIBUTION** northern Africa, Afghanistan, where possibly introduced, and Pakistan to India, Sri Lanka, Nepal, and central Myanmar; up to 1600m, mostly below 900m; abundant; introduced to Mauritius, South and East Africa, Egypt, Arabia, and Middle East, southeastern China, Taiwan, Japan, and Singapore; feral populations in southern Britain, western Europe, U.S.A., West Indies, and Venezuela. **SUBSPECIES** yellowish-green African and darker green Asiatic subspecies well differentiated, but within each population only minor geographical differences. 1. *P. k. krameri* yellowish green; ♂ black stripe across lower cheeks; nape suffused blue; upper mandible red, lower mandible black; ♀ lacks all head markings. *Range* north-central Africa from southern Mauritania to southern Sudan and northern Uganda. 2. *P. k. parvirostris* like *krameri*, but head and cheeks less yellowish. *Range* eastern Sudan through Eritrea to Djibouti and rarely northern Somalia. 3. *P. k. borealis* greener, less yellowish, larger, all-red bill; ♂ nape to hindneck suffused blue, darker rose-pink collar on hindneck. *Range* easternmost Afghanistan, northern Pakistan and northern India, to Nepal and central Myanmar. 4. *P. k. manillensis* like *borealis*, but lower mandible black; ♂ more prominent rose-pink collar. *Range* Sri Lanka, Rameswaram Island, and peninsular India. **SIMILAR SPECIES** Mauritius Parakeet *P. echo* (see below) darker green; ♂ rose-pink band on sides of neck not forming collar on hindneck; ♀ all-black bill. Alexandrine Parakeet *P. eupatria* (plate 71) maroon "shoulder-patch"; larger with much larger bill. **LOCALITIES** Mole National Park, northern Ghana, and Yankari National Park, Nigeria. Chitwan National Park, Nepal. Corbett National Park, Uttar Pradesh, and Buxa Tiger Reserve, West Bengal, northern India. Easily seen in much of peninsular India.

MAURITIUS PARAKEET *Psittacula echo* 42cm

Larger and more stocky than similar *P. krameri*, with shorter, broader tail and darker green coloration; ♂ rose-pink band on sides of neck not continuing as collar encircling hindneck, and occiput to nape and lower cheeks suffused mauve-blue; ♀ yellowish band on sides of neck, and all-black bill; nasal *chaa-choa…chaa-choa*, high-pitched *chee…chee*. **DISTRIBUTION** Mauritius, Mascarene Islands, Indian Ocean; critically endangered, CITES I. **LOCALITY** Black River Gorges National Park, southwestern Mauritius.

♂

♀

Juv

P. k. krameri

ROSE-RINGED PARAKEET

♂

P. k. borealis

♂

♀

MAURITIUS PARAKEET

PLATE 71 *PSITTACULA* **PARROTS (in part)**

162

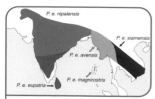

ALEXANDRINE PARAKEET
Psittacula eupatria 58cm

Large parrot with massive red bill giving "top-heavy" appearance and only green-headed *Psittacula* parrot with red "shoulder-patch"; sexes differ, JUV like ♀; screaming *kee-aar*, resonant *gr-raak…gr-raak*.
DISTRIBUTION easternmost Afghanistan and western Pakistan to Indochina and northern Thailand; also Sri Lanka and Andaman Islands; up to 800m, occasionally 1600m; locally common; introduced to Japan. **SUBSPECIES** five identifiable subspecies. 1. *P. e. eupatria* ♂ wide rose-pink collar encircling hindneck; black stripe across lower cheeks; occiput and nape to cheeks mauve-blue; maroon-red "shoulder-patch"; ♀ lacks head markings. *Range* Sri Lanka and peninsular India. 2. *P. e. nipalensis* more grayish green; larger; ♂ wider black stripe across lower cheeks. *Range* easternmost Afghanistan, eastern and southern Pakistan, and central India to Nepal, Bhutan, and Assam to Bangladesh. 3. *P. e. magnirostris* larger, heavier bill; ♂ narrow blue band on hindneck and brighter red "shoulder-patch." *Range* Andaman and Coco Islands, Bay of Bengal. 4. *P. e. avensis* like *nipalensis*, but neck and underparts more yellowish and smaller bill; ♂ narrow blue line on hindneck. *Range* Cachar district of Assam to southern Myanmar at about lat. 16°N. 5. *P. e. siamensis* face and neck yellowish; ♂ occiput and nape washed blue and paler red "shoulder-patch." *Range* northern and western Thailand to Cambodia, central Laos, rarely in north, and central Vietnam. **SIMILAR SPECIES** Rose-ringed Parakeet *P. krameri* (plate 70) smaller and much smaller bill; no maroon "shoulder-patch." **LOCALITIES** Chitwan National Park, Nepal. Gorumara National Park and Buxa Tiger Reserve, West Bengal, northern India. Pench Tiger Reserve, Tadoba National Park, and Borivali National Park, Maharashtra, peninsular India.

MALABAR PARAKEET
Psittacula columboides 38cm

Smaller *Psittacula* parrot with distinctive gray-and-green plumage coloration, with narrow black collar encircling hindneck; ♂ greenish-blue band below black collar, and red bill; ♀ without greenish-blue band, and with all-black bill; JUV mainly green with narrow black collar and pale orange bill; discordant *screek*, sharp *cheet…cheet*. **DISTRIBUTION** Western Ghats, peninsular India, from about lat. 19°N south to southern Kerala at about lat. 8°30'N; mostly 450 to 1000m, occasionally 1600m; locally common. **SIMILAR SPECIES** Plum-headed Parrot *P. cyanocephala* (gray-headed ♀, plate 72) brighter yellowish green; longer central tail-feathers; no black collar on hindneck; palest yellow upper mandible; different calls. **LOCALITIES** Wynaad Wildlife Sanctuary, Kerala, Anamalai Wildlife Sanctuary, Tamil Nadu, and Blackwoods Camp at Molem Wildlife Sanctuary (Bhagwan Mahavir National Park), Goa, southern India.

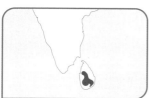

EMERALD-COLLARED PARAKEET
Psittacula calthorpae 29cm

Smallest *Psittacula* parrot, and another species with distinctive gray-and-green plumage coloration; shorter central tail-feathers; head and upperparts gray with broad green collar encircling hindneck, and gray-mauve rump; ♂ bicolored red/brownish bill; ♀ all-black bill; JUV mainly green, and pale orange bill; chattering *ak-ak-ak-ak*.
DISTRIBUTION southwestern to central-southern Sri Lanka; up to 1700m, occasionally 2000m; locally common, but generally scarce. **SIMILAR SPECIES** Plum-headed Parakeet *P. cyanocephala* (gray-headed ♀, plate 72) see above. **LOCALITIES** Horton Plains National Park, Sinharaja Rainforest National Heritage Wilderness Area, Bodhinagala Forest Reserve, Peak Wilderness Sanctuary, and Peradeniya Botanical Gardens, Kandy, Sri Lanka.

ALEXANDRINE PARAKEET

♂

P. e. eupatria

♀

P. e. avensis

♂

♂

♀

MALABAR PARAKEET

Juv

♂

♀

EMERALD-COLLARED PARAKEET

Juv

PLATE 72 *PSITTACULA* **PARROTS (in part)**

164

PLUM-HEADED PARAKEET
Psittacula cyanocephala 33cm

One of two similar smaller *Psittacula* parrots with sexually dimorphic head coloration—♂ red and bluish purple, ♀ gray; long central tail-feathers broadly tipped white; ♂ black line encircling hindneck bordered below by bluish-green collar, maroon "shoulder-patch," and bicolored orange/gray bill; ♀ variable yellow collar encircling hindneck, no maroon "shoulder-patch," and bicolored yellowish-white/pale gray bill; JUV head dusky green with orange frontal band, and pale orange bill; shrill *too-ik...too-ik*, rapid *pe-pe-pe-pe-pe*, musical *queeah-quah...kwink-kwink-queeeah* (♂). **DISTRIBUTION** Sri Lanka, peninsular India north to northeastern Pakistan, and Nepal, Bhutan, and Bangladesh; up to 1500m; common. **SIMILAR SPECIES** Blossom-headed Parakeet *P. roseata* (see below) paler head coloration in both sexes; red "shoulder-patch" in both sexes; central tail-feathers blue tipped yellow. Malabar Parakeet *P. columboides* (plate 71) gray head in both sexes; black collar around hindneck; no maroon "shoulder-patch." Slaty-headed Parakeet *P. himalayana* (plate 73) darker gray head in both sexes; red upper mandible in both sexes; larger; different calls. **LOCALITIES** Chitwan National Park, Nepal. Rajaji National Park, Uttar Pradesh, northern India, and Wynaad Wildlife Sanctuary, Kerala, southern India.

BLOSSOM-HEADED PARAKEET
Psittacula roseata 30cm

Smaller *Psittacula* parrot similar to *P. cyanocephala*, but paler head coloration in both sexes, and long central tail-feathers blue tipped yellow; musical *twee-too...twee...twee...too...too...twee-too*, shrill *too-ik...too-ik* in flight. **DISTRIBUTION** Bhutan, northeastern India, and Bangladesh east to Indochina and southern China; formerly eastern Nepal; up to 1000m; uncommon. **SUBSPECIES** two poorly differentiated subspecies.
1. *P. r. roseata* ♂ forecrown and face pink, remainder of head pale bluish lilac; fine black line encircling hindneck; brownish-red "shoulder-patch"; bicolored orange/gray bill. ♀ head dull bluish gray; yellow collar encircling hindneck; smaller brownish-red "shoulder-patch"; bill brownish yellow. JUV green head; no brownish-red "shoulder-patch"; bill pale yellow. *Range* Bhutan, northeastern India from Sikkim and Assam south to West Bengal and Tripura, Bangladesh, and northern Myanmar; intergrades with *juneae* in Tripura region; old specimen records from eastern Nepal.
2. *P. r. juneae* more yellowish green, and more extensive red "shoulder-patch" in both sexes. *Range* Tripura region and southernmost Assam, northeastern India, where merges with *roseata*, and northern Myanmar east through Thailand, north of Prachaup, to central and southern Laos, Cambodia, Vietnam, and southeastern China, in southern Guangxi and western Guandong; reports from southern Yunnan and southeastern Xizang, southern China, could refer to *roseata*. **SIMILAR SPECIES** Plum-headed Parakeet *P. cyanocephala* (see above) darker head coloration in both sexes; ♀ without red "shoulder-patch"; long central tail-feathers tipped white. Gray-headed Parakeet *P. finschii* (plate 73) darker gray head in both sexes; red upper mandible in both sexes; different calls. Red-breasted Parakeet *P. alexandri* (plate 71) rose-pink underparts; no red "shoulder-patch"; bicolored red-and-black (♂) or all-black (♀) bill. **LOCALITIES** Saw district, southern Chin Hills, western Myanmar. Xé Pian National Biodiversity Conservation Area, Champasak and Attapu Provinces, southern Laos. Ky Anh district, Nghe Tinh Province, north Vietnam.

INTERMEDIATE PARAKEET
"Psittacula intermedia"

Described as separate species, and live birds said to be trapped in Uttar Pradesh, northern India, but known to be *P. cyanocephala* × *P. himalayana* hybrid.

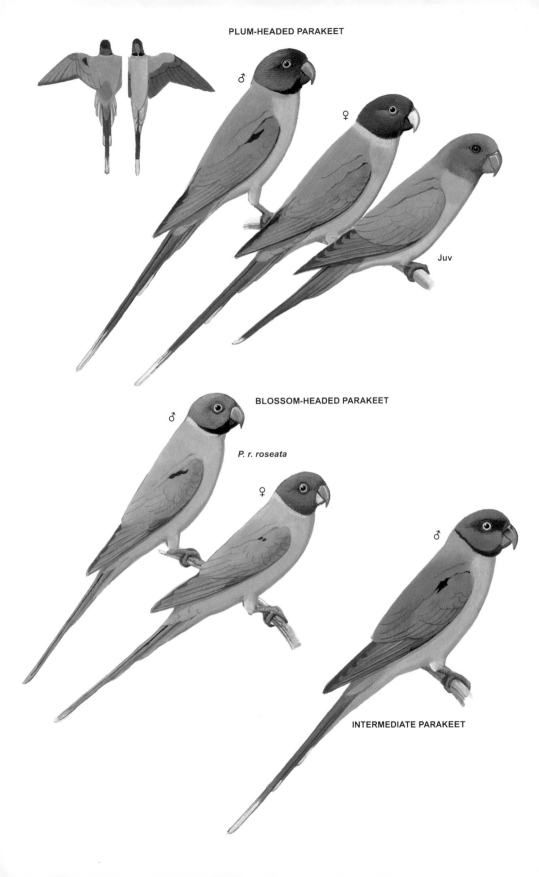

PLUM-HEADED PARAKEET

♂

♀

Juv

BLOSSOM-HEADED PARAKEET

♂

P. r. roseata

♀

♂

INTERMEDIATE PARAKEET

PLATE 73 *PSITTACULA* PARROTS (in part)

166

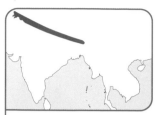

SLATY-HEADED PARAKEET
Psittacula himalayana 40cm

Larger *Psittacula* parrot, and one of two very similar species with gray head and red bill in both sexes; central tail-feathers blue broadly tipped yellow; ♂ with maroon "shoulder-patch," ♀ without maroon "shoulder-patch"; JUV grayish-green head and orange bill; high-pitched *too-i...too-i*, shrill *scree-eet*, prolonged *wee-eenee*.
DISTRIBUTION Himalayas from eastern Afghanistan and northern Pakistan east through northern India and Nepal to Bhutan, Arunachal Pradesh, where probably overlaps with *P. finschii*, and northern Assam, mostly north of Brahmaputra River; mostly 460 to 2400m; common. **SIMILAR SPECIES** Gray-headed Parakeet *P. finschii* (see below) paler yellowish green; narrower central tail-feathers violet-blue tipped yellowish white; mostly allopatric. Plum-headed Parakeet *P. cyanocephala* (gray-headed ♀, plate 72) paler gray head; no black on lower cheeks; no maroon "shoulder-patch"; yellowish-white upper mandible; smaller; different calls.
LOCALITIES Rajaji and Corbett National Parks, Uttar Pradesh, northern India. Chitwan National Park, Nepal.

GRAY-HEADED PARAKEET
Psittacula finschii 40cm

More yellowish green than *P. himalayana*, and with narrower central tail-feathers violet-blue tipped yellowish white. **DISTRIBUTION** northeastern India, in Arunachal Pradesh, where probably overlaps with *P. himalayana*, and southern Assam, mostly south of Brahmaputra River, east through Bangladesh, Myanmar, northern and southwestern Thailand, and Indochina to southern China, in Yunnan, southeastern Xizang, and southwestern Sichuan; up to 2700m; common. **SIMILAR SPECIES** Slaty-headed Parakeet *P. himalayana* (see above) darker green with bluish tinge; central tail-feathers broadly tipped yellow; mostly allopatric. Blossom-headed Parakeet *P. roseata* (gray-headed ♀, plate 72) paler gray head; no black on lower cheeks; smaller; different calls. **LOCALITIES** Bonzon district, western Myanmar. Phou Xang He National Biodiversity Conservation Area, Savannakhét Province, central Laos. Ky Anh district, Nghe Tin Province, north Vietnam.

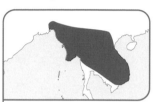

DERBYAN PARAKEET *Psittacula derbiana* 50cm

Unmistakable; large *Psittacula* parrot with lavender-purple head and breast, bold black facial markings, and greenish-yellow "wing-patch"; ♀ with pink band behind black cheeks; JUV head and underparts grayish-green suffused dull blue, and orange-red bill; metallic *cree-eeo...cree-eeo*, corvid-like *kraaa...kraaa*. **DISTRIBUTION** eastern Himalayas and Tibetan Plateau, from Arunachal Pradesh, northeastern India, and southeastern Tibet to southwestern China, in southeastern Xizang, southwestern Sichuan, and western Yunnan; isolated population some 50km from Simao, Yunnan; mostly 2700 to 4000m. **LOCALITIES** Mishmi Hills and Hotspring district, Arunachal Pradesh, northeastern India. Simao district, Yunnan, southern China.

NICOBAR PARAKEET *Psittacula caniceps* 56cm

Unmistakable; large *Psittacula* parrot with yellowish-gray head, bold black facial markings, and bicolored red/black (♂) or all-black (♀) bill; JUV grayish-green head and throat; corvid-like *kraan...kraan*. **DISTRIBUTION** Great Nicobar, Little Nicobar, Menschal, and Kondul, in Nicobar Islands, Bay of Bengal; near-threatened. **LOCALITIES** probably widespread on small islands.

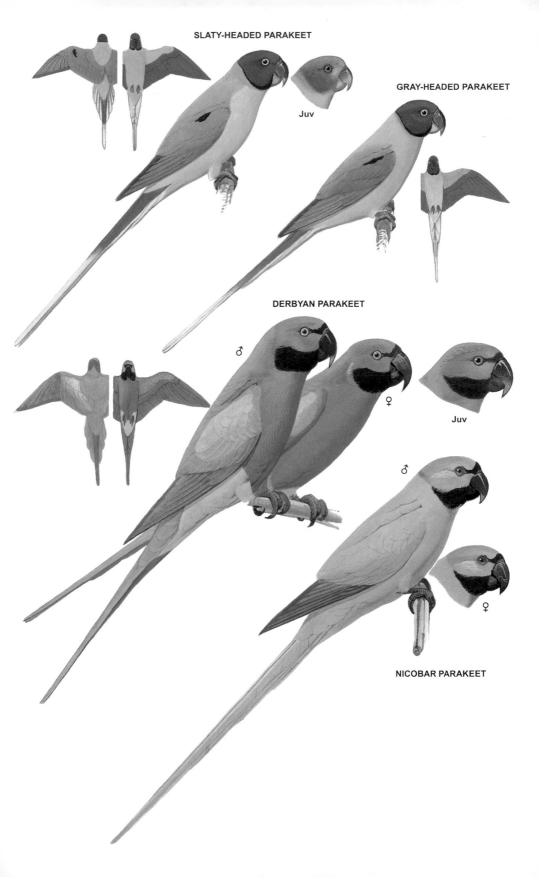

SLATY-HEADED PARAKEET

GRAY-HEADED PARAKEET

Juv

DERBYAN PARAKEET

♂

♀

Juv

♂

♀

NICOBAR PARAKEET

PLATE 74 *PSITTACULA* **PARROTS (in part)**

168

RED-BREASTED PARAKEET
Psittacula alexandri **33cm**

Midsized rose-breasted *Psittacula* parrot with black facial markings and greenish-yellow "wing-patch"; slight sexual dimorphism; JUV face brownish gray, entire underparts green, shorter tail, and bill brownish red; shrill *kek-kek-kek-kek-kek*, plaintive *kewn*, nasal *kaink*.
DISTRIBUTION northern India and Nepal to Indochina, southeastern China, and Andaman Islands in Bay of Bengal; also Java, Bali, islands in Java Sea, and islands off western Sumatra, Indonesia; up to 1500m; common; introduced to southeastern Borneo, Singapore, Japan, Hong Kong, and cities in southeastern China. **SUBSPECIES** eight subspecies in two groupings differentiated by red bill in both sexes (1–3) or black bill in ♀ (4–8). 1. *P. a. alexandri* head gray variably tinged bluish; lower cheeks and line from forehead to eyes black; throat to abdomen rose-pink, and lower abdomen to undertail-coverts green suffused bluish; red bill. *Range* Java and Bali, Indonesia; introduced to southeastern Borneo and Singapore. 2. *P. a. kangeanensis* like *alexandri*, but less bluish suffusion to gray head; more extensive greenish-yellow "wing-patch." *Range* Kangean Islands in Java Sea. 3. *P. a. dammermani* larger than *alexandri*, and crown more bluish. *Range* Karimunjawa Islands in Java Sea. 4. *P. a. fasciata* head grayish blue with greenish suffusion around eyes; throat to abdomen dark pink washed lilac-blue on foreneck and sides of breast, less prominent in ♀; ♂ upper mandible red, lower mandible black; ♀ all-black bill. *Range* lower Himalayas from northern Uttar Pradesh east to Arunachal Pradesh and Assam, northern India, and through Myanmar, Thailand, south to Ranong, and Indochina to southeastern China, in southeastern Xizang, Yunnan, southwestern Guangxi, and Hainan Island; introduced to Singapore, Hong Kong, and nearby cities in southern Guandong, southeastern China. 5. *P. a. abbotti* like *fasciata*, but paler plumage coloration; larger. *Range* Andaman Islands in Bay of Bengal. 6. *P. a. cala* like *fasciata*, but upperparts paler green; throat to upper abdomen less suffused bluish, especially in ♀; larger. *Range* Simeulue Island, off northwest Sumatra. 7. *P. a. major* larger than *cala*, and less bluish suffusion on lower abdomen to undertail-coverts. *Range* Lasia and Babi Islands, off northwest Sumatra. 8. *P. a. perionca* like *major*, but smaller and lower abdomen to undertail-coverts brighter green. *Range* Nias Island, off west Sumatra.
SIMILAR SPECIES Derbyan Parakeet *P. derbiana* (plate 73) darker lavender-purple underparts; larger; different calls. Other *Psittacula* parrots in range have green underparts. **LOCALITIES** Ragunan Zoo grounds, Jakarta, Java, Indonesia. Nan Bai Cat Tien National Park, southern Vietnam. Pakhui Wildlife Sanctuary, Arunachal Pradesh, and Buxa Tiger Reserve, West Bengal, northeastern India. Chitwan National Park, Nepal.

RED-BREASTED PARAKEET

♂ ♀ *P. a. alexandri*

♂

♀ *P. a. fasciata*

♂ Juv

P. a. cala

PLATE 75   *PSITTACULA* (in part) AND *PSITTINUS* PARROTS

170

LONG-TAILED PARAKEET
Psittacula longicauda 42cm

Midsized *Psittacula* parrot with very long central tail-feathers and distinctive facial pattern featuring rose-red cheeks to ear-coverts; sexually dimorphic, JUV duller; discordant *kiak…kiak*, high-pitched *pee-yo…pee-yo*, quavering *kraaak…kraaak*, scolding *cheet-cheet-cheet*. **DISTRIBUTION** Malay Peninsula, Singapore, Borneo, Sumatra and adjacent islands, and Andaman and Nicobar Islands in Bay of Bengal; up to 300m; near-threatened. **SUBSPECIES** four well-marked and one poorly differentiated subspecies. 1. *P. l. longicauda* ♂ crown dark green; sides of head to nape and hindneck rose-red; lower cheeks black; mantle and upper back yellowish-green suffused bluish gray; rump to upper tail-coverts pale blue; upper mandible red, lower mandible brown. ♀ lower cheeks dark green; nape and hindneck green; bill dark brown; shorter tail. JUV head green suffused rose-red on upper cheeks; very short tail. *Range* Malay Peninsula south from Kedah, and Anambas Islands, and on Singapore, including offshore islets, Borneo, and Sumatra, including offshore Nias and Bangka Islands. 2. *P. l. defontainei* like *longicauda*, but crown more yellowish green; upper cheeks deep red in both sexes. *Range* Natuna Islands, Riau Archipelago, Bintan, Karimata, and Belitung Islands, Indonesia. 3. *P. l. modesta* larger; ♂ crown dull red tinged greenish, and sides of head deep crimson; ♀ crown greenish brown tinged red on occiput. *Range* Enggano Island, off southwestern Sumatra. 4. *P. l. tytleri* larger; ♂ nape to mantle and upper back yellowish green strongly suffused grayish mauve, and lower back to rump green; ♀ crown to hindneck and mantle yellowish green. *Range* Coco and Andaman Islands, Bay of Bengal. 5. *P. l. nicobarica* larger; ♂ crown to nape and hindneck yellowish green, and upperparts yellowish green slightly suffused bluish on mantle; ♀ upperparts, including crown and nape bright yellowish green. *Range* Nicobar Islands, Bay of Bengal. **SIMILAR SPECIES** Other *Psittacula* parrots in range lack red face. **LOCALITIES** Botanical Gardens, Reservoir Parks, and Bukit Timah Nature Reserve, Singapore. Tanjung Puting and Kutai National Parks, Kalimantan, Indonesia. Way Kambas National Park, south Sumatra, Indonesia.

BLUE-RUMPED PARROT
Psittinus cyanurus 18cm

Small stocky green parrot with short, rounded tail and proportionately large bill; sexually dimorphic, JUV differs from adults. Forests, secondary growth, plantations; arboreal; pairs, small flocks; noisy and conspicuous in swift, direct flight; whistling *tee-link*, trisyllabic *wee-chi-chi*, high-pitched *peep*. **DISTRIBUTION** southernmost Myanmar to Borneo and Sumatra; up to 700m; near-threatened. **SUBSPECIES** two well-marked and one poorly differentiated subspecies. 1. *P. c. cyanurus* ♂ head blue; underparts pale brown; mantle and upper back black; lower back to upper tail-coverts blue; red underwing-coverts diagnostic in flight; upper mandible red, lower mandible brown. ♀ head brown; upperparts green with blue on lower back; bill brown. JUV like ♀, but head green. *Range* Tenasserim, southernmost Myanmar, and southwestern Thailand south through Malay Peninsula to Singapore, Sumatra, Bangka Island, and Borneo. 2. *P. c. pontius* like *cyanurus*, but larger. *Range* Mentawai Islands, Indonesia. 3. *P. c. abbotti* ♂ crown green and remainder of head blue; underparts yellowish green; upperparts green; larger. ♀ head green; larger. *Range* Simeulue and Siumat Islands, off west coast of Sumatra, Indonesia. **SIMILAR SPECIES** Vernal Hanging Parrot *Loriculus vernalis* and Blue-crowned Hanging Parrot *L. galgulus* (plate 76) only other small parrots in range; different head markings; blue underwings; much smaller. **LOCALITIES** Khao Pra-Bang Khran Wildlife Sanctuary, Krabi-Trang, peninsular Thailand. Danum Valley Conservation Area, Sabah, Malaysia. Kutai National Park, Kalimantan Timur, and Bukit Tigapulih area, Sumatra, Indonesia.

LONG-TAILED PARAKEET

P. l. modesta

♂

♀

♂

P. l. longicauda

♀

Juv

♂

♀

P. l. nicobarica

♂

♀

P. l. tytleri

♂

♀

P. c. abbotti

BLUE-RUMPED PARROT

♂

♀

♂

♀

Juv

P. c. cyanurus

PLATE 76 **HANGING PARROTS (in part)**

172 Small green parrots with very short, rounded tail, very fine, sharply-pointed bill, and blue underwings; red rump and upper tail-coverts in all but one species; sexes differ, JUV like ♀. Forests, secondary growth, village gardens, plantations; arboreal; pairs or small parties in swift flight; lorikeet-like behavior, coming to flowering trees and shrubs to feed on nectar. Two species in Sulawesi, but elsewhere species replace each other geographically. Other hanging parrots occur in the Australasian Distribution (see plates 26–28).

VERNAL HANGING PARROT
Loriculus vernalis 13cm

One of two red-billed hanging parrots with pale eye; all-green head in both sexes; ♂ blue "throat-patch"; ♀ & JUV little or no blue on throat; squeaky *tee-sip* or *de-zee-zeet*, sharp *chee-chee-chee*, shrill *tsit-tsit*. **DISTRIBUTION** southwestern peninsular India, south from lat. 19°N, and up east coast to West Bengal and northeastern India, Nepal, and Bhutan, but not Sikkim, to Bangladesh and east through Myanmar to southernmost China, in southwestern Yunnan, and Indochina to Thailand, south on peninsula to Satul and Songkla; also Andaman Islands and Mergui Archipelago; up to 1800m; common. **SIMILAR SPECIES** Blue-crowned Hanging Parrot L. galgulus (see below, possible sympatry in southernmost peninsular Thailand) black bill and dark eye; ♂ blue crown and red "throat-patch." **LOCALITIES** Wynaad Wildlife Sanctuary, Kerala, and Bondla Wildlife Sanctuary, Goa, southern India. Gorumara National Park and Buxa Tiger Reserve, West Bengal, northern India. Xé Pian National Diversity Conservation Area, southern Laos. Nam Bai Cat Tien National Park, southern Vietnam. Khao Yai National Park, Thailand.

SRI LANKAN HANGING PARROT
Loriculus beryllinus 13cm

Another red-billed hanging parrot with pale eye, and only small green parrot in range; forehead and crown scarlet, becoming golden orange on nape to hindneck; rump and upper tail-coverts dark red; sexes alike; JUV green head, brownish bill, and pale brown eye; sharp *twit-twit-twit*, warbling notes. **DISTRIBUTION** Sri Lanka; up to 1600m; common. **LOCALITIES** Gardens at Ratnaloka Tour Inn and Pompekele Forest, Ratnapura, Gilimale Forest, and Sinharaja National Heritage Wilderness Area, Sri Lanka.

BLUE-CROWNED HANGING PARROT
Loriculus galgulus 12cm

Small black-billed hanging parrot with mantle suffused orange-yellow, and only small green parrot in much of range; ♂ blue crown and red "throat-patch"; ♀ little or no blue on crown, and throat green; JUV like ♀, but rump green, feathers edged red; squeaky *tsee* or *dzi*, repeated *ti-ti-ti-ti* or ringing *ti...ti...ti*, shrill *trrirt*, disyllabic *tir-rit* or *squeak-it*. **DISTRIBUTION** Malay Peninsula, south from lat. 7°N, Singapore, Anambas Islands, Riau Archipelago, Bangka, Belitung, and Mendanau Islands, Sumatra and outlying islands, Indonesia, Borneo, and offshore Labuan and Maratua Islands; vagrant to westernmost Java, and feral population near Jakarta; up to 1300m; common. **SIMILAR SPECIES** possibly sympatric with other hanging parrots at extremities of range; Vernal Hanging Parrot L. vernalis (see above) red bill and pale eye; no red on throat or blue on crown. Yellow-throated Hanging Parrot L. pusillus (plate 26) red bill and pale eye; no blue on crown or red on throat. **LOCALITIES** Kerau Wildlife Reserve, Pahang, Malay Peninsula, and Danum Valley Conservation Area, Sabah, Malaysia. Tanjung Puting National Park, Kalimantan Tengah, and Way Kambas National Park, south Sumatra, Indonesia.

VERNAL HANGING PARROT

♂ ♀

SRI LANKAN HANGING PARROT

Juv

BLUE-CROWNED HANGING PARROT

♂ ♀

Juv

PLATE 77 BLUE MACAWS

174

HYACINTH MACAW
Anodorhynchus hyacinthinus 100cm

Unmistakable; larger of two blue macaws with massive black bill, yellow bare eye-ring and lappet at base of lower mandible, and long, strongly graduated tail. A third species—Glaucous Macaw A. *glaucus* from Río Paraguay region—now extinct (see plate 146).
Differentiated from smaller A. *leari* by narrower, crescent-shaped lappet encircling base of lower mandible; sexes alike, JUV paler yellow eye-ring and lappet; discordant *kraaa-aaa*, shrill *kraa-ee...kraa-ee*. Open woodland and gallery forest with palm trees; pairs, family trios, small flocks; feeds mainly on palm nuts taken in trees and on ground; attracted to recently burned areas where palm nuts split open by fire; conspicuous and noisy; effortless, buoyant flight with shallow wingbeats; long, streamerlike tail gives distinctive appearance in flight.
DISTRIBUTION three apparently isolated populations in central South America: (i) northeastern Brazil, locally north of lower Amazon River in Amapá, and south of river between Rio Tapajos and Rio Tocatins; (ii) central-eastern Brazil, from southern Piauí and southern Maranhão to northern Goiás and northwestern Bahia, centered on Chapada das Mangabeiras; (iii) south-central Brazil, in upper Rio Paraguay basin and southwestern Mato Grosso to neighboring eastern Bolivia and northernmost Paraguay: endangered, CITES I. **LOCALITIES** BioBrasil Reserve, southern Piauí, northern Brazil. Caiman Lodge Wildlife Refuge and Pousada Aguapé at São Jose Ranch, Mato Grosso do Sul, southern Brazil. Paso Bravo and San Luis National Parks, northern Paraguay.

LEAR'S MACAW *Anodorhynchus leari* 75cm

Unmistakable; smaller than similar, allopatric A. *hyacinthinus*, and differentiated by "tear-drop" shaped lappets at base of lower mandible; paler blue head and neck contrasting with darker blue back and wings; sexes alike, JUV paler yellow eye-ring and lappets; croaking *greee-ah*, and *ara-ara...trrahra* cries. On strongly fissured plateau in dry scrubland or caatinga woodland with thorny bushes and *Syagrus* palms, the main food tree; pairs, small groups; feeds in trees and on ground; conspicuous and noisy; very shy; roosts and nests in burrows in cliff-face; steady flight with strong, deliberate wingbeats. **DISTRIBUTION** northeastern Bahia, central-eastern Brazil, where two colonies: (i) Toca Velha and Serra Branca, south of Raso da Catarina plateau; (ii) possibly distinct subpopulation some 200km to east, at Senta Sé-Campo Formoso: critically endangered, CITES I. **LOCALITY** Estação Ecológica do Rosa da Catarina, northern Bahia, Brazil.

SPIX'S MACAW *Cyanopsitta spixii* 56cm

Unmistakable; only small blue macaw; forehead, cheeks, and ear-coverts grayish blue; bare lores to eye-ring dark gray; sexes alike, JUV pale bare lores to eye-ring; rolling *kraa-aark*, screeches. Formerly favored *Tabebuia*-dominated gallery woodlands along ephemeral watercourses in dry caatinga scrubland; formerly pairs or small groups; timid; flapping flight with distinctive silhouette produced by long tail. **DISTRIBUTION** Formerly in Rio São Francisco valley, northern Bahia, northeastern Brazil; extinct in wild, survives in captivity; CITES I.

HYACINTH MACAW

LEAR'S MACAW

Juv

SPIX'S MACAW

PLATE 78 **LARGE *ARA* MACAWS (in part)**

176 **Large parrots with massive bill and long, strongly graduated tail; bare face white, becoming deep pink when excited or alarmed, and with lines of small feathers; three "species pairs," each of two much-alike species; sexes alike, JUV resembles adults. One species—Cuban Macaw *Ara tricolor* from Cuba and possibly Hispaniola—now extinct (see plate 146). Forests, secondary growth, woodlands, remnant forest patches in cultivation; pairs, small groups, large numbers at nighttime roost and traditional clay licks; arboreal, but will come to ground to pick up palm nuts; direct flight with shallow wingbeats.**

BLUE AND YELLOW MACAW
Ara ararauna 85cm

One of two large similar blue-and-yellow macaws; forecrown green; throat black; bare face with fine lines of greenish-black feathers; guttural *rraa-aaar* or *kurr-raak*. **DISTRIBUTION** eastern Panama to Colombia, except Río Cauca valley and Nariño, eastern and possibly western Ecuador, and eastern Peru east through Venezuela, in far northeast from northern Monagas to central Delta Amacuro and in south mostly south of Río Orinoco in central Amazonas, to Guianas, and south through Brazil, except much of southeast where locally extinct, to Bolivia, and extralimitally to northeastern Paraguay; reintroduced to Trinidad; up to 500m; locally common. **SIMILAR SPECIES** Blue-throated Macaw *A. glaucogularis* (see below) blue forecrown, throat, and sides of neck; bare face with greenish-blue feathers; different calls. **LOCALITIES** Ducke Forest Reserve, near Manaus, Amazonas, BioBrasil Reserve, southern Piauí, and Caiman Lodge Wildlife Refuge, Mato Grosso do Sul, Brazil. Kapawi Lodge, Pastaza, Ecuador. Manú Biosphere Reserve, Madre de Dios, southeastern Peru. Noel Kempff Mercado National Park, Santa Cruz, Bolivia.

BLUE-THROATED MACAW
Ara glaucogularis 85cm

Like A. *ararauna*, but forecrown blue; throat to sides of neck and forecheeks greenish blue; bare face with lines of greenish-blue feathers; calls softer and more high pitched than calls of A. *ararauna*. **DISTRIBUTION** east of upper Río Mamoré in Llanos de Mojos. El Beni, central Bolivia; up to 300m; critically endangered, CITES I. **SIMILAR SPECIES** Blue and Yellow Macaw A. *ararauna* (see above). **LOCALITY** localized north and south of Trinidad city, El Beni; local guide required.

BLUE AND YELLOW MACAW

BLUE-THROATED MACAW

PLATE 79 LARGE *ARA* MACAWS (in part)

178

SCARLET MACAW *Ara macao* 85cm

Slightly smaller of two similar red macaws with bicolored black-and-white bill; strident *rraa-aaarr* or *kurr-rraak*, growling *kree-eet*. **DISTRIBUTION** Central America and northern South America; up to 500m; scarce, locally common in remote areas, CITES I. **SUBSPECIES** two poorly differentiated subspecies. 1. A. m. *macao* scarlet; flight feathers, back, rump, and tail-coverts blue; median and secondary wing-coverts yellow tipped green; tail scarlet tipped blue; bare face with inconspicuous lines of minute red feathers. *Range*
locally in Costa Rica, mostly Pacific slope, and southern Panama, on Azuero Peninsula and Isla Coiba, then disjunctly through northern South America, east of Andes, from Río Magdalena valley, Colombia, to Guianas, and south to eastern Bolivia and northern Mato Grosso, Brazil. 2. A. m. *cyanopterus* median and secondary wing-coverts yellow tipped blue. *Range* Oaxaca and southern Tamaulipas, southeastern Mexico, disjunctly to northeastern Nicaragua. **SIMILAR SPECIES** Green-winged Macaw A. *chloropterus* (see below) darker red; green upper wing-coverts; bare face with conspicuous feathered lines. **LOCALITIES** Cockscomb Basin Wildlife Sanctuary, southern Belize. Corcovado National Park, Peninsula de Osa, and Lapa Ríos Nature Reserve, Golfo Dulce, Costa Rica. Tinigua National Park, Meta, Colombia. Manú Biosphere Reserve, Madre de Dios, southeastern Peru. Ducke Forest Reserve, near Manaus, Amazonas, and Cristalino Jungle Lodge, Cristalino State Park, northern Mato Grosso, Brazil.

GREEN-WINGED MACAW
Ara chloropterus 90cm

Slightly larger than A. *macao*, and darker red; flight feathers, back, rump, and tail-coverts blue; median wing-coverts, scapulars, and tertials green; tail dark red tipped blue; bare face with conspicuous lines of red feathers; raucous *raw-aawk...raw-aawk*, prolonged *aar-aaark*, screeching *ree-eeah*, corvid-like *kraaah*. **DISTRIBUTION** eastern Panama to Río Atrato region, northwestern Colombia, and northern to central South America, east of the Andes, from eastern Colombia, Venezuela, and Guianas, south to Paraná and Mato Grosso, Brazil, eastern Peru, and eastern Bolivia to Paraguay, and northern Argentina in Formosa; up to 1000m; more thinly dispersed and generally less numerous than other large macaws. **SIMILAR SPECIES** Scarlet Macaw A. *macao* (see above) paler scarlet; yellow upper wing-coverts; minutely feathered lines across bare face not noticeable. **LOCALITIES** Darién Biosphere Reserve, Darién, eastern Panama. Los Katios National Park, Chocó, Colombia. BioBrasil Reserve, southern Piauí, and Caiman Lodge Wildlife Refuge, Mato Grosso do Sul, Brazil. Manú Biosphere Reserve, Madre de Dios, southeastern Peru.

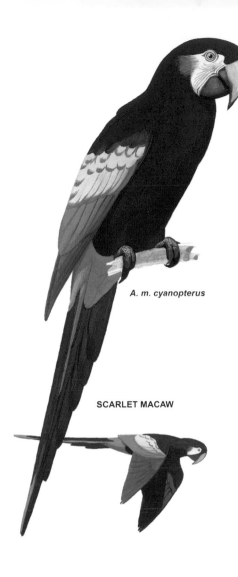

A. m. cyanopterus

SCARLET MACAW

GREEN-WINGED MACAW

PLATE 80   LARGE *ARA* MACAWS (in part)

180

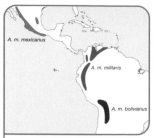

MILITARY MACAW *Ara militaris* 75cm

Smaller of two similar green macaws with red forehead, blue wings, and blue-tipped red tail; raucous *kraa-aaak*, shrill *kee-eeah*, or *kaa-ee-aah*. **DISTRIBUTION** Mexico, and disjunctly in western South America; mostly 500 to 3000m; vulnerable, CITES I. **SUBSPECIES** three doubtfully distinct subspecies. 1. *A. m. militaris* feathered lines on bare face red in front of eyes and black underneath eyes; throat slightly suffused olive-brown; tail dark red broadly tipped blue. *Range* northwestern Venezuela and western Colombia south to middle Urubamba region, Cuzco, southeastern Peru; rare visitor to Pacific slope of Andes in northwestern Peru. 2. *A. m. mexicanus* larger than *militaris*. *Range* Pacific slope in Mexico, from southern Sonora south to Jalisco, and Caribbean slope from eastern Nuevo Leon to San Luis Potosí. 3. *A. m. bolivianus* throat reddish brown; reddish bases to feathers of ear-coverts; tail tipped darker blue. *Range* east slope of Andes in Bolivia, south from Beni, and extreme northwestern Argentina, in Jujuy and Salta. **SIMILAR SPECIES** Great Green Macaw *A. ambiguus* (see below) probably indistinguishable in field, but mostly allopatric; more yellowish green; tail paler orange-red. Thick-billed Parrot *Rhynchopsitta pachyrhyncha* (plate 83) much smaller with shorter tail; feathered face; red from forehead to above and behind eye. **LOCALITIES** El Cielo Biosphere Reserve, Tamaulipas, Mexico. Cueva de los Guácharos National Park, Huila, Colombia. Guatopo National Park, Miranda, Venezuela. Amboró National Park, Santa Cruz, Bolivia.

GREAT GREEN MACAW *Ara ambiguus* 85cm

Larger than similar *A. militaris*; raucous shouts and squawks, growling *aa-aahk* or *aowrk*. **DISTRIBUTION** Central America and northwestern South America; up to 1500m; vulnerable, CITES I. **SUBSPECIES** two poorly differentiated subspecies. 1. *A. a. ambiguus* yellowish green; forehead scarlet; feathered lines on bare face scarlet in front of eyes, black underneath eyes; throat tinged olive-brown; tail orange-red tipped dull blue. *Range* Caribbean lowlands of eastern Honduras and easternmost Nicaragua to Panama and northwestern Colombia. 2. *A. a. guayaquilensis* like *ambiguus*, but smaller, narrower bill. *Range* two populations in western Ecuador, in north in Esmeraldas, and in south in Chongon Hills, near Guyaquil, Guayas. **SIMILAR SPECIES** Military Macaw *A. militaris* (see above) probably indistinguishable in field, but mostly allopatric; darker green and tail darker red. **LOCALITIES** Río Plátano Biosphere Reserve, eastern Honduras. Sarapiquís Neotropical Center and La Trimbina Rainforest Reserve, Costa Rica. Darién Biosphere Reserve, Darién, Panama. Los Katíos National Park, Chocó, Colombia. Cerro Blanco Reserve, near Guayaquil, western Ecuador.

RED-FRONTED MACAW *Ara rubrogenys* 60cm

Unmistakable; smaller green macaw, and only green macaw in range; orange-red crown, ear-coverts, and "shoulder-patch"; red-and-yellow underwings; shrill ringing notes, raucous *raa-aaah*, dueting from pairs. Woodlands and dry forest in arid intermontane valleys; pairs, family groups, larger flocks at nighttime roost; noisy and conspicuous; feeds in trees and on ground; strong flight with shallow wingbeats. **DISTRIBUTION** central-southern Bolivia, from southern Cochabamba and western Santa Cruz to eastern Potosí; mostly 1100 to 2500m; endangered, CITES I. **LOCALITIES** valleys of Río Caine and Río Mizque, Cochabamba.

MILITARY MACAW

A. m. militaris

GREAT GREEN MACAW

A. a. ambiguus

Juv

RED-FRONTED MACAW

PLATE 81   SMALLER GREEN MACAWS (in part)

182

CHESTNUT-FRONTED MACAW
Ara severus 46cm

Midsized green macaw with chestnut-brown forehead and blue-tipped brownish-red tail; thighs and bend of wing red; brownish-black feathered lines on bare face; brownish-red undersides of wings and tail diagnostic in flight; sexes alike, JUV like adults. harsh *ahh-aarra*, high-pitched *ghehh*. Forests, secondary growth, plantations; pairs, small groups, larger flocks at nighttime roost; arboreal; fast, direct flight.
DISTRIBUTION Darién, eastern Panama, and Pacific slope of Andes in western Colombia to southern Ecuador, and east of Andes from Colombia south to Santa Cruz, eastern Bolivia; north of Amazon River east to western and southern Venezuela and Guianas; south of Amazon River east to Mato Grosso, south-central Brazil, but apparently avoiding much of Amazon River basin; up to 1500m; common; introduced to southern Florida, U.S.A. **SIMILAR SPECIES** Blue-winged Macaw *Primolius maracana* (see below) red forehead and abdominal patch; yellow underwings; much smaller. Red-bellied Macaw *Orthopsittaca manilata* (plate 82) bare face yellow; red abdominal patch and yellow underwings; much smaller. **LOCALITIES** Darién Biosphere Reserve, Darién, Panama. Tinigua National Park, Meta, Colombia. Cristalino Jungle Lodge, Cristalino State Park, northern Mato Grosso, Brazil. Manú Biosphere Reserve, Madre de Dios, southeastern Peru.

BLUE-HEADED MACAW
Primolius couloni 43cm

Only midsized green macaw with blue head; blue-tipped red tail; bare face gray; sexes alike, JUV duller; rasping *purr* or *raaah*, high-pitched shrieks. Foothill forests, tall secondary growth, clumps of *Mauritia* palms; pairs, small groups; arboreal; inconspicuous and fairly quiet; swift, direct flight. **DISTRIBUTION** eastern Peru, south from Río Huallaga valley, Loreto, and westernmost Brazil, only in Acre, to northern Bolivia, east of Andes south to about lat. 14°S; mostly 150 to 1300m; near-threatened, CITES I. **SIMILAR SPECIES** Blue-headed Parrot *Pionus menstruus* (plate 126) much smaller with very short, squarish tail; red undertail-coverts; feathered face; different flight. **LOCALITIES** Tingo María National Park, Huánuco, and Manú Biosphere Reserve, Madre de Dios, southeastern Peru.

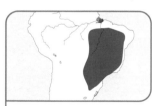

BLUE-WINGED MACAW
Primolius maracana 43cm

Only midsized green macaw with red forehead and red lower back; red abdominal patch; bare face pale yellow; sexes alike, JUV duller, less extensive red markings; shrill *gheh* or *krek...krek...krek*. Forests, mangroves, gallery woodland; pairs, small parties; not conspicuous or noisy; distinctive flight with jerky, upward pitches.
DISTRIBUTION eastern Brazil, at Ilha de Marajó, Pará, and southern Pará and Maranhão south to Mato Grosso do Sul and São Paulo, and eastern Paraguay to northeastern Argentina, in Misiones and northern Corrientes, where now probably extirpated; up to 1000m; near-threatened, CITES I.
SIMILAR SPECIES Chestnut-fronted Macaw *Ara severus* (see above) chestnut-brown forehead; no red abdominal patch; red underwings; larger. Red-bellied Macaw *Orthopsittaca manilata* (plate 82) see above. **LOCALITIES** Serra Negra, Pernambuco, and Serra do Cachimbo, Pará, Brazil. Rio Doce State Park, Minas Gerais, Brazil. Fazenda Paraíso Reserve and Caetetus Ecological Station, São Paulo, Brazil.

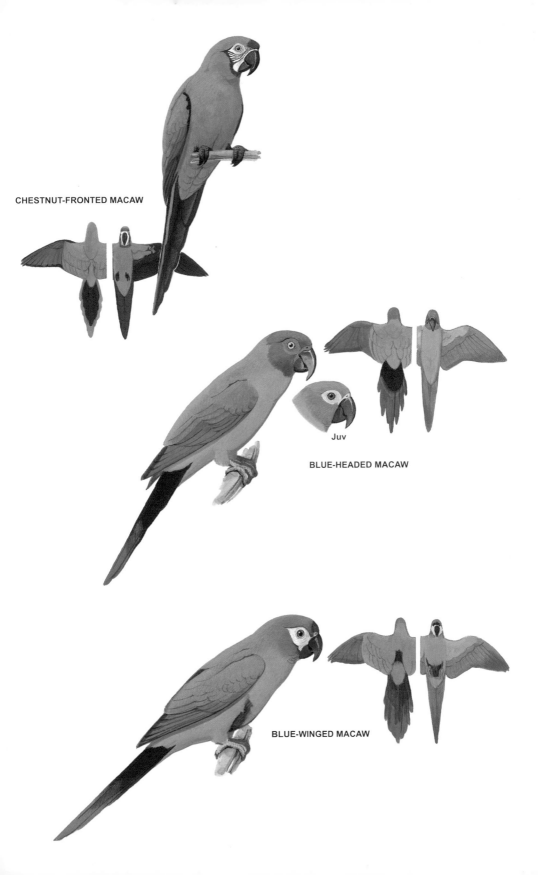

CHESTNUT-FRONTED MACAW

BLUE-HEADED MACAW

Juv

BLUE-WINGED MACAW

PLATE 82 **SMALLER GREEN MACAWS (in part)**

184

YELLOW-COLLARED MACAW
Primolius auricollis 38cm

Unmistakable; only macaw with black head and yellow collar on hindneck; tail brownish red broadly tipped blue; bare face cream-white; sexes alike, JUV like adults; gull-like *scree-eeat…scree-eeat*. **DISTRIBUTION** northern and eastern Bolivia to southwestern Mato Grosso, Brazil, and northern Paraguay, mainly west of Río Paraguay, to northwestern Argentina, in eastern Jujuy and northern Salta; also on and around Ilha do Bananal in western Goiás, northeastern Brazil, where possibly isolated; up to 600m, locally to 1700m; common. **LOCALITIES** readily seen in farmlands on outskirts of Corumba, Mato Grosso do Sul, Brazil, and Santa Cruz city, Bolivia.

RED-BELLIED MACAW
Orthopsittaca manilata 41cm

Midsized olive-green macaw with carunculated yellow skin on bare face; fine scalloping on neck and underparts; brownish-red abdominal patch; green tail; sexes alike, JUV like adults; wailing *choii-aa*, loud *wrr-rake…wrr-rake*, rhythmical *screeet…screeet*. Dependent on *Mauritia* palms for food and nesting, so rarely found far from palm groves in swampy lands; often in large, noisy flocks. **DISTRIBUTION** Trinidad, Guianas, and eastern Venezuela to southern Colombia, eastern Ecuador, and eastern Peru, and northern Bolivia to Amazonian Brazil, but apparently absent from Rio Negro drainage in upper Amazon River basin; up to 500m; common. **SIMILAR SPECIES** Chestnut-fronted Macaw *Ara severus* (plate 81) chestnut-brown forehead; white bare face; no red abdominal patch; red tail and red underwings; larger. Blue-winged Macaw *Primolius maracana* (plate 81) red forehead; white bare face; red tail. **LOCALITIES** Nariva Swamp Reserve, Trinidad. Botanic Gardens, Georgetown, Guyana. Tinigua National Park, Meta, Colombia. Cristalino Jungle Lodge, Cristalino State Park, northern Mato Grosso, Brazil.

RED-SHOULDERED MACAW
Diopsittaca nobilis 30cm

Smallest macaw, resembling green *Aratinga* conures, but identified by bare face; sexes alike, JUV duller; shrill *kreek-kreek-kreek* in flight, harsh *ark-ark-ark-ark*. **DISTRIBUTION** Guianas and eastern Venezuela to northeastern Brazil, and eastern Brazil to southeastern Peru and northern Bolivia; up to 1400m; common. **SUBSPECIES** two well marked and one poorly differentiated subspecies. 1. *D. n. nobilis* forecrown blue; bend of wing, carpal edge, and lesser underwing-coverts red; bare face white; all-black bill. *Range* north of Amazon River in eastern Venezuela, in Delta Amacuro to northern Monagas and northern to eastern Bolívar, Guianas, and northeastern Brazil, in Roraima, northern Pará, and Amapá. 2. *D. n. cumanensis* (Noble Macaw, possibly separate species) larger than *nobilis*, and more massive bill with pale upper mandible. *Range* northeastern Brazil south of Amazon River, from Alagoas and Bahia west to Maranhão and southeastern Pará south to central Goiás, where intergrades with *longipennis*. 3. *D. n. longipennis* larger than *cumanensis*. *Range* inland Brazil, from central Goiás and western Minas Gerais to northwestern São Paulo and Mato Grosso do Sul, through southeastern Peru to eastern and central Bolivia; apparently no recent records from Espírito Santo or Rio de Janeiro, but introduced population in São Paulo city. **SIMILAR SPECIES** White-eyed Conure *Aratinga leucophthalma* (plate 87) bare white eye-ring, but feathered face; no blue on forecrown; all-pale bill. **LOCALITIES** Botanic Gardens, Georgetown, Guyana. Caiman Lodge Wildlife Refuge, Mato Grosso do Sul, Brazil.

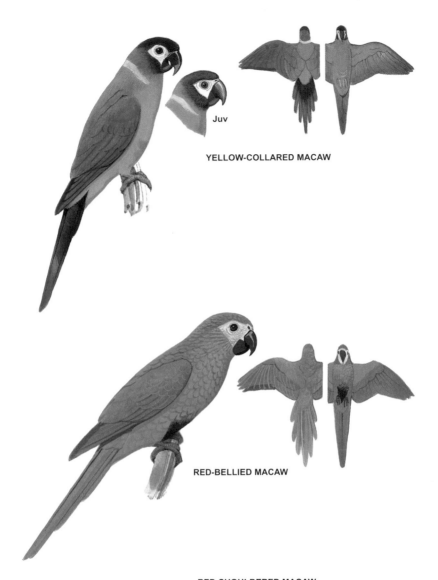

YELLOW-COLLARED MACAW

Juv

RED-BELLIED MACAW

RED-SHOULDERED MACAW

D. n. nobilis

Juv

D. n. cumanensis

PLATE 83 MACAW ALLIES

186

GOLDEN CONURE *Guaruba guarouba* 34cm

Unmistakable; midsized yellow parrot with green flight feathers and massive bill; sexes alike, JUV green on head, breast, and back; discordant *kray* in flight, repeated *keek-keek-keek*. Forests and forest clearings; small flocks, roosting and nesting communally. **DISTRIBUTION** northern Brazil, south from Amazon River to about lat. 5°N between Rio Tapajós, western Pará, and Rio Turiacu, western Maranhão, also upper Rio Madeira and tributaries, northern Rondônia and northwestern Mato Grosso; up to 500m; endangered, CITES I. **LOCALITIES** Amazônia National Park, Pará, and Jamari National Forest, Rondônia, Brazil.

YELLOW-EARED CONURE
Ognorhynchus icterotis 42cm

Large green parrot with long, graduated tail; forehead, lores, and elongated ear-coverts yellow; sexes alike, JUV like adults; goose-like *raanh-raanh*. Associated with *Ceroxylon* palms in montane forests; mostly small flocks; strong, direct flight. **DISTRIBUTION** northwestern Ecuador, north from Pichincha and western Cotopaxi, and western Colombia, north to Antioquia and northwestern Norte de Santander; recent records only from few localities in Cordilleras Central and Oriental, Colombia, and western Cotopaxi, Ecuador; mostly 1200 to 3500m; critically endangered, CITES I. **SIMILAR SPECIES** Golden-plumed Conure *Aratinga branickii* (plate 84) orange frontal band; narrow yellow "tufts" behind ear-coverts; smaller. **LOCALITIES** Cueva de los Guacharos National Park, Huila, Colombia. Cerro Golondrinas Reserve, near Ibarra, Imbabura, and Caripero district, western Cotopaxi, Ecuador.

THICK-BILLED PARROT
Rhynchopsitta pachyrhyncha 38cm

One of two similar large green parrots with short, wedge-shaped tail and large, strongly compressed black bill; forecrown, broad band above eye, bend of wing, and thighs red; greater underwing-coverts yellow; sexes alike, JUV red forehead only; raucous *scronk*, repeated *haw-haw-haw*, sharp *kuk-kuk-kuk*, rolling *aa-ahr*. Highland *Pinus* and *Pinus-Quercus* forests; arboreal; wanders in search of *Pinus* seeds, the staple food; strong, direct flight. **DISTRIBUTION** northwestern to central Mexico, along Sierra Madre Occidental from western Chihuahua and eastern Sonora to Michoacán or occasionally Jalisco; formerly north to southwestern U.S.A.; mostly 1500 to 3400m; endangered, CITES I. **SIMILAR SPECIES** Military Macaw *Ara militaris* (plate 80) larger with longer tail; bare face; no red above eye or on bend of wing. **LOCALITY** Cebadillas de Yaguirachic, and Tutuaca Forest Reserve, Chihuahua, Mexico.

MAROON-FRONTED PARROT
Rhynchopsitta terresi 40cm

Darker green than similar, possibly conspecific *R. pachyrhyncha*, with deep maroon forecrown and broad band above eye; olive greater underwing-coverts; habits and calls like *R. pachyrhyncha*. Highland *Pinus-Abies-Quercus* forests near roosting and nesting cliff-face. **DISTRIBUTION** northeastern Mexico, along 300km of Sierra Madre Oriental, in southeastern Coahuila, central-western Nuevo León, and southwestern Tamaulipas; mostly 1300 to 3700m; vulnerable, CITES I. **SIMILAR SPECIES** Military Macaw *Ara militaris* (plate 80) see above. **LOCALITY** El Taray Sanctuary, near Monterrey, Nuevo León, Mexico.

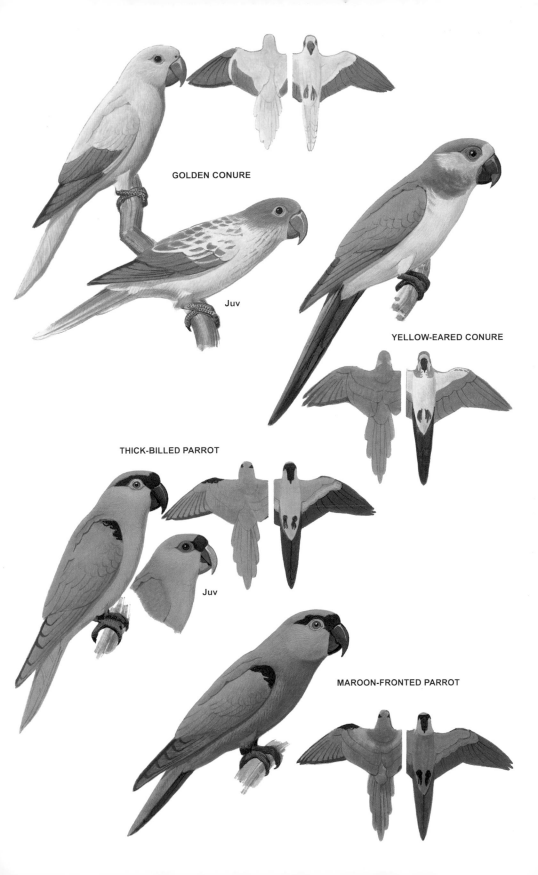

GOLDEN CONURE

Juv

YELLOW-EARED CONURE

THICK-BILLED PARROT

Juv

MAROON-FRONTED PARROT

PLATE 84 *ARATINGA* **CONURES (in part)**

188 Small to midsized, mostly green parrots with long, graduated tail; sexes alike, JUV often lacks features of adults, so can be difficult to identify; includes several distinctive groupings often treated as separate genera, but here considered subgenerically differentiated. Forests, woodlands, dry scrublands, farmlands; mostly arboreal, but some ground feeding; pairs, small to large flocks; noisy and conspicuous in flight, but can be hidden amidst foliage; shrill screeching calls, and chatter of feeding flocks.

GOLDEN-PLUMED CONURE
Aratinga branickii (formerly *Leptosittaca branickii*) 35cm

Midsized green conure with orange frontal band and yellow stripe beneath eye continuing to tuft of elongated feathers behind ear-coverts; abdomen suffused orange; undertail brownish red; shrill *kree-ah…kree-ah*, high-pitched *rhaaaa-aa*. Attracted to fruiting *Podocarpus*. **DISTRIBUTION** disjunctly in Andes highlands of Colombia to southern Peru; mostly 2400 to 3400m; vulnerable. **SIMILAR SPECIES** Yellow-eared Conure *Ognorhynchus icterotis* (plate 83) forehead, upper cheeks, and elongated ear-coverts yellow; no orange suffusion on abdomen; larger. **LOCALITIES** Puracé National Park, Cauca, and Los Nevados National Park, Tolima, Colombia. Podocarpus National Park, and Tapichalaca Biological Reserve, Loja/Zamora-Chinchipe, southern Ecuador.

NANDAY CONURE
Aratinga nenday (formerly *Nandayus nenday*) 30cm

Unmistakable; midsized green conure with black head; throat and upper breast suffused blue; thighs orange-red; bare eye-ring pink; strident *kree-ah…kree-ah*, *krehh* and *kriie-kriie*. **DISTRIBUTION** upper Río Paraguay basin, from Santa Cruz, southeastern Bolivia, and Mato Grosso do Sul, Brazil, through central Paraguay to northern Argentina; up to 800m; common; feral populations near Buenos Aires, Argentina, in Florida and southern California, U.S.A., and Puerto Rico, West Indies. **LOCALITIES** Caiman Lodge Wildlife Refuge, Pousada Aguapé at São Jose Ranch, and Passo do Lontra, Mato Grosso do Sul, Brazil. Outskirts of Asunción and at Estancia Santa Asunción, Presidente Hayes, Paraguay.

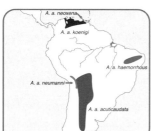

BLUE-CROWNED CONURE
Aratinga acuticaudata 37cm

Unmistakable; only green conure with blue forecrown and red undertail; prominent white eye-ring; loud *cheeah-cheeah*, musical *krraa*. **DISTRIBUTION** disjunctly east of Andes in northeastern Colombia, northern Venezuela, and northeastern Brazil to northern Argentina; up to 2600m; common; introduced to Florida and southern California, U.S.A. **SUBSPECIES** three discernible and two poorly differentiated subspecies. 1. *A. a. acuticaudata* crown to cheeks and ear-coverts dull blue; pale upper mandible, gray lower mandible. *Range* lowlands in eastern Bolivia and southwestern Mato Grosso do Sul, Brazil, to Paraguay, western Uruguay, and northern Argentina, south to La Pampa and southwestern Buenos Aires. 2. *A. a. neumanni* only forehead to nape blue; underparts washed blue. *Range* eastern slopes of Andes (1500 to 2650m) in Cochabamba, Santa Cruz, Chuquisiaca, and probably Tarija, southern Bolivia. 3. *A. a. haemorrhous* like *neumanni*, but paler blue forehead and forecrown; no blue on underparts; all-pale bill. *Range* northeastern Brazil, in Piauí and northern Bahia. 4. *A. a. koenigi* like *neumanni*, but undertail less brownish red; smaller. *Range* northeastern Colombia and northern Venezuela, east to Monagas. 5. *A. a. neoxena* like *haemorrhous*, but underparts suffused bluish; smaller. *Range* Margarita Island, northern Venezuela. **LOCALITIES** Catatumbo-Barí National Park, Norte de Santander, and El Tuparro National Park, Vichada, Colombia. Caiman Lodge Wildlife Refuge, Mato Grosso do Sul, Brazil.

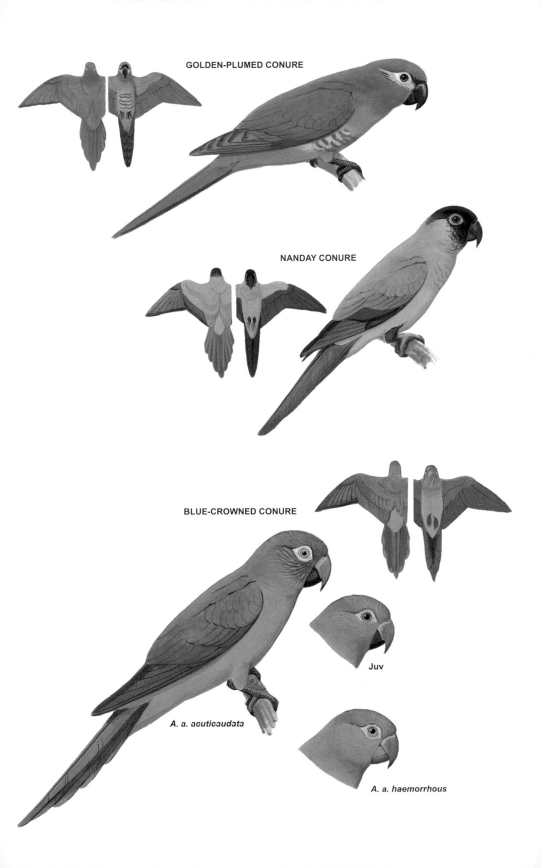

GOLDEN-PLUMED CONURE

NANDAY CONURE

BLUE-CROWNED CONURE

A. a. acuticaudata

Juv

A. a. haemorrhous

PLATE 85 *ARATINGA* **CONURES (in part)**

190

GREEN CONURE *Aratinga holochlora* 30cm

One of three closely allied, probably conspecific, all-green midsized conures in Central America; shrill *screek…screek* and *cree-ik…crii-crii-criir*, deeper *kreh-kreh…kteh-kreh* or *ruh-ruh-ruh*. **DISTRIBUTION** northwestern and eastern Mexico; up to 2000m; common; introduced to southern Texas and southern Florida, U.S.A. **SUBSPECIES** two discernible subspecies. 1. *A. h. holochlora* yellowish-green underparts and olive-yellow underwings; some birds with few scattered red feathers on head. *Range* disjunctly in eastern Mexico, from eastern Nuevo León and Tamaulipas to central Veracruz, and southeastern Veracruz and eastern Oaxaca to eastern Chiapas, ranging seasonally to adjacent Pacific slope on Isthmus of Tehuantepec. 2. *A. h. brewsteri* darker green with bluish suffusion on crown. *Range* highlands of Sonora, northern Sinaloa, and southwestern Chihuahua, northwestern Mexico. **SIMILAR SPECIES** Pacific Conure *A. strenua* (see below) indistinguishable in field. Olive-throated Conure *A. nana* (plate 94) olive-brown underparts; white eye-ring; smaller. Orange-fronted Conure *A. canicularis* (plate 94) distinctive head markings; olive throat; smaller. **LOCALITIES** El Naranjo to El Meco road, San Luis Potosí, and Francisco I. Madero National Park, Chiapas, Mexico.

PACIFIC CONURE *Aratinga strenua* 33cm

Differentiated only at close quarters from very similar *A. holochlora* by larger size, heavier bill, and stouter legs; calls like *A. holochlora*. **DISTRIBUTION** Pacific slope from eastern Oaxaca, southern Mexico, to inland Guatemala and central Nicaragua; up to 2500m; common. **SIMILAR SPECIES** Green Conure *A. holochlora* (see above) indistinguishable in field. Red-throated Conure *A. rubritorquis* (see below) red throat; smaller. Olive-throated Conure *A. nana* (plate 94) see above. Orange-fronted Conure *A. canicularis* (plate 94) see above. **LOCALITIES** Nighttime roosts at Antiguo Cuscatlán, a suburb of San Salvador, El Salvador. Volcán Masaya National Park, Nicaragua.

SOCORRO CONURE *Aratinga brevipes* 32cm

Unmistakable; only parrot on island; like *A. holochlora*, but darker green and different wing pattern—p10 shorter than p7, vice versa in *A. holochlora* and *A. strenua*; high-pitched *kree-kree…kree-kree*, shorter *kee-kee-kee*. **DISTRIBUTION** Socorro Island in Islas Revillagigado, off west coast of Mexico. **LOCALITY** slopes of Cerro Evermann, Socorro Island.

RED-THROATED CONURE
Aratinga rubritorquis 28cm

Smaller green conure with throat and foreneck red, often intermixed yellow; JUV throat and foreneck green; screaming *kreeah-kreeah* and *krri-krreea*, persistent *krieh-krieh…krieh-krieh*. **DISTRIBUTION** central highlands and adjacent Pacific slope in eastern Guatemala to western Nicaragua; mostly 600 to 1800m; common. **SIMILAR SPECIES** Pacific Conure *A. strenua* (see above) no red on throat; larger. Olive-throated Conure *A. nana* (plate 94) see above. Orange-fronted Conure *A. canicularis* (plate 94) see above. **LOCALITIES** Guisayote Reserve and Celaque National Park, Honduras. Volcan Masaya National Park and Volcan Mombacho Nature Reserve, Nicaragua.

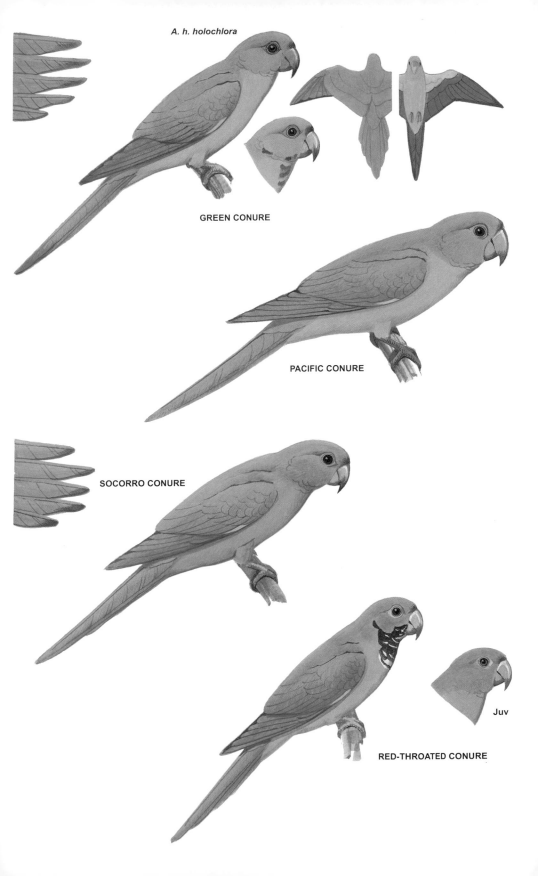

A. h. holochlora

GREEN CONURE

PACIFIC CONURE

SOCORRO CONURE

Juv

RED-THROATED CONURE

PLATE 86 *ARATINGA* CONURES (in part)

192

CRIMSON-FRONTED CONURE
Aratinga finschi 28cm

Smaller green conure with red forecrown not extending to eyes; bend of wing, carpal edge, and outer lesser underwing-coverts red; outer greater underwing-coverts yellow; some birds with few scattered red feathers on neck; white eye-ring; JUV little or no red on forecrown or underwing-coverts, and outer greater underwing-coverts olive; raucous *kaa-kaa-kaa*, guttural *keerr-keerr* or *kew-lee-kee-kee…kew-keerr*. **DISTRIBUTION** southwestern Nicaragua to Costa Rica, throughout Caribbean slope, regularly visiting Guanacaste, and on southern Pacific slope, and east to Azuero Peninsula, western Panama; up to 1650m; increasing. **SIMILAR SPECIES** Olive-throated Conure *A. nana* (plate 94) and Brown-throated Conure *A. pertinax* (plate 92) no red on forecrown or underwing-coverts; olive or brown throat; smaller. Sulphur-winged Conure *Pyrrhura hoffmanni* (plate 107) no red on forecrown or underwing-coverts; yellow on upper side of wings; smaller. **LOCALITIES** Bosque del Río Tigre Sanctuary and Lodge, Dos Brazos, and Lapa Ríos Nature Reserve, Golfo Dulce, Costa Rica.

CUBAN CONURE *Aratinga euops* 26cm

Unmistakable, only small green parrot in range; scattered red feathers on head and underparts, lacking in JUV; carpal edge and lesser underwing-coverts red; white eye-ring; loud *crick-crick-crick*, low whispering while perched. **DISTRIBUTION** Cuba, and formerly Isla de Pinos; uncommon. **LOCALITIES** Zapata Peninsula, Sierra de Najasa, and Guantánamo, Cuba.

A. c. chloroptera

A. c. maugei
(extinct)

HISPANIOLAN CONURE
Aratinga chloroptera 32cm

Larger green conure with red bend of wing, carpal edge, and outer underwing-coverts, mostly lacking in JUV; white eye-ring; shrill screeches. **DISTRIBUTION** Haiti and Dominican Republic, Hispaniola, West Indies; up to 3000m; uncommon: introduced to Guadeloupe and Puerto Rico. **SUBSPECIES** doubtfully distinct A. c. maugei formerly on Mona Island and possibly Puerto Rico, but now extinct. **SIMILAR SPECIES** Olive-throated Conure *A. nana* (plate 94) no red markings; olive-brown underparts; smaller. On Puerto Rico can be confused with other introduced parakeets. **LOCALITIES** Sierra de Baoruco and Sierra de Neiba, Dominican Republic. Massif de la Selle and Massif du Nord, Haiti.

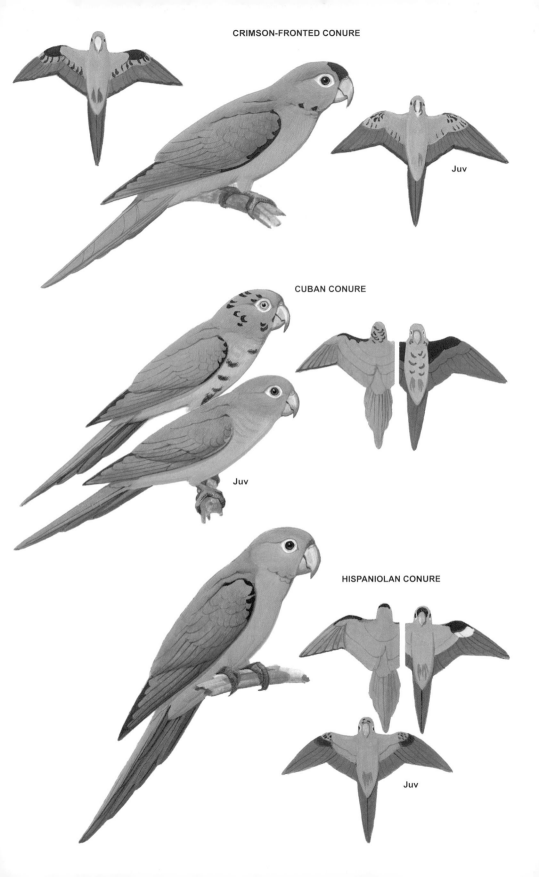

CRIMSON-FRONTED CONURE

Juv

CUBAN CONURE

Juv

HISPANIOLAN CONURE

Juv

WHITE-EYED CONURE
Aratinga leucophthalma 32cm

Midsized pale-billed green conure with distinctive red-and-yellow pattern on underwing-coverts and scattered red feathers on head and neck; only "all-green" conure in much of range; sexes alike, JUV duller; metallic *che-chek*, grating *scraaah…scraaah*, trilling notes. **DISTRIBUTION** much of South America, east of Andes; up to 2500m; common. **SUBSPECIES** three poorly differentiated subspecies. 1. *A. l. leucophthalma* red on carpal edge and edge of forewing; outer lesser underwing-coverts red and outer greater underwing-coverts yellow; white eye-ring. *Range* Guianas, northeastern Venezuela, from southeastern Sucre, Monagas, and northern Anzoátegui to northeastern Bolívar, and eastern Colombia, south from Meta, south through Brazil, except upper Amazon River basin and dry northeast, to eastern Bolivia, Paraguay, northern Uruguay, and northern Argentina, south to Catamarca, northern Santa Fé, and Entre Ríos. 2. *A. l. callogenys* like *leucophthalma*, but larger and with heavier, more robust bill; westernmost birds darker green. *Range* upper Amazonia, near foothills of Andes, in southeastern Colombia, eastern Ecuador, and northeastern Peru to extreme northwestern Brazil; intergrades with *leucophthalma* throughout central Amazonia. 3. *A. l. nicefori* like *callogenys*, but paler, more yellowish green; red band across forehead. *Range* known only from type specimen collected at Guaicaramo, Río Guavio, Meta, Colombia; possibly aberrant *callogenys* or *A. leucophthalma* × *A. wagleri* hybrid. **SIMILAR SPECIES** can be confused with "red-fronted" green conures in western part of range, but lacks red forecrown and with red-and-yellow pattern on underwing-coverts. **LOCALITIES** Tinigua National Park, Meta, and Amacayacu National Park, Amazonas, Colombia. Amazonia National Park, Pará, Cristalino Jungle Lodge, Cristalino State Park, northern Mato Grosso, and Caiman Lodge Wildlife Refuge, Mato Grosso do Sul, Brazil. Manú Biosphere Reserve, Madre de Dios, southeastern Brazil.

MITRED CONURE *Aratinga mitrata* 38cm

One of four very similar midsized pale-billed, "red-fronted" green conures of uncertain taxonomic status; red extending from forecrown to lores and below eyes; white eye-ring; sexes alike, JUV red restricted to forecrown and lores; strident *weee-weee…queiiee-queiiee…weee-weee*, loud *cheeah…cheeah*, calls deeper and harsher than Red-fronted Conure *A. wagleri*. Favors dry forest and adjoining open woodland (Arndt, pers. comm.). **DISTRIBUTION** northern Peru south to northwestern Argentina; mostly 1000 to 3500m; common; introduced to Florida, California, and Maui, Hawaii, U.S.A. **SUBSPECIES** three subspecies, but probably north–south cline of increasing red on head. 1. *A. m. chlorogenys* red frontal band extending to lores and narrow, incomplete ring around eye; upper cheeks and ear-coverts green. *Range* northern and central Peru, in Río Utcubamba valley and surrounding highlands in Amazonas and northern Cajamarca, and eastern Andean valleys in Huánuco and Junín. 2. *A. m. mitrata* red forecrown and lores extending to irregular band underneath eye to ear-coverts and upper cheeks; some scattered red feathers on sides of neck and breast; thighs red. *Range* southern Peru, at about lat. 12°50'S in Ayacucho, south to Salta, northwestern Argentina. 3. *A. m. tucumana* red forecrown and lores extending to cheeks and ear-coverts. *Range* northwestern Argentina, in Tucumán and Córdoba, and probably also Catamarca and La Rioja. **SIMILAR SPECIES** easily confused with sympatric "red-fronted" green conures (plate 88), but only species with red extending underneath eyes; juveniles probably indistinguishable in field. **LOCALITIES** Cerro Palmarcito, Chuquisaca, and Río Caine valley, Cochabamba, Bolivia. Calilegua National Park, Jujuy, and El Rey National Park, Salta, northwestern Argentina.

WHITE-EYED CONURE

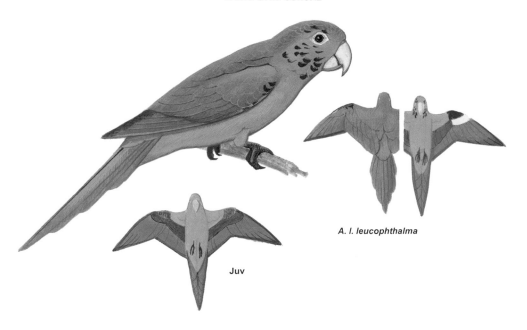

A. l. leucophthalma

Juv

MITRED CONURE

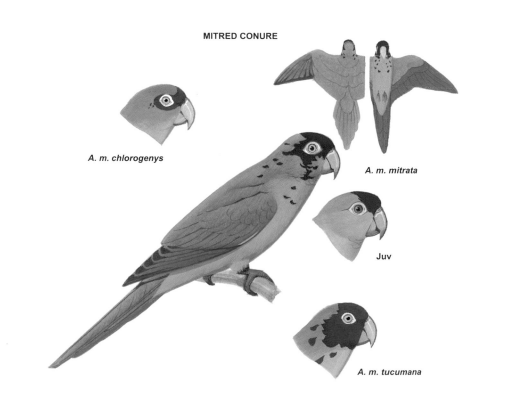

A. m. chlorogenys

A. m. mitrata

Juv

A. m. tucumana

PLATE 88 *ARATINGA* **CONURES (in part)**

196

MOUNTAIN CONURE *Aratinga alticola* 35cm

Midsized red-fronted green conure probably indistinguishable in field from A. *hockingi*; dull green upperparts suffused glaucous; red frontal band not exceeding 16mm in width, and not extending to forecrown or to eyes; variable scattered red feathers on lores and around eyes; thighs green; cream-white eye-ring; JUV undescribed. **DISTRIBUTION** known only from Huancavelica and Cuzco, southern Peru, and Cochabamba, Bolivia. **SIMILAR SPECIES** probably cannot be distinguished in field from Hocking's Conure A. *hockingi* (see below) and JUV Mitred Conure A. *mitrata* (plate 87). **LOCALITIES** vicinity of Anco, Huancavelico, southern Peru. Carrasco Ichilo National Park, Cochabamba, Bolivia.

HOCKING'S CONURE *Aratinga hockingi* 35cm

Possibly conspecific with Red-fronted Conure A. *wagleri* (plate 89) and distinguishable from very similar A. *alticola* only at close quarters; "half-moon" shaped red frontal band exceeding 17.5mm in width, and extending to forecrown but not to eyes; sometimes scattered red feathers on lores, around eyes, and on cheeks; thighs green; cream-white eye-ring; JUV uniformly green without red frontal band. Possibly favors cloudforests, visits cornfields. **DISTRIBUTION** mountains east of Río Utcubamba valley, Amazonas, and eastern Andean valleys in Huánuco and Cuzco, Peru; recorded between 1760m and 3000m; locally common. **SIMILAR SPECIES** probably cannot be distinguished in field from Mountain Conure A. *alticola* (see above) and JUV Mitred Conure A. *mitrata* (plate 87). **LOCALITIES** Chosgon district and Rioja to Pedro Ruiz road, Río Utcubamba valley, Amazonas, northern Peru (Arndt pers com.).

CORDILLERAN CONURE
Aratinga frontata 36cm

Only pale-billed, "red-fronted" conure with red on bend of wing to carpal edge; sexes alike, JUV without red markings; nasal *keh-keh* in flight. Favors dry and semihumid forests, visits orchards and cornfields. **DISTRIBUTION** southern Ecuador and western Peru; up to 3000m; locally common; feral population in Lima. **SUBSPECIES** two discernible subspecies. 1. A. *f. frontata* red forecrown to lores; red bend of wing to carpal edge, and red thighs; cream-white eye-ring. *Range* Pacific slope of Andes in southwestern Ecuador and western Peru, south to Tacna. 2. A. *f. minor* darker green; more extensive, paler red on thighs; smaller. *Range* central Andes of Peru from Río Marañón valley, Amazonas, and possibly Zumba district in neighboring southernmost Ecuador, south to Ayacucho and Apurímac. **SIMILAR SPECIES** Red-masked Conure *Aratinga erythrogenys* (plate 89) entirely red crown and face; red on underwing-coverts; JUV also with red on underwing-coverts. Other "red-fronted" green conures lack red on bend of wing to carpal edge. **LOCALITY** Machu Picchu Historical Reserve, Cuzco, southern Peru.

MOUNTAIN CONURE

HOCKING'S CONURE

CORDILLERAN CONURE

A. f. frontata

PLATE 89 *ARATINGA* **CONURES (in part)**

198

RED-MASKED CONURE
Aratinga erythrogenys 33cm

Midsized pale-billed, green conure with entirely red crown and face; red bend of wing to carpal edge and outer lesser underwing-coverts; sexes alike, JUV crown and face green, and less red on bend of wing to carpal edge and outer lesser underwing-coverts; screeching *screee-screeah* or *scrah-scrah-scra-scra*, rasping *screet-screet-screet*.

DISTRIBUTION lowlands and Pacific slope of western Ecuador and northwestern Peru, south to Lambayeque and Cajamarca; up to 2500m; near-threatened; feral population in Lima, and introduced to Florida and California, U.S.A., and Grand Cayman Island, West Indies. **SIMILAR SPECIES** Cordilleran Conure A. *frontata* (plate 88) red restricted to forecrown and lores. **LOCALITIES** Machalilla National Park, Manabí, and Loma Alta Ecological Reserve, Guayas, Ecuador. Tumbes National Forest, Tumbes, northwestern Peru.

RED-FRONTED CONURE
Aratinga wagleri 36cm

Midsized pale-billed, green conure with red forecrown; only "red-fronted" green conure in range, though possibly conspecific with Hocking's Conure A. *hockingi* (plate 88); sexes alike, JUV little or no red on forecrown; strident *skreek*. Local presence associated with cliff-face roosting and nesting sites. **DISTRIBUTION** northern Venezuela and western Colombia; mainly 700 to 2800m; locally common. **SUBSPECIES** two poorly differentiated subspecies. 1. A. *w. wagleri* red forecrown not extending to eyes; sometimes red band or scattered red feathers on throat. *Range* northwestern Venezuela, from Sierra de Perijá, Zulia, and both slopes of Andes in Merida east to mountains of Yaracuy and Distrito Federal, and western Colombia, from Santa Marta Mountains and Sierra de Perijá south along all Andean cordilleras to Nariño. 2. A. *w. transilis* darker green, and darker red on forecrown less extensive posteriorly. *Range* coastal mountains of Sucre and northern Monagas, northeastern Venezuela. **SIMILAR SPECIES** White-eyed Conure A. *leucophthalma* (plate 87) possible sympatry in Sucre, northeastern Venezuela; no red forecrown; distinctive red-and-yellow pattern on outer underwing-coverts. **LOCALITIES** Henri Pittier National Park, Aragua, Venezuela. Catatumbo-Barí National Park, Norte de Santander, and Cueva de los Guácharos National Park, Huila, Colombia.

DUSKY-HEADED CONURE
Aratinga weddellii 28cm

Unmistakable; only black-billed, green conure with bluish-gray head; tail green tipped dark blue; wide bare eye-ring white; sexes alike, JUV head more greenish, less gray, and narrower white eye-ring; series of *jee-jeek* notes. Favors várzea forest along watercourses, and attracted to *Erythrina* blossoms; tame. **DISTRIBUTION** east of Andes in southeastern Colombia, south from Caquetá and Vaupés, to eastern Ecuador, and eastern Peru to northwestern Brazil in western Amazonas to northwestern Mato Grosso, and northern Bolivia, south to Cochabamba; up to 500m, locally 900m; common. **LOCALITIES** Amacayacu National Park, Amazonas, southern Colombia. Manú Biosphere Reserve, Madre de Dios, southeastern Peru.

RED-MASKED CONURE

Juv

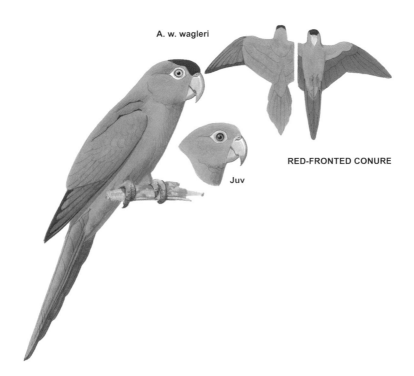

A. w. wagleri

Juv

RED-FRONTED CONURE

DUSKY-HEADED CONURE

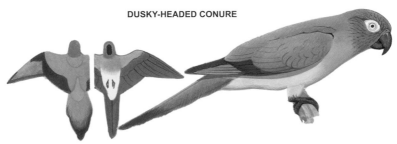

PLATE 90 *ARATINGA* CONURES (in part)

200

GOLDEN-CAPPED CONURE
Aratinga auricapillus 30cm

Only black-billed green conure with golden-yellow crown and orange face; sexes alike, JUV duller than adults; calls presumably like *A. jandaya*. **DISTRIBUTION** eastern Brazil; up to 2200m; near-threatened. **SUBSPECIES** two discernible subspecies with intermediates in range overlap. 1. *A. a. auricapillus* forehead and lores to around eyes orange-red; crown golden-yellow; abdomen, lower breast, and lower back to rump brownish red; gray eye-ring; JUV less yellow on crown, and less red on underparts. *Range* northern and central Bahia, eastern Brazil; birds from southern Bahia intermediate between this subspecies and *aurifrons*. 2. *A. a. aurifrons* forecrown orange-red; lower back to rump green. *Range* southeastern Brazil, from Minas Gerais and southern Goiás to Santa Catarina. **SIMILAR SPECIES** Jandaya Conure *A. jandaya* (see below) possibly locally sympatric in marginal contact zone; entirely yellow head and neck. Peach-fronted Conure *A. aurea* and Cactus Conure *A. cactorum* (plate 93) olive-brown throat and breast; lower underparts yellow; pale bare or feathered eye-ring. **LOCALITIES** Monte Pascual National Park, Bahia, and Rio Doce State Park and Serra da Canastra National Park, Minas Gerais, Brazil.

JANDAYA CONURE *Aratinga jandaya* 30cm

One of two similar black-billed, yellow-and-green conures; differentiated from *A. solstitialis* by green back, wings, and lower underparts; head and neck yellow suffused orange on face; breast, abdomen, and underwing-coverts orange-red; gray or white eye-ring; sexes alike; JUV head and neck paler yellow variably marked green, and breast to abdomen paler orange; screeching *kink-kink-kank*.

DISTRIBUTION northeastern Brazil, from southeastern Pará, and near São Luis, Maranhão, east to Rio Grande do Norte and south to Alagoas, northernmost Bahia, and eastern Goiás; feral population at Belem, Pará; common. **SIMILAR SPECIES** Golden-capped Conure *A. auricapillus* (see above) possibly sympatric in marginal contact zone; green cheeks, neck, and breast. Peach-fronted Conure *A. aurea* and Cactus Conure *A. cactorum* (plate 93) see above. **LOCALITY** BioBrasil Reserve, near Barreiras, southern Piauí.

SUN CONURE *Aratinga solstitialis* 30cm

Black-billed yellow conure with green primaries and secondary-coverts; differentiated from *A. jandaya* by yellow mantle, back, upper wing-coverts, and lower underparts; face, and breast to abdomen suffused orange; grayish-white eye-ring; sexes alike, JUV upperparts and neck variably mixed green; screeching *screek-screek-screek* in flight, chuckling notes while perched. **DISTRIBUTION** far northeastern Brazil, Guyana, and possibly northwestern and southern Suriname, and extreme southeastern Venezuela; up to 1200m; locally common. **SUBSPECIES** two well marked, apparently isolated subspecies. 1. *A. s. solstitialis* deep yellow; mantle and upper wing-coverts yellow; olive-green tail; grayish-white eye-ring. *Range* central Guyana and neighboring Roraima, far northeastern Brazil; sight records from northwestern Suriname and extreme southeastern Venezuela. 2. *A. s. maculata* paler yellow; mantle and upper wing-coverts green edged yellow. *Range* northeastern Brazil, in northern Pará and lower Rio Canuma in northeastern Amazonas; unconfirmed sight records from southern Suriname. **LOCALITIES** Contão and Boa Vista districts, Roraima, and Monte Alegre district, Pará, northeastern Brazil.

GOLDEN-CAPPED CONURE

A. a. auricapillus

Juv

A. a. aurifrons

JANDAYA CONURE

Juv

A. s. solstitialis

Juv

SUN CONURE

PLATE 91 *ARATINGA* CONURES (in part)

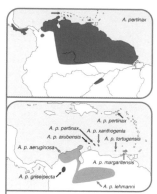

BROWN-THROATED CONURE
Aratinga pertinax 25cm

Polytypic black-billed, green conure with strong geographical variation in plumage coloration; identified by varying combinations of yellow and brown on face and throat to upper breast; sexes alike, JUV duller than adults; repeated *crik-crik…crak-crak* in flight, *cheer-cheedit* when perched. **DISTRIBUTION** easternmost Costa Rica to central Panama, northern South America, mostly north of Amazon River, and islands off north coast of Venezuela; up to 1600m; common; introduced to St. Thomas and Saba, in Virgin Islands, West Indies. **SUBSPECIES** (in part, see plate 92) 14 subspecies differentiated mainly by extent and intensity of yellow on head and brown on throat to upper breast. 1. A. *p. pertinax* forehead, lores and sides of head orange-yellow; throat to upper breast pale olive-brown; variable orange marking on center of abdomen; iris yellow. *Range* Curaçao, in Netherlands Antilles, off north coast of Venezuela; introduced to St. Thomas and Saba in Virgin Islands. 2. A. *p. xanthogenia* brighter orange-yellow extending to crown. *Range* Bonaire, in Netherlands Antilles. 3. A. *p. arubensis* orange-yellow restricted to forehead and around eyes; crown dull greenish blue; lores, cheeks, and ear-coverts light brown faintly streaked dull yellow. *Range* Aruba, in Netherlands Antilles. 4. A. *p. aeruginosa* forecrown darker blue; dull yellow only encircling eye; lores to cheeks, ear-coverts, and upper breast grayish brown. *Range* northern Colombia from Caribbean coast south in middle Río Magdalena valley to northern Santander, and east through Guajira Peninsula to northwestern Zulia, northwestern Venezuela. 5. A. *p. griseipecta* like *aeruginosa*, but cheeks, throat, and upper breast olive-gray; no yellow encircling eye; forehead white. *Range* known only from Río Sinú valley, northeastern Colombia. 6. A. *p. lehmanni* similar to *aeruginosa*, but more extensive orange-yellow feathered eye-ring; less extensive greenish blue on forecrown. *Range* central and eastern Colombia, east of Andes, from Casanare south to northernmost Vaupés and east to Río Orinoco in western Amazonas, southern Venezuela. 7. A. *p. tortugensis* like *aeruginosa*, but more extensive orange-yellow on sides of head; larger. *Range* Tortuga Island, off north coast of Venezuela. 8. A. *p. margaritensis* forehead white; forecrown dull greenish blue; throat and upper breast pale olive-brown; little or no orange on darker green abdomen. *Range* Margarita and Los Frailes Islands, off north coast of Venezuela.

BROWN-THROATED CONURE

A. p. pertinax

Juv

A. p. xanthogenia

A. p. arubensis

A. p. aeruginosa

A. p. griseipecta

A. p. margaritensis

PLATE 92 *ARATINGA* **CONURES (in part)**

204

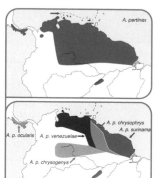

BROWN-THROATED CONURE
Aratinga pertinax (cont.)

SUBSPECIES (in part, see also plate 91) 9. A. *p. venezuelae* like *margaritensis*, but paler, more yellowish green; no white frontal band; abdomen suffused orange. *Range* generally throughout Venezuela, except extreme northwest where replaced by *aeruginosa*, Delta Amacuro and southeastern Monagas occupied by *surinama*, southeastern Bolívar inhabited by *chrysophrys*, and western Amazonas within range of *lehmanni*. 10. A. *p. chrysophrys* similar to *venezuelae*, but darker green; lores, cheeks, and ear-coverts darker brown shaft-streaked dull yellow; forehead pale brownish yellow. *Range* interior of Guyana, Cerro Roraima district to upper Río Caroní in southeastern Bolívar, Venezuela, and extreme northern Roraima in northeastern Brazil. 11. A. *p. surinama* like *chrysophrys*, but orange-yellow extending from below eyes to cheeks and base of lower mandible; narrow orange-yellow frontal band. *Range* French Guiana, Suriname, and along coast of Guyana to Delta Amacuro and southeastern Monagas, northeastern Venezuela. 12. A. *p. chrysogenys* darker coloration than other subspecies; no pale frontal band; crown dark greenish blue; center of abdomen extensively tinged dark orange. *Range* recorded from lower Rio Negro and Rio Solimões region, west to Igarapé Belem, Amazonas, northwestern Brazil, but exact range undetermined. 13. A. *p. paraensis* forehead and crown bluish green; outer webs of primaries and secondaries blue; iris red. *Range* apparently isolated south of Amazon River, where known only from Rio Tapajós and its tributary, Rio Cururu, western Pará, northern Brazil. 14. A. *p. ocularis* crown and forehead green; orange-yellow in front of and below eye; sides of head buff-brown, darker on ear-coverts; throat and upper breast orange-brown. *Range* easternmost Costa Rica, where recorded along Río Coto, and Pacific lowlands of Panama from western Chiriquí east to western Panama province, in vicinity of Panama City and Tocumen. **SIMILAR SPECIES** Peach-fronted Conure A. *aurea* (plate 93) orange forecrown and orange feathered eye-ring; dark blue in flight feathers. **LOCALITIES** along Pan-American Highway, west from Playa Coronado, Panama. Botanic Gardens, Georgetown, Guyana. Henri Pittier National Park, Aragua, Venezuela. Tayrona National Park, Magdalena, Paramillo National Park, Córdoba, and Tinigua National Park, Meta, Colombia. Amazonia National Park, Pará, Brazil.

A. p. venezuelae

A. p. surinama

A. p. paraensis

BROWN-THROATED CONURE

A. p. ocularis

PEACH-FRONTED CONURE
Aratinga aurea 26cm

Small black-billed, green conure with orange frontal band and feathered eye-ring; sexes alike, JUV duller with bare eye-ring pale gray; shrill screeching in flight. Favors open woodland, woodlots in cultivation or pasturelands, and low scrubby vegetation; often feeds on ground; some mutual exclusion with Cactus Conure A. *cactorum*, where one species common the other normally absent or scarce. **DISTRIBUTION** southernmost Suriname and far northeastern Brazil south to northwestern Argentina; up to 600m; common. **SUBSPECIES** two poorly differentiated subspecies, probably north–south cline of increasing size. 1. A. *a. aurea* forehead and forecrown yellow-orange; hindcrown to occiput dull blue; throat, cheeks, and breast pale brownish olive. *Range* southernmost Suriname and neighboring extreme northeastern Brazil, and south of Amazon River from northeastern and inland Brazil to eastern Bolivia, southeastern Peru, and northwestern Argentina; up to 600m; common. 2. A. *a. major* slightly darker green; paler yellow-orange frontal band; larger. *Range* exact range undetermined, but has been recorded from northern Paraguay. **SIMILAR SPECIES** Cactus Conure A. *cactorum* (see below) no orange frontal band; brown crown and face with contrasting green ear-coverts; bare white eye-ring, and pale bill. Brown-throated Conure A. *pertinax* (plate 91, 92) no orange forecrown; cheeks and ear-coverts olive-brown shaft-streaked paler. Golden-capped Conure A. *auricapillus* (plate 90) red frontal band, lores, and around eyes; golden-yellow forecrown; lower underparts brownish red; dark eye-ring. **LOCALITIES** Serra da Canastra National Park, Minas Gerais, and Caiman Lodge Wildlife Refuge, Mato Grosso do Sul, Brazil. Los Ferros Lodge, Noel Kempff Mercado National Park, Santa Cruz, Bolivia. Cerro Corá National Park, Amambay, Paraguay.

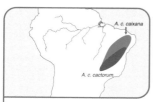

CACTUS CONURE *Aratinga cactorum* 25cm

Small pale-billed, green conure with olive-brown crown and throat to breast, and yellow abdomen; sexes alike, JUV duller, with green crown and olive breast to abdomen; strident *cri-cri-cri* or *screet…screet*. Closely associated with caatinga scrubland where dominated by low, thorny shrubs and low trees, also in semidesert country and degraded pastures; some mutual exclusion with A. *aurea*, where one species common the other normally absent or scarce. **DISTRIBUTION** interior of northeastern Brazil; up to 600m; common. **SUBSPECIES** two poorly differentiated subspecies. 1. A. *c. cactorum* crown, face, and upper breast olive-brown with contrasting green ear-coverts; lower breast to abdomen orange-yellow; bare eye-ring white. *Range* restricted to Bahia, south of Rio São Francisco, and adjacent northeastern Minas Gerais. 2. A. *c. caixana* paler green; throat and breast buff-brown; abdomen less orange, more yellow. *Range* west and north of Rio São Francisco in northwestern Bahia and western Pernambuco north to Ceará, Piauí, and southeastern Maranhão. **SIMILAR SPECIES** Peach-fronted Conure A. *aurea* (see above) orange forecrown and feathered eye-ring; black bill. Golden-capped Conure A. *auricapillus* (plate 90) see above. **LOCALITIES** Serra da Capivara National Park, Piauí, and Chapada Diamantina National Park, Bahia, Brazil.

PEACH-FRONTED CONURE

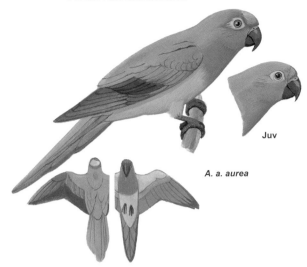

Juv

A. a. aurea

CACTUS CONURE

A. c. cactorum

A. c. caixana

PLATE 94 *ARATINGA* **CONURES (in part)**

208

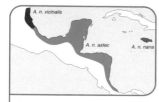

OLIVE-THROATED CONURE
Aratinga nana 26cm

Only *Aratinga* conure with all-green head and olive-brown throat to abdomen; sexes alike, JUV like adults; screeching *screek…screek*, higher-pitched *krrieh-krrie-krreah*. **DISTRIBUTION** Jamaica and Caribbean slope of tropical Central America; up to 1100m; common; introduced to Dominican Republic, but subspecies not determined. **SUBSPECIES** three slightly differentiated subspecies; Jamaican and Central American populations often treated as separate species. 1. *A. n. nana* throat to upper abdomen dark olive-brown; bare eye-ring white. *Range* Jamaica, West Indies. 2. *A. n. astec* (Aztec Conure) smaller than *nana*, but with proportionately longer wings; throat to upper abdomen paler olive-brown; smaller bill. *Range* Caribbean slope from Veracruz, southeastern Mexico, to Almirante Bay region, westernmost Panama. 3. *A. n. vicinalis* like *astec*, but brighter green; underparts more olive, less brownish. *Range* northeastern Mexico, from central Tamaulipas south to northeastern Veracruz, where intergrades with *astec*. **SIMILAR SPECIES** Red-throated Conure *A. rubritorquis* (plate 85) orange-red throat and foreneck; breast to lower abdomen green; bare eye-ring brownish gray. Other *Aratinga* conures in range have all-green underparts and are larger. **LOCALITIES** Ria Lagartos Natural Park, Yucatán, southern Mexico. Tikal National Park, El Petén, Guatemala. Río Bravo Conservation and Management Area, Orange Walk District, and Chin Chich Lodge, Belize. Pico Bonito National Park, northern Honduras.

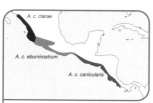

ORANGE-FRONTED CONURE
Aratinga canicularis 24cm

Small, pale-billed, green conure with orange frontal band and olive throat; sexes alike, JUV like adults; screeching *kreer…kreei-krrei* or *rreek…ree-reeh*, rapidly repeated *can-can-can*. Presence dependent on availability of arboreal termitaria for nesting. **DISTRIBUTION** Pacific slope of Central America; up to 1500m; common; introduced to Puerto Rico, West Indies. **SUBSPECIES** three subspecies differentiated primarily on extent of orange frontal band. 1. *A. c. canicularis* broad orange frontal band extending to lores; crown dull blue; throat and upper breast pale brownish olive; bare eye-ring dull orange-yellow. *Range* Pacific slope from Chiapas, southern Mexico, to Honduras and northwestern Costa Rica. 2. *A. c. eburnirostrum* narrower orange frontal band; blue restricted to forecrown; brownish spot on each side of base of lower mandible. *Range* southwestern Mexico, from easternmost Michoacán south to Oaxaca. 3. *A. c. clarae* orange frontal band much reduced; dark gray at base of lower mandible. *Range* western Mexico, from Sinaloa south to Colima. **SIMILAR SPECIES** Red-throated Conure *A. rubritorquis* (plate 85) see above. Crimson-fronted Conure *A. finschi* (plate 86) red forecrown, bend of wing to carpal edge, and outer underwing-coverts. Other *Aratinga* conures in range have all-green head. **LOCALITIES** Mazatlan to La Noria road, Sinaloa, and Sierra de Atoyac, Guerrero, Mexico. Tarrales Ecolodge, on lower slopes of Atitlan Volcano, Guatemala. El Impossible and Walter Thilo Deininger National Parks, El Salvador. Volcan Masaya National Park and Volcan Mombacho Nature Reserve, Nicaragua. Santa Rosa National Park, northwestern Costa Rica.

OLIVE-THROATED CONURE

A. n. nana

A. n. astec

ORANGE-FRONTED CONURE

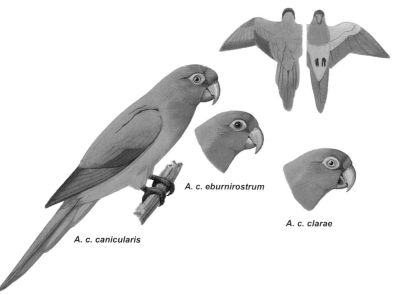

A. c. eburnirostrum

A. c. clarae

A. c. canicularis

PLATE 95 *CYANOLISEUS* AND *MYIOPSITTA* PARROTS

PATAGONIAN CONURE
Cyanoliseus patagonus 45cm

Unmistakable; large olive-brown parrot with long, strongly graduated tail and prominent bare eye-ring; sexes alike, JUV like adults; shrill *scree-ah…scree-ah*; grating *graaap…graaap* interspersed with sharp *keeew* notes (*bloxami*). Open country, especially near watercourses; highly conspicuous, noisy, and gregarious; very large flocks at some cliff-face nesting and roosting sites; swift, direct flight with distinctive streamlined silhouette and exposed yellow rump. **DISTRIBUTION** central Chile, much of Argentina and southern Uruguay; up to 2000m; endangered in Chile, common elsewhere. **SUBSPECIES** one isolated, well-marked subspecies, two discernible subspecies, and one doubtfully distinct subspecies. 1. *C. p. patagonus* head, neck, back, and breast olive-brown; lower back and abdomen to undertail-coverts yellow; center of abdomen orange-red; whitish marking on each side of breast. *Range* breeds in southern Argentina, ranging north in winter to central Argentina and southern Uruguay. 2. *C. p. conlara* like *patagonus*, but darker breast. *Range* resident in central Argentina. 3. *C. p. andinus* duller than *patagonus* with less yellow on underparts and center of abdomen washed orange-red. *Range* northwestern Argentina, from Salta and Catamarca south to central Mendoza and northern San Luís, where some intergradation with *patagonus*. 4. *C. p. bloxami* irregular whitish pectoral band; brighter yellow and orange-red on underparts; larger. *Range* formerly central Chile, from Atacama south to Valdiva, but now restricted to few localities in central provinces, including Bío Bío; endangered. **LOCALITIES** El Cóndor Beach, near Viedma, Río Negro, and Estancia la Esperanza, Valdés Peninsula, Chubut, southern Argentina. Río de los Cipreses National Reserve, O'Higgins, Chile.

MONK PARAKEET *Myiopsitta monachus* 29cm

Common cagebird with color mutations; midsized green parrot with long, graduated tail and distinctive coloration featuring gray face and breast; sexes alike, JUV like adults; shrill screeches, squawks, and chatter. In highlands of Bolivia frequents thorn steppe or woodland in dry intermontane valleys, elsewhere in open country, including cultivation and pasturelands; noisy, conspicuous flocks at or near conspicuous communal nests of twigs in treetops or (*luchsi*) crevices in cliff-face; often feeds on ground; swift, direct flight. **DISTRIBUTION** central Bolivia and southern Brazil south to central Argentina; up to 1000m, or *luchsi* at 1300 to 3000m; common; feral populations at Santiago, Chile, and Rio de Janeiro, Brazil, on Puerto Rico and Grand Cayman Island, West Indies, at numerous urban centers in North America and Europe, and in some Mediterranean localities. **SUBSPECIES** three poorly differentiated subspecies, and one isolated, well-marked subspecies; ranges of lowland subspecies not accurately determined. 1. *M. m. monachus* crown and face bluish gray; breast brownish-gray barred grayish white; olive-yellow abdominal band; underwings blue. *Range* southern Rio Grande do Sul, extreme southeastern Brazil, through Uruguay to northeastern Argentina, south to eastern Buenos Aires. 2. *M. m. calita* more yellowish green; breast dull olive barred grayish white; abdomen dull olive-yellow. *Range* western Argentina, from Salta and western Buenos Aires south to Río Negro; probably not differentiated from *cotorra*. 3. *M. m. cotorra* like *calita*, but less yellow on abdomen. *Range* southern Mato Grosso, Brazil, and eastern Bolivia through Paraguay to northern Argentina, in Formosa, Chaco, and Corrientes. 4. *M. m. luchsi* forecrown paler gray, almost white; breast uniformly pale gray without barring. *Range* highlands of Bolivia, from southeastern La Paz, southern Cochabamba and western Santa Cruz to northern Chuquisaca; isolated altitudinally, and sometimes treated as separate species. **LOCALITIES** Caiman Lodge Wildlife Refuge, Mato Grosso do Sul, Brazil. Between Peña Colorado and Ele Ele, along Río Mizque, Cochabamba, Bolivia. Los Esteres del Ibera, Corrientes, and Costanera Sur Reserve, Buenos Aires, Argentina.

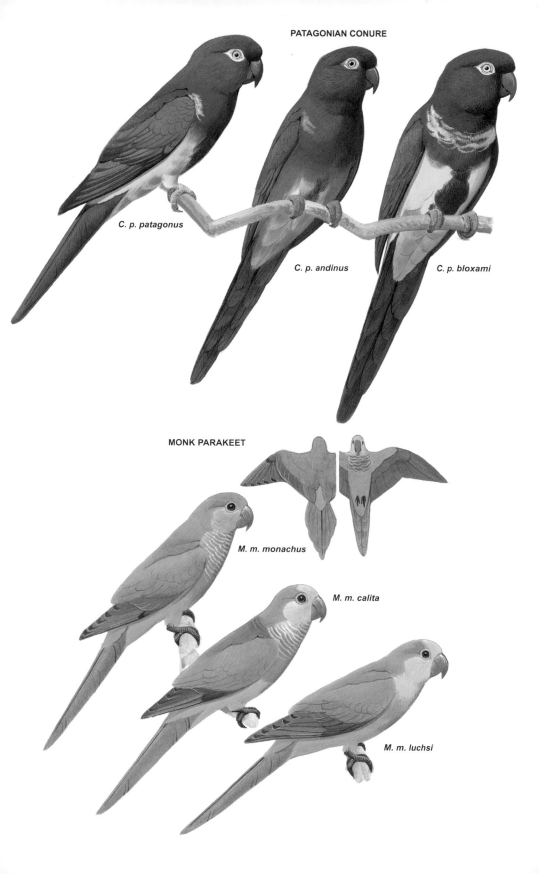

PATAGONIAN CONURE

C. p. patagonus

C. p. andinus

C. p. bloxami

MONK PARAKEET

M. m. monachus

M. m. calita

M. m. luchsi

PLATE 96 *ENICOGNATHUS* CONURES

212 Two very similar midsized green conures with long, graduated tail and dark margins to feathers producing overall barred appearance; sexes alike, JUV resembles adults. Forests, particularly *Nothofagus* and *Araucaria* forests, woodlands, cultivation with remnant woodlots or scattered trees; noisy flocks; feeds in treetops and on ground; swift, direct flight.

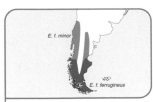

AUSTRAL CONURE
Enicognathus ferrugineus 33cm

Smaller than *E. leptorhynchus*, and with short upper mandible; shrill *grie...grie* in flight, more strident *grieee-grieee* when alarmed. **DISTRIBUTION** central Chile and Neuquén, Argentina, south to Tierra del Fuego; most southerly distributed of parrots; up to 2000m; common. **SUBSPECIES** two slightly differentiated subspecies; size difference may be clinal. 1. *E. f. ferrugineus* dull green, feathers edged dusky black; dark brownish frontal band not extending beneath or behind eye; brownish-red patch in center of abdomen; tail brownish red. *Range* southernmost Chile, in Aisén and Magallanes, and southern Argentina, from southwestern Chubut south to Tierra del Fuego. 2. *E. f. minor* darker green; darker brownish-red abdominal patch less extensive and sometimes lacking; slightly smaller. *Range* central and southern Chile, from O'Higgins south to Aisén, and eastern slopes of Andes in southwestern Argentina, from Neuquén south to western Chubut. **SIMILAR SPECIES** Slender-billed Conure *E. leptorhynchus* (see below) very similar in field, but appears brighter green and less "barred," particularly on underparts; brighter, crimson-red frontal band extending beneath and behind eye (adults); elongated upper mandible. **LOCALITIES** Villarrica National Park, Cautín, Chiloé National Park, Chiloé, Palermo Birding Reserve, Laguna Blanca district, Magallanes, and Torres del Paine National Park, Magallanes, southern Chile. Tierra del Fuego National Park, Tierra del Fuego, southern Argentina.

SLENDER-BILLED CONURE
Enicognathus leptorhynchus 40cm

Larger than *E. ferrugineus*, and with elongated, less-curved upper mandible; crimson-red lores and frontal band extending beneath and behind eye, restricted to forehead and lores in JUV; brownish-red patch on center of abdomen; tail brownish red; eye-ring pink-gray, pale gray in JUV; screeching in flight, harsh *scraart...scraart* while perched. Favors *Nothofagus* forests, and in late summer attracted to ripening *Araucaria* cones. **DISTRIBUTION** central Chile, from Aconcagua south to Isla de Chiloé and occasionally northern Aisén; up to 2000m; common. **SIMILAR SPECIES** Austral Conure *E. ferrugineus* (see above) very similar in field, but appears duller green and more heavily "barred"; dull brownish frontal band not extending beneath or behind eye; short upper mandible. **LOCALITIES** Chiloé National Park, Chiloé, and Cerro Ñielol National Park, Temuco City, Arauania, Chile.

AUSTRAL CONURE

E. f. ferrugineus

Juv

SLENDER-BILLED CONURE

PLATE 97 *PYRRHURA* CONURES (in part)

214 Small to midsized green parrots with long, graduated tail; coloration often features barring or scalloping on breast; sexes alike, JUV like or duller than adults. Lowland and mountain forests, secondary growth, woodlands; arboreal; usually small flocks; noisy and conspicuous in flight, but well concealed amidst foliage; swift, undulating or erratic flight.

A: SPECIES WITH BARRING ON BREAST (in part)

A-1 (in part, see plate 98) Three similar green-cheeked species without red on primary-coverts; brown, yellow, and white barring on breast; variable red abdominal patch; dark ear-coverts; prominent white eye-ring.

MAROON-BELLIED CONURE
Pyrrhura frontalis 26cm

Identified by green crown and wholly or mostly olive upper tail; sharp *aack-aack-aack*, screeches in flight. **DISTRIBUTION** Uruguay and southeastern Brazil to eastern Paraguay and northern Argentina; up to 1400m; common. **SUBSPECIES** three slightly differentiated subspecies. 1. *P. f. frontalis* narrow rufous frontal band; crown and cheeks to hindneck and mantle green; ear-coverts grayish brown; breast olive-brown barred dull yellow and margined darker brown; center of abdomen brownish red; tail olive broadly tipped brownish red. *Range* southeastern Brazil, from southern Bahia and eastern Minas Gerais to Rio de Janeiro. 2. *P. f. kriegi* tail very narrowly tipped brownish red. *Range* southeastern Brazil, from western Minas Gerais to Rio Grande do Sul. 3. *P. f. chiripepe* tail entirely olive; bend of wing orange-red. *Range* Uruguay, northern Argentina, and southern Paraguay. **SIMILAR SPECIES** White-eared Conure *P. leucotis* (plate 102) maroon face and brown-gray eye-ring; brownish-red back to rump. **LOCALITIES** Lombardia Biological Reserve, Espírito Santo, and Intervales State Park and Morro do Diabo Reserve, São Paulo, Brazil. Iguazú National Park, Misiones, Argentina.

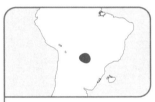

BLAZE-WINGED CONURE
Pyrrhura devillei 26cm

Like *P. frontalis*, but scarlet bend of wing, carpal edge, and lesser underwing-coverts; crown ash-brown; cheeks green; breast olive-brown barred yellowish brown; upper tail olive; white eye-ring; calls like *P. frontalis*. **DISTRIBUTION** southeastern Bolivia, northern Paraguay, and far southwestern Mato Grosso, Brazil; hybridization with *P. frontalis* in northern Paraguay and far southwestern Mato Grosso, so species possibly conspecific; locally common. **SIMILAR SPECIES** Green-cheeked Conure *P. molinae* (plate 98) green bend of wing and underwing-coverts; upper tail brownish red. **LOCALITIES** Caiman Lodge Wildlife Refuge and Pousada Aguapé at São Jose Ranch, Mato Grosso do Sul, Brazil. Along Río Apa, northern Paraguay.

MAROON-BELLIED CONURE

P. f. frontalis

P. f. chiripepe

BLAZE-WINGED CONURE

PLATE 98 *PYRRHURA* CONURES (in part)

216

A: SPECIES WITH BARRING ON BREAST (cont.)

A-1 (cont., see plate 97) Three similar green-cheeked species without red on primary-coverts; brown, yellow, and white barring on breast; variable red abdominal patch; dark ear-coverts; prominent white eye-ring.

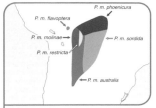

GREEN-CHEEKED CONURE
Pyrrhura molinae 26cm

Differentiated from *P. frontalis* by brown crown, and brownish-red tail; repeated *kree-aat*, sharp *kreet* or *quee*. **DISTRIBUTION** southern Brazil, eastern Bolivia, northwestern Argentina and northwestern Paraguay; possibly also southernmost Peru; up to 3000m; common. **SUBSPECIES** five discernible and one well-marked subspecies.

1. *P. m. molinae* crown to nape brown tinged green; cheeks bright green; some blue feathers on hindneck rarely forming indistinct collar; breast pale brown barred grayish white, or dull yellow toward center of breast, feathers tipped dusky brown; white eye-ring. *Range* highlands of eastern Bolivia, approaching to within 30km of range of *flavoptera*. 2. *P. m. phoenicura* tail green toward base. *Range* southern Brazil and northeastern Bolivia. 3. *P. m. restricta* breast brownish-gray barred white; cheeks green tinged blue; blue collar on hindneck; flanks and undertail-coverts suffused blue. *Range* known only from Santa Cruz, Bolivia. 4. *P. m. sordida* GREEN MORPH like *restricta*, but indistinct barring on breast, center of breast suffused yellowish; little or no bluish suffusion on flanks or undertail-coverts. YELLOW MORPH throat and upper breast yellowish-white barred brown; lower underparts yellow indistinctly barred green and brown; undertail-coverts yellowish-white tinged blue. *Range* southern Brazil and northwestern Paraguay. 5. *P. m. australis* paler than *molinae*; center of breast suffused yellow; more extensive brownish-red abdominal patch. *Range* southern Bolivia and northwestern Argentina. 6. *P. m. flavoptera* like *molinae*, but bend of wing and carpal edge reddish orange; alula intermixed blue and yellowish-white feathers. *Range* known only from Cochabamba–La Paz border, northern Bolivia. **SIMILAR SPECIES** Maroon-bellied Conure *P. frontalis* (*P. f. chiripepe* possibly sympatric in Salta, northern Argentina; plate 97) crown green and upper tail olive. Blaze-winged Conure *P. devillei* (plate 97) scarlet bend of wing, carpal edge, and lesser underwing-coverts; upper tail olive. Black-capped Conure *P. rupicola* (plate 106) black crown and green abdomen; red primary-coverts and green tail. **LOCALITIES** Morro do Urucum, near Corumbá, Mato Grosso do Sul, southern Brazil. Caliegua National Park, Jujuy, and El Rey National Park, Salta, northern Argentina.

P. m. molinae

P. m. sordida

yellow morph

P. m. australis

P. m. flavoptera

PLATE 99 *PYRRHURA* CONURES (in part)

218

A: SPECIES WITH BARRING ON BREAST (cont.)

A-2. One or possibly more species with red primary-coverts and dark maroon to blackish tail; white and brown barring on breast; white or gray eye-ring.

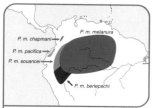

MAROON-TAILED CONURE
Pyrrhura melanura 24cm

Small green conure with green cheeks contrasting with dark crown and prominently barred breast sometimes extending around neck; red primary-coverts and forewing conspicuous; shrill *kree* repeated rapidly. **DISTRIBUTION** western Amazon River basin, from central Colombia and southern Venezuela to northern Peru and northwestern Brazil; up to 3200m; common. **SUBSPECIES** one poorly differentiated and four discernible subspecies, including two distinctive isolates that could be separate species. 1. *P. m. melanura* crown brown, feathers margined green; breast barred buff-white and dusky brown; primary-coverts red broadly tipped orange-yellow; tail dark brownish red with green at base. *Range* upper Amazon River basin, from southeastern Colombia, in extreme southeastern Guainía, Vaupés, and Amazonas, and southern Venezuela, in Amazonas and eastern Bolívar, to Rio Negro and Rio Solimões and tributaries, northwestern Brazil, and middle to lower Río Napo drainage in eastern Ecuador and northeastern Peru. 2. *P. m. souancei* breast centrally suffused dusky brown and more broadly barred grayish white; primary-coverts all-red; indistinct reddish-brown abdominal patch in some birds; tail more broadly green at base. *Range* eastern base of Andes in southern Colombia, south from Macarena Mountains, and in eastern Ecuador and possibly northeastern Peru. 3. *P. m. berlepschi* like *souancei*, but breast more broadly barred grayish white, sometimes appearing almost uniformly grayish white; abdomen tinged reddish brown. *Range* lower eastern slopes of Andes in Morona-Santiago, southeastern Ecuador, and eastern Peru, south to Río Huallaga valley; much individual plumage variation in *berlepschi* and *souancei* could negate subspecific differentiation. 4. *P. m. pacifica* (possibly separate species) forecrown green; carpal edge red; no reddish-brown tinge on abdomen; breast dusky brown narrowly barred grayish white; eye-skin gray and eye-ring feathered. *Range* only *Pyrrhura* conure on Pacific slope of Andes in northwestern Ecuador, south to northern Los Rios, and in Nariño, extreme southwestern Colombia. 5. *P. m. chapmani* (possibly separate species) like *souancei*, but breast and entire neck brown broadly barred buff-white; crown brown without green margins to feathers; primary-coverts intermixed red and green; larger. *Range* subtropical zone (1600 to 2800m) in upper Río Magdalena valley on eastern slopes of Cordillera Central, in Huila and southern Tolima, central Colombia. **SIMILAR SPECIES** Painted Conure *P. picta* and Deville's Conure *P. lucianii* (plate 103) and Rose-fronted Conure *P. roseifrons* (plate 104) distinctive head patterns with sharply defined ear-coverts; "chevroned" breast markings; back and rump brownish red. Wavy-breasted Conure *P. peruviana* (plate 104) dark maroon face with buff-white ear-coverts; "chevroned" breast markings; back and rump brownish red. White-necked Conure *P. albipectus* (plate 106) white collar encircling neck; yellow breast; orange-yellow ear-coverts. **LOCALITIES** Sangay National Park, Morona-Santiago, and La Selva Lodge, Sucumbios, Ecuador. Cueva de los Guácharos National Park, Huila, and Tinigua National Park, Meta, Colombia.

P. m. melanura

P. m. souancei

P. m. berlepschi

P. m. pacifica

P. m. chapmani

PLATE 100 *PYRRHURA* CONURES (in part)

220

A: SPECIES WITH BARRING ON BREAST (cont.)

A-3. Two closely allied, possibly conspecific, conures with blue on underparts and cheeks; brown and buff barring on breast; prominent white eye-ring; species replace each other geographically.

CRIMSON-BELLIED CONURE
Pyrrhura perlata **24cm**

Unmistakable; adults readily identified by crimson abdomen; breast brown barred pale buff and darker brown; cheeks yellowish green, becoming blue on lower cheeks; bend of wing and lesser underwing-coverts red; thighs to undertail-coverts blue; tail brownish red with green at base; JUV abdomen green, often with some red markings; shrill *tieww...kritieww*. **DISTRIBUTION** central and southern Amazon River basin, south of Amazon River in northern Brazil, from Rio Tapajós drainage, western Pará, west to Rio Madeira, eastern Amazonas, south to central-western Mato Grosso, at Rio Jaurú drainage, and neighboring northeastern Bolivia; up to 600m; locally common. **LOCALITIES** Amazonia National Park, Pará, and Cristalino Jungle Lodge, Cristalino State Park, northern Mato Grosso, northern Brazil. Noel Kempff Mercado National Park, Santa Cruz, Bolivia.

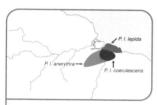

PEARLY CONURE *Pyrrhura lepida* **24cm**

Differentiated from *P. perlata* by green abdomen; sexes alike, JUV like adults; calls like *P. perlata*. **DISTRIBUTION** northeastern Brazil, south of Amazon River; up to 500m; near-threatened. **SUBSPECIES** two discernible and one well-marked subspecies. 1. *P. l. lepida* breast brown barred buff and dark brown, suffused blue on lower breast; bend of wing, carpal edge, and lesser underwing-coverts red; undertail-coverts blue; tail brownish red. *Range* northeastern Pará and northwestern Maranhão. 2. *P. l. coerulescens* breast grayish brown with strong blue suffusion extending up to throat. *Range* central and eastern Maranhão. 3. *P. l. anerythra* no blue suffusion on breast; little or no red on green bend of wing and lesser underwing-coverts. *Range* eastern Pará, except northeast where replaced by *lepida*. **SIMILAR SPECIES** Santarem Conure *P. amazonum* (plate 103) "chrevroned" breast markings; brownish-red abdominal patch, and back to upper tail-coverts brownish red; green underwing-coverts. **LOCALITIES** National Forests in the Serra dos Carajás, Pará, and Gurupi Biological Reserve, Maranhão, northern Brazil.

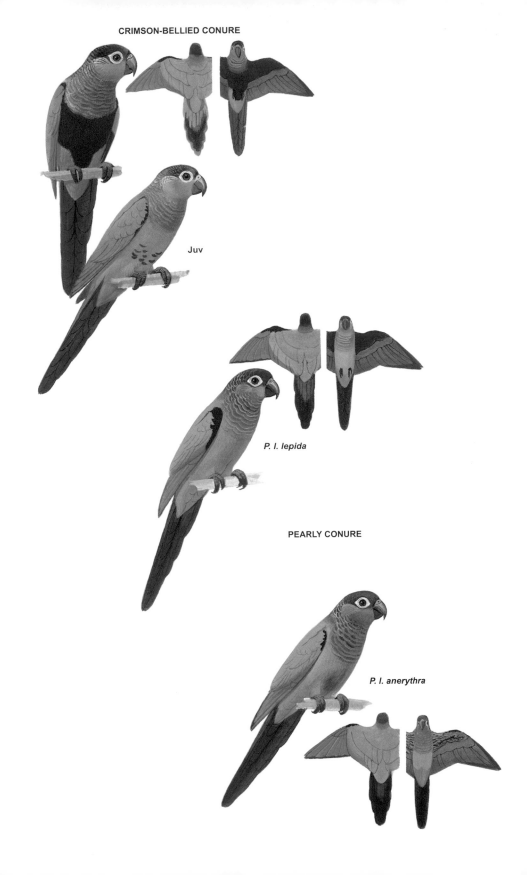

CRIMSON-BELLIED CONURE

Juv

P. l. lepida

PEARLY CONURE

P. l. anerythra

PLATE 101 *PYRRHURA* CONURES (in part)

222

A: SPECIES WITH BARRING ON BREAST (cont.)

A-4. Three allopatric green conures with indistinctly barred breast, reddish-brown ear-coverts, and brownish-red tail; white eye-ring.

BROWN-BREASTED CONURE
Pyrrhura calliptera 22cm

Only *Pyrrhura* conure with yellow on primary-coverts to carpal edge; breast rufous barred paler brown and dusky black; pale bill; JUV like adults, but primary-coverts to carpal edge green; harsh, far-carrying *screeyr...screeyr*. **DISTRIBUTION** central Colombia, on both slopes of Cordillera Oriental from extreme northeastern Boyacá south to southwestern Cundinamarca, at about lat. 4°20'N, but apparently very rare or absent on western slope; mostly 1500 to 3400m; scarce and declining. **LOCALITIES** Chingaza National Park and Carpanta Biological Reserve, Cundinamarca, and Guanentá-Alto Río Fonce Fauna and Flora Sanctuary, Santander, Colombia.

FIERY-SHOULDERED CONURE
Pyrrhura egregia 26cm

Only *Pyrrhura* conure with reddish orange on bend of wing, carpal edge, and lesser underwing-coverts; sexes alike, JUV little or no reddish orange on bend of wing and carpal edge; grating *jjaa-eek*. **DISTRIBUTION** Pantepui region of southeastern Venezuela, western Guyana, and northernmost Roraima, Brazil; mostly 700 to 1800m; locally common. **SUBSPECIES** two poorly differentiated subspecies. 1. *P. e. egregia* breast green barred yellowish white and dusky brown; center of abdomen suffused brownish red; tail brownish red with green at base. *Range* highlands of western Guyana, mainly in Pakaraima Mountains, south-central Venezuela, and northernmost Roraima, Brazil. 2. *P. e. obscura* upperparts darker green, and little or no brownish red on abdomen. *Range* highlands of southeastern Venezuela; doubtfully distinct from *egregia*. **SIMILAR SPECIES** Painted Conure *P. picta* (plate 103) brownish-red face with buff-white ear-coverts; strongly "chevroned" breast markings; brownish-red abdominal patch, and brownish-red back to upper tail-coverts. **LOCALITY** Vicinity of La Escalera, in Canaima National Park, Bolívar, Venezuela.

RED-EARED CONURE
Pyrrhura hoematotis 25cm

Plainly colored *Pyrrhura* conure with indistinct barring on breast and neck, but conspicuous brownish-red ear-coverts; sexes alike, JUV like adults; staccato *ca-ca-ca-ca-ca* repeated very rapidly. **DISTRIBUTION** coastal cordillera in northwestern Venezuela; mostly 600 to 2400m; common. **SUBSPECIES** two apparently isolated subspecies. 1. *P. h. hoematotis* feathers of sides of neck dusky brown margined gray; breast green tinged olive and barred dusky gray. *Range* coastal cordillera in Aragua to Miranda at Curupao and Guarenas. 2. *P. h. immarginata* no gray margins to feathers of sides of neck; breast with light subterminal bands to feathers, but not barred dusky gray. *Range* cordillera in southeastern Lara, where known only from Cubito, with sight record from Yacambú. **SIMILAR SPECIES** Emma's Conure *P. emma* (plate 102) dark brownish-maroon face with white ear-coverts; more strongly barred breast; back to upper tail-coverts brownish red. Red-fronted Conure *Aratinga wagleri* (plate 89) red forecrown, but no red ear-coverts; uniformly green underparts; tail green; pale bill; larger. **LOCALITY** Rancho Grande Biological Station, Henri Pittier National Park, Aragua, northern Venezuela.

BROWN-BREASTED CONURE

FIERY-SHOULDERED CONURE

P. e. egregia

Juv

RED-EARED CONURE

P. h. hoematotis

P. h. immarginata

PLATE 102 *PYRRHURA* CONURES (in part)

224

A: SPECIES WITH BARRING ON BREAST (cont.)

A-5. Four allopatric *Pyrrhura* conures with brownish-red face and brownish-red patch on back to rump.

WHITE-EARED CONURE
Pyrrhura leucotis 23cm

Identified by narrower barring on green breast and blue on forecrown; ear-coverts buff; gray-brown eye-ring; sharp *teer-teer* in flight, high-pitched *chee-cheet-chee* or *ki-ki*. **DISTRIBUTION** eastern Brazil, from southern Bahia south to Espírito Santo and eastern Minas Gerais; up to 600m; generally scarce. **SIMILAR SPECIES** Maroon-bellied Conure *P. frontalis* (plate 97) green face with dark ear-coverts; green back and rump; larger. Blue-throated Conure *P. cruentata* (plate 107) distinctive facial pattern; no barring on blue breast. **LOCALITIES** Sooretama Biological Reserve and Linhares Reserve, Espírito Santo. Rio Doce State Park, Minas Gerais.

GRAY-BREASTED CONURE
Pyrrhura griseipectus 23cm

Like *P. leucotis*, but breast dull gray barred white and dark brown; ear-coverts white; no blue on brown forecrown; calls like *P. leucotis*. **DISTRIBUTION** northeastern Brazil, in Ceará, possibly disjunctly in Alagoas, and at Serra Negra, Pernambuco, and only *Pyrrhura* species in range; up to 600m; critically endangered. **SIMILAR SPECIES** Cactus Conure *Aratinga cactorum* and Peach-fronted Conure *A. aurea* (plate 93) no barring on olive-brown breast; no white ear-coverts; back and rump green. **LOCALITIES** Serra Negra Biological Reserve, Pernambuco, and Serra do Baturité, Ceará.

MAROON-FACED CONURE
Pyrrhura pfrimeri 22cm

Only *Pyrrhura* conure with wholly chestnut-red face, including ear-coverts; crown to hindneck dull blue; breast dull bluish-green barred white; eye-ring gray-white; calls like *P. leucotis*. Closely associated with dry, deciduous forest on rocky terrain. **DISTRIBUTION** only between Serra Geral and Rio Parana, Goiás, north-central Brazil, and only *Pyrrhura* species in range; up to 600m; vulnerable. **SIMILAR SPECIES** Peach-fronted Conure *Aratinga aurea* (plate 93) see above. **LOCALITIES** Mata Grande National Forest and Terra Ronca State Park, Goiás.

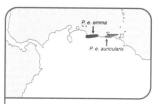

EMMA'S CONURE *Pyrrhura emma* 23cm

Identified by blue crown to nape and sides of neck; loud *kik-kik-kik-kik*, nasal *wa-ke-ke-ke-ka*. **DISTRIBUTION** mountains of northern Venezuela; up to 1700m; fairly common. **SUBSPECIES** two isolated subspecies differentiated primarily by color of cere and eye-ring. 1. *P. e. emma* ear-coverts dirty white; breast brownish green barred yellowish white; bare cere and eye-ring white. *Range* coastal mountains from Yaracuy and Carabobo east to Distrito Federal, and interior mountains of southern Aragua, Miranda, and western Anzoátegui. 2. *P. e. auricularis* clearer white ear-coverts; bare cere and eye-ring gray. *Range* mountains in eastern Anzoátegui, Sucre, and northern Monagas. **SIMILAR SPECIES** Red-eared Conure *P. haematotis* (plate 101) green crown and face; red ear-coverts; little barring on breast; green back and rump. **LOCALITIES** Guatopo National Park, Miranda, and Cueva del Guácharo National Park, northern Monagas.

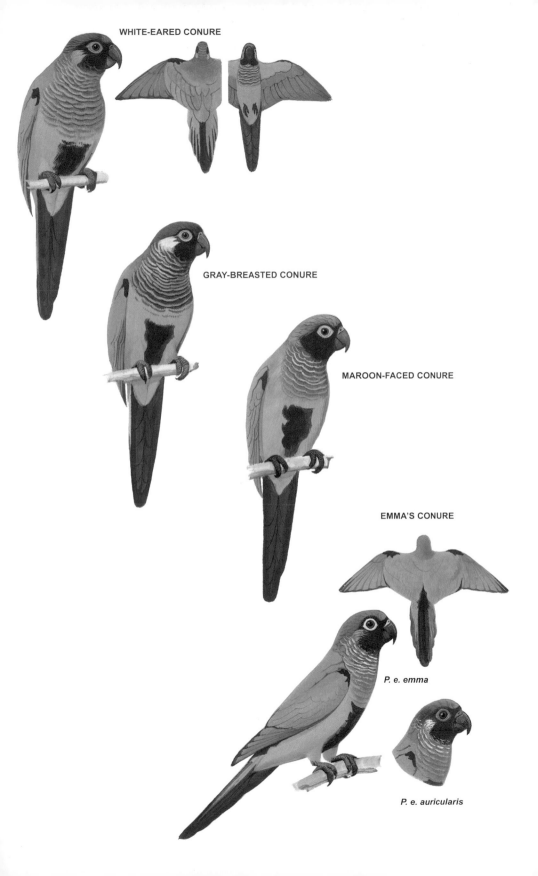

WHITE-EARED CONURE

GRAY-BREASTED CONURE

MAROON-FACED CONURE

EMMA'S CONURE

P. e. emma

P. e. auricularis

PLATE 103 *PYRRHURA* CONURES (in part)

B: SPECIES WITH SCALLOPING ON BREAST (in part)

B-1 (in part, see plates 104, 105). Closely allied species with chevroned scalloping on breast, brownish red on abdomen and on lower back to rump, and prominent yellowish or pale buff to brownish ear-coverts; tail brownish red with green at base; white or grayish eye-ring; gray bill.

PAINTED CONURE *Pyrrhura picta* 22cm

Identified by dark maroon face, pale buff ear-coverts, and red on bend of wing; crown to nape brown, suffused blue on forecrown; upper breast dusky brown scalloped buff, lower breast greenish brown scalloped yellowish buff; bare eye-ring gray; JUV bend of wing green with few scattered red feathers, and grayish-white eye-ring; coarse *pik-pik* or *pik…pik…pik*. **DISTRIBUTION** Venezuela, south of the Río Orinoco in southern Delta Amacuro, Bolívar and eastern Amazonas, through Guianas to northernmost Brazil, where southern limits in Amapá, northern Pará and northern Amazonas are uncertain. **SIMILAR SPECIES** Fiery-shouldered Conure *P. egregia* (plate 101) green face with reddish-brown ear-coverts; lightly barred breast; green back and rump; orange-yellow bend of wing; pale bill and white eye-ring. Maroon-tailed Conure *P. melanura* (plate 99) green face and ear-coverts; barred breast; red primary-coverts; green back and rump; white eye-ring. **LOCALITIES** Nouragues Field Station, northwest of Ipoucin Crique, French Guiana. Canaima National Park, southeastern Bolívar, Venezuela. Uaçá district and outskirts of Oiapoque, Amapá, northern Brazil.

SANTAREM CONURE *Pyrrhura amazonum* 22cm

Like *P. picta*, but with narrow blue band in front of eyes, darker reddish-brown face, and green bend of wing; calls similar to *P. picta*. **DISTRIBUTION** eastern Amazonia, northern Brazil; up to 600m; common. **SUBSPECIES** two poorly differentiated subspecies with ranges north and south of Amazon River. 1. *P. a. amazonum* crown to nape dark brown with blue band in front of eyes; ear-coverts brownish buff; upper breast brownish gray scalloped grayish buff, lower breast brownish gray scalloped yellowish green; bare eye-ring gray. *Range* north bank of Amazon River from about Obidos east to Monte Alegre, Pará. 2. *P. a. microtera* narrow frons and face darker maroon-brown; upper breast dusky brown scalloped grayish buff; smaller. *Range* south bank of Amazon River at Santarem, and Rio Tapajós, Pará, east to westernmost Maranhão and northernmost Mato Grosso. **SIMILAR SPECIES** Crimson-bellied Conure *P. perlata* (plate 100) green-and-blue cheeks; barred breast; abdomen crimson and undertail-coverts blue; back and rump green. **LOCALITIES** Amazonia National Park, Pará, and Cristalino Jungle Lodge, Cristalino State Park, northern Mato Grosso, Brazil.

DEVILLE'S CONURE *Pyrrhura lucianii* 22cm

Identified by almost wholly grayish-brown head, including dark ear-coverts; little or no blue on forecrown; upper breast brownish gray scalloped pale gray, lower breast brownish gray scalloped yellowish green; green bend of wing; bare eye-ring cream-white; calls not recorded. **DISTRIBUTION** known only from Tefé district and Rio Purús, Amazonas, northern Brazil; up to 800m; little known, but locally common. **SIMILAR SPECIES** Maroon-tailed Conure *P. melanura* (plate 99) green face and ear-coverts; barred breast; red primary-coverts; green back and rump. **LOCALITY** Mamirauá Reserve, near Tefé, Amazonas.

PAINTED CONURE

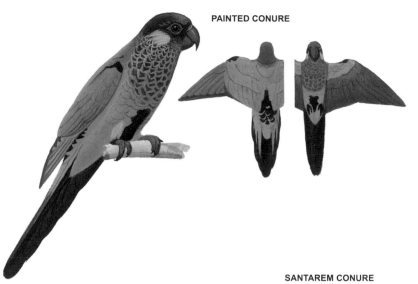

SANTAREM CONURE

P. a. amazonum

DEVILLE'S CONURE

PLATE 104 *PYRRHURA* CONURES (in part)

228

B: SPECIES WITH SCALLOPING ON BREAST (cont.)

B-1 (cont., see plates 103, 105). Closely allied species with chevroned scalloping on breast, brownish red on abdomen and on lower back to rump, and prominent yellowish or pale buff to brownish ear-coverts; tail brownish red with green at base; white or gray eye-ring; gray bill.

ROSE-FRONTED CONURE
Pyrrhura roseifrons 22cm

Adult identified by rose-red crown and face with yellowish ear-coverts; breast dull greenish brown scalloped greenish yellow; bend of wing and carpal edge red; bare eye-ring cream-white; JUV crown and face dark maroon-brown with off-white ear-coverts and green bend of wing to carpal edge; rolling *prrrt…prrrt*, also screeches.

DISTRIBUTION disjunctly in western Amazonia, in north from vicinity of São Paulo de Olivença and Rio Javarí, western Amazonas, northern Brazil, and Requena district, Río Ucayali, Loreto, northeastern Peru, south to Yurinaqui Alto and Conchapen, Junín, eastern Peru, and in south from Itahuania, Madre de Dios, southeastern Peru, to Teoponte, La Paz, northern Bolivia; up to 1200m; common. **SIMILAR SPECIES** Black-capped Conure *P. rupicola* (plate 106) green face and ear-coverts; broadly scalloped breast; red primary-coverts and carpal edge; no red on abdomen or on back and rump; green tail. Maroon-tailed Conure *P. melanura* (plate 99) green face and ear-coverts; barred breast; red primary-coverts; green back and rump. Crimson-bellied Conure *P. perlata* (plate 100) green-and-blue cheeks; barred breast; crimson abdomen and blue undertail-coverts; green back and rump. **LOCALITIES** Palmari Lodge, Rio Javarí, Amazonas, northern Brazil. Manú Biosphere Reserve, Madre de Dios, southeastern Peru.

AMAZON RED-FRONTED CONURE
Pyrrhura parvifrons 22cm

Differentiated from similar, and possibly conspecific *P. roseifrons* by having rose-red restricted to narrow frons; face and crown dark grayish brown with variable blue suffusion and scattered red feathers on crown; ear-coverts buff-white; upper breast brownish gray scalloped grayish white, lower breast greenish gray scalloped greenish yellow; bare eye-ring white; calls not recorded, but probably like *P. roseifrons*. **DISTRIBUTION** northern Peru, where recorded north and northeast of range of *P. roseifrons* in Loreto and San Martín. **SIMILAR SPECIES** Maroon-tailed Conure *P. melanura* (plate 99) see above. **LOCALITIES** vicinity of Santa Cecilia and Yurmaguas, Loreto, and Shanusi, San Martín.

WAVY-BREASTED CONURE
Pyrrhura peruviana 22cm

Resembles *P. parvifrons*, but with blue, instead of red, on forehead; calls like *P. roseifrons*. **DISTRIBUTION** disjunctly in southeastern Ecuador to northern Peru and in southern Peru; up to 1200m; common. **SUBSPECIES** two poorly differentiated, isolated subspecies. 1. *P. p. peruviana* forehead blue; crown to nape dark brown; buff-white ear-coverts; upper breast dark grayish brown scalloped grayish buff, lower breast dark green scalloped yellowish white; bare eye-ring grayish white. *Range* Río Macuma, northern Morona-Santiago, southeastern Ecuador, south to about lat. 5°30'S in northern Loreto, Peru. 2. *P. p. dilutissima* blue restricted to narrow frons; darker buff ear-coverts; upper breast paler brown scalloped pale buff. *Range* upper Río Apurimac valley, southern Peru. **SIMILAR SPECIES** Maroon-tailed Conure *P. melanura* (plate 99) see above. Black-capped Conure *P. rupicola* (plate 106) see above. **LOCALITY** Sangay National Park, Morona-Santiago, Ecuador.

ROSE-FRONTED CONURE

AMAZON RED-FRONTED CONURE

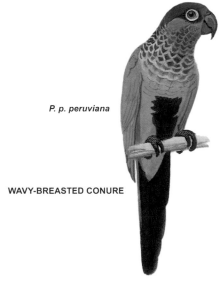

P. p. peruviana

WAVY-BREASTED CONURE

PLATE 105 *PYRRHURA* CONURES (in part)

230

B: SPECIES WITH SCALLOPING ON BREAST (cont.)

B-1 (cont., see plates 103, 104). Closely allied species with chevroned scalloping on breast, brownish red on abdomen and on lower back to rump, and prominent yellowish or pale buff to brownish ear-coverts; tail brownish red with green at base; white or gray eye-ring; gray bill.

MADEIRA CONURE *Pyrrhura snethlageae* 22cm

Identified by distinctive brown-and-buff "arrow-head" scalloping on breast; ear-coverts buff-white; little blue on brown forecrown; bare eye-ring brown; calls like *P. picta*. **DISTRIBUTION** Rio Madeira drainage, from Amazonas, northern Brazil, south to extreme northeastern Bolivia, in eastern Pando to extreme northeastern Santa Cruz; up to 300m; common. **SIMILAR SPECIES** Black-capped Conure *P. rupicola* (plate 106) green face and ear-coverts; broadly scalloped breast; red primary-coverts and carpal edge; no red on abdomen or back and rump; green tail. Crimson-bellied Conure *P. perlata* (plate 100) green and blue cheeks; barred breast; crimson abdomen and blue undertail-coverts; green back and rump. **LOCALITIES** forests near Ariquemes, Rondônia, Brazil. Noel Kempff Mercado National Park, Santa Cruz, Bolivia.

SINÚ CONURE *Pyrrhura subandina* 21cm

Like *P. picta*, but brighter maroon-red face; dull red and blue frontal band; ear-coverts russet-brown; breast dark brown scalloped grayish buff; bend of wing green; bare eye-ring gray; calls not recorded, but probably like *P. picta*. **DISTRIBUTION** lower Río Sinú valley, Córdoba, northwestern Colombia, and only *Pyrrhura* species in range; up to 1300m; uncommon. **LOCALITY** Paramillo National Park, Córdoba, Colombia.

PERIJÁ CONURE *Pyrrhura caeruleiceps* 22cm

Identified by blue forecrown, with brown hindcrown suffused blue; blue nuchal collar; breast dusky brown scalloped grayish buff; bend of wing red; bare eye-ring dull gray; calls not described. **DISTRIBUTION** Sierra de Perijá on Colombia–Venezuela border, and western slopes of Cordillera Oriental in Magdalena south to Santander, northern Colombia, and only *Pyrrhura* species in range; mostly 450 to 2000m; scarce. **LOCALITY** possibly Tamá National Park, Norte de Santander, Colombia.

B-2. Single species with more rounded, less chevroned scalloping on breast, brownish red on abdomen and lower back to rump.

AZUERO CONURE *Pyrrhura eisenmanni* 25cm

Like *P. caeruleiceps*, but forehead to nape sooty brown; ear-coverts buff-white; breast dark brown scalloped buff and white; bend of wing green; bare eye-ring grayish brown; short *eek* in flight, loud *peea*, and harsh *kleek-kleek*. **DISTRIBUTION** southwestern Azuero Peninsula, central Panama, and only *Pyrrhura* species in range; up to 1660m; common. **SIMILAR SPECIES** Crimson-fronted Conure *Aratinga finschi* (plate 86) red forecrown, carpal edge, and outermost underwing-coverts; no scalloping on green breast. **LOCALITY** Cerro Hoya National Park, Azuero Peninsula, central Panama.

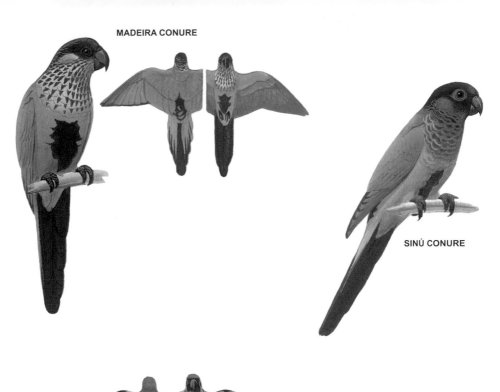

MADEIRA CONURE

SINÚ CONURE

PERIJÁ CONURE

AZUERO CONURE

PLATE 106 *PYRRHURA* CONURES (in part)

232

B: SPECIES WITH SCALLOPING ON BREAST (cont.)

B-3. Single species with more rounded, less chevroned scalloping on breast, but without brownish red on abdomen or on lower back and rump; green tail.

BLACK-CAPPED CONURE
Pyrrhura rupicola 25cm

Only green-cheeked *Pyrrhura* conure with rounded scalloping on breast, black crown, and red primary-coverts; rolling *jiree* in flight, high-pitched yapping notes. **DISTRIBUTION** western Amazon River basin; up to 300m; common. **SUBSPECIES** two doubtfully distinct subspecies. 1. *P. r. rupicola* forehead to occiput black; cheeks and ear-coverts green; hindneck and upper breast dark gray-brown scalloped white, lower breast dark brown scalloped dull yellow; primary-coverts and carpal edge red; eye-ring white. *Range* central-eastern Peru. 2. *P. r. sandiae* narrower buff-white scalloping on breast and almost absent on hindneck. *Range* southeastern Peru to northern Bolivia, south to La Paz and Beni, and westernmost Brazil, near Rio Branco, Acre. **SIMILAR SPECIES** Rose-fronted Conure *P. roseifrons* (plate 104) rose-red face and forecrown; yellowish ear-coverts; brownish red on abdomen and lower back to rump; brownish-red tail. Wavy-breasted Conure *P. peruviana* (plate 104) and Madeira Conure *P. snethlageae* (plate 105) dark maroon-red face; brownish-red abdomen and lower back to rump; green primary-coverts and carpal edge; brownish-red tail. **LOCALITIES** Manú Biosphere Reserve and Explorer's Inn, Madre de Dios, southeastern Peru.

C: SPECIES WITHOUT BARRING OR SCALLOPING ON BREAST (in part)

Not closely allied to each other or to other *Pyrrhura* species, these distinctively plumaged species lack barring or scalloping on breast.

EL ORO CONURE *Pyrrhura orcesi* 22cm

Only green-headed *Pyrrhura* conure lacking colored ear-coverts and with red frontal band; bend of wing, carpal edge, and primary-coverts red; tail green broadly tipped brownish red; pale bill; trilling *tchreeet...tchreeet*. Favors cloudforest. **DISTRIBUTION** west slope of Andes in southwestern Ecuador, in Azuay north to above Manta Real, and in El Oro south to Buenaventura; mainly 600 to 1200m; endangered. **SIMILAR SPECIES** Maroon-tailed Conure *P. melanura* (plate 99) barring on breast; no red on forehead; crown brown. **LOCALITY** Buenaventura forest, near Piñas, El Oro, Ecuador.

WHITE-NECKED CONURE
Pyrrhura albipectus 24cm

Unmistakable; distinctive coloration features white neck and yellow breast; crown to nape dark gray-brown; ear-coverts yellow-orange; primary-coverts and carpal edge red; dark bill; calls like *P. melanura*. **DISTRIBUTION** east slope of Andes in southeastern Ecuador, in northwestern Morona-Santiago and southern Zamora-Chinchipe, and in Cordillera del Condor, northernmost Peru; mostly 900 to 1700m; vulnerable. **LOCALITY** Río Bombuscara valley, Podocarpus National Park, southern Ecuador.

BLACK-CAPPED CONURE

P. r. rupicola

EL ORO CONURE

WHITE-NECKED CONURE

PLATE 107 *PYRRHURA* CONURES (in part)

234

C: SPECIES WITHOUT BARRING OR SCALLOPING ON BREAST (cont.)

Not closely allied to each other or to other *Pyrrhura* species, these distinctively plumaged species lack barring or scalloping on breast.

SANTA MARTA CONURE
Pyrrhura viridicata 25cm

Unmistakable; identified by broad orange-red abdominal band, purplish-brown ear-coverts, and yellow and orange-red forewing; lesser underwing-coverts red marked yellow; tail brownish red with green at base; pale bill; bare eye-ring white; calls like *P. picta* (plate 103). **DISTRIBUTION** Sierra Nevada de Santa Marta, northern Colombia, and only *Pyrrhura* species in range; mostly 1800 to 2800m; endangered. **SIMILAR SPECIES** Red-fronted Conure *Aratinga wagleri* (plate 89) uniformly green underparts; green bend of wing and carpal edge; green upper tail; larger. **LOCALITIES** Sierra Nevada de Santa Marta National Park and San Lorenzo district, Magdalena, northern Colombia.

ROSE-CROWNED CONURE
Pyrrhura rhodocephala 24cm

Unmistakable; only *Pyrrhura* conure with red crown and white primary-coverts; ear-coverts maroon; carpal edge yellow tinged white; brownish-red tail; pale bill; bare eye-ring white; clear *clee*, sharp *kik* or rapidly repeated *kik-kik-kik-kik*. **DISTRIBUTION** northwestern Venezuela, from northern Tachira and northwestern Barinas north to southern Lara, and only *Pyrrhura* species in range; mostly 1500 to 2500m; near-threatened. **SIMILAR SPECIES** Red-fronted Conure *Aratinga wagleri* (plate 89) see above. **LOCALITY** Yacambú National Park, southern Lara.

P. h. hoffmanni

P. h. gaudens

SULPHUR-WINGED CONURE
Pyrrhura hoffmanni 24cm

Only *Pyrrhura* conure with yellow on flight feathers; red ear-coverts; piping *toweet-deet-deet-toweet*, grating *zee-wheet*. **DISTRIBUTION** southern Costa Rica and western Panama, and only *Pyrrhura* species in range; mostly 1300 to 3000m; common. **SUBSPECIES** two subspecies differentiated by head markings. 1. *P. h. hoffmanni* feathers of head and throat shaft-streaked and tipped yellow. *Range* southern Costa Rica, mainly in Cordillera de Talamanca and outlying ranges, and reaching Caribbean approaches to Cordillera Central. 2. *P. h. gaudens* feathers of crown and occiput shaft-streaked greenish-yellow to orange-red and tipped dull red. *Range* western Panama, in western Chiriquí, and adjacent Bocas del Toro to Santa Fé region, northwestern Veraguas. **SIMILAR SPECIES** Crimson-fronted Conure *Aratinga finschi* (plate 89) red forecrown, carpal edge, and outermost underwing-coverts; green face and ear-coverts; no yellow on wings; larger. **LOCALITY** Genesis II Lodge, Tapanti Region, Talamanca Mountains, Costa Rica.

BLUE-THROATED CONURE
Pyrrhura cruentata 30cm

Unmistakable; identified by distinctive head pattern featuring brownish-red face with brownish orange on sides of neck; upper breast and collar encircling hindneck blue; abdomen and lower back to rump brownish red; gray bill; large size; high-pitched chattering. **DISTRIBUTION** eastern Brazil, from southern Bahia south to Rio de Janeiro; up to 400m, occasionally 900m; vulnerable, CITES I. **SIMILAR SPECIES** Maroon-bellied Conure *P. frontalis* (plate 97) green face and brownish ear-coverts; barring on breast; green back to lower rump; smaller. White-eared Conure *P. leucotis* (plate 102) maroon face and pale ear-coverts; barring on breast; smaller. **LOCALITIES** Sooretama Biological Reserve and adjoining Linhares Reserve, Espirito Santo. Estação Vera Cruz Reserve, Bahia.

SANTA MARTA CONURE

ROSE-CROWNED CONURE

Juv

BLUE-THROATED CONURE

SULPHUR-WINGED CONURE

P. h. hoffmanni

P. h. gaudens

PLATE 108 MOUNTAIN PARAKEETS (in part)

236 Small green parakeets with broad, blunt bill and long, strongly graduated (subgenus *Psilopsiagon*) or short, wedge-shaped to rounded (subgenus *Bolborhynchus*) tail; no bare eye-ring; sexes alike, JUV resembles adults. Highland forests, woodlands, dry scrublands, and farmlands or cultivation with remnant trees; normally small to large noisy flocks maintaining tight formation in fast flight; altitudinal and local movements; high-pitched twittering calls.

SIERRA PARAKEET
Bolborhynchus aymara 20cm

Unmistakable; only small parakeet with grayish-brown head and pale gray throat to breast; pink bill and legs; twittering *cheer-psi-psi-cheer-psi*. Characteristic of dry scrubby hillsides and valleys where presence dependent on available water; communal roosting and nesting in burrows in earth-bank. **DISTRIBUTION** eastern slopes of Andes from La Paz and Cochabamba, central Bolivia, south to Mendoza and western Córdoba, northwestern Argentina; unconfirmed reports from northernmost Chile; mostly 1800 to 3000m, seasonally to 4000m; common. **SIMILAR SPECIES** Mountain Parakeet *B. aurifrons* (see below) all-green with yellow on face. **LOCALITIES** Condorito National Park, Córdoba, northern Argentina. Aconquija Mountains and outskirts of Tafi del Valle city, Tucumán, and Quebrada de Humahuaca, Jujuy, northern Argentina.

MOUNTAIN PARAKEET
Bolborhynchus aurifrons 18cm

Only *Bolborhynchus* parakeet with deep blue on primaries; piping *tchee-tchee-tchee-tchee*, twittering *trreet, tzirr-zirr* or *zit*, and rolling *preeet* or *priiiit*. Communal roosting and nesting in burrows in earth-bank; fond of *Lepidophyllum* seeds and buds. **DISTRIBUTION** disjunctly from northwestern Peru to central Chile and northwestern Argentina; mostly 1000 to 3500m, seasonally 4100m; common. **SUBSPECIES** four subspecies differentiated by intensity of yellow in plumage and bill color. 1. *B. a. aurifrons* face, throat and sides of breast yellow (♂) or green (♀), cheeks emerald green; bill horn-colored. *Range* central Peru. 2. *B. a. robertsi* underparts darker green; ♂ with yellow restricted to forehead and chin. *Range* Río Marañón valley, northwestern Peru. 3. *B. a. margaritae* both sexes like ♀ *aurifrons*; bill horn-colored (♂) or dusky gray (♀); legs gray. *Range* southern Peru, central-western Bolivia, and northern Chile to northwestern Argentina. 4. *B. a. rubrirostris* like *margaritae*, but underparts darker green tinged blue; bill flesh-pink (♂) or dusky gray (♀). *Range* northwestern Argentina and central Chile. **SIMILAR SPECIES** Sierra Parakeet *B. aymara* (see above) gray-brown head and pale gray throat to breast. Andean Parakeet *B. orbygnesius* (plate 109) darker green; no yellow on face; shorter, rounded tail; favors more humid habitats. **LOCALITIES** Lomas de Lachay National Reserve, north of Lima, western Peru. Agua Blanca y Las Salinas Reserve, above Arequipa, and near Río Marañón on road from Sihuas, Ancash, to Huacrachuco, Huánuco, Peru. Dry valleys outside La Paz and Cochabamba cities, Bolivia. Huancar area on road to Abrapampa, Jujuy, northern Argentina.

SIERRA PARAKEET

MOUNTAIN PARAKEET

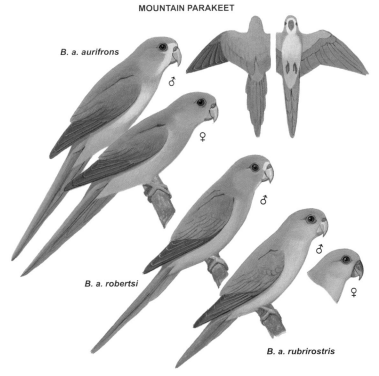

B. a. aurifrons

♂

♀

B. a. robertsi

♂

♂

♀

B. a. rubrirostris

PLATE 109 MOUNTAIN PARAKEETS (in part)

238

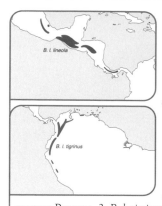

BARRED PARAKEET
Bolborhynchus lineola 16cm

Unmistakable; only small all-green, short-tailed parrot with strongly barred plumage; far-carrying *shree-eet* or *churr-ee*, slurred *cheer-churr* and nasal *jur-jur-jur*. Montane forests, secondary growth and woodlands; unpredictable, erratic movements and possibly long-distance migration in search of seeding bamboo. **DISTRIBUTION** Central America and disjunctly in northwestern South America; mostly 1500 to 3000m, occasionally 600m; locally common. **SUBSPECIES** two subspecies differentiated by intensity of barring. 1. *B. l. lineola* upperparts barred black; bend of wing black; underwings bluish green; central tail-feathers broadly (♂) or narrowly (♀) edged black. *Range* southern Mexico, south from central Veracruz and northern Oaxaca, south to northern Veraguas, western Panama. 2. *B. l. tigrinus* darker green more heavily barred black. *Range* disjunctly from Colombia and westernmost Venezuela south to central Peru, in Huánuco, Ayacucho, and Cuzco. **SIMILAR SPECIES** other small green parrots in range lack black barring. **LOCALITIES** Biotopo del Quetzal (Mario Dary Riviera Biological Reserve), Guatemala. La Amistad National Park, Cordillera de Talamanca, southern Costa Rica. Sierra Nevada de Santa Marta National Park, Magdalena, and Cueva de los Guácharos National Park, Huila, Colombia. Guaramacal National Park, Trujillo-Portuguesa, northwestern Venezuela. Podocarpus National Park, Zamora-Chinchipe, southern Ecuador.

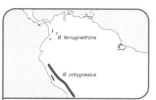

ANDEAN PARAKEET
Bolborhynchus orbygnesius 17cm

Stocky, all-green parakeet with bluish-green underwings; short green tail; chattering *dy-dy-dy-gy…dy-dy-dy-gy*, series of *gurk* notes, repeated *chi-teet-teet…chi-teet-teet*, rolling *rrueet-ee…rrueetee*. **DISTRIBUTION** eastern slopes of Andes from Cajamarca and La Libertad, northern Peru, south to northern Cochabamba and Santa Cruz, central-western Bolivia, and western slopes only in Lima, central Peru; mostly 3000 to 4000m, locally 1500 to 6000m; common. **SIMILAR SPECIES** Mountain Parakeet *B. aurifrons* (plate 108) brighter yellowish green; yellow face; longer tail; mostly on western slope, so rarely in contact. *Forpus* parrotlets (plate 112) blue on rump and wings; normally in low vegetation or grasslands. *Touit* parrotlets (plates 114, 115) prominently spotted wings in *T. stictopterus*; blue face and red underwing-coverts in *T. huetii*. **LOCALITIES** Machu Picchu Hotel (Hotel Ruinas) and mountain trails above Machu Picchu, Cuzco, and around San Mateo on main road from Lima to Huancayo, Peru. Near Pillahuata, in Manu Biosphere Reserve, Cuzco, Peru. *Polylepis* forests of San Miguel, Cochabamba, Bolivia (not common).

RUFOUS-FRONTED PARAKEET *Bolborhynchus ferrugineifrons* 18cm

Stocky green parakeet with rufous facial marking; bluish-green underwings; chattering calls like *B. orbygnesius*. Characteristically on cold, scrubby or sparsely wooded mountain slopes near or above treeline. **DISTRIBUTION** (see map above) Cordillera Central, central Colombia, where recorded mostly from Nevado del Tolima and eastern slope of Páramo del Ruiz, on Tolima–Quindío border, and slopes of Volcán du Puracé, Cauca, but may occur in intervening regions; mostly 2400 to 4000m; endangered. **SIMILAR SPECIES** none; only small parrot in very restricted, high-altitude range. **LOCALITIES** Los Nevados National Park, Tolima, and Puracé National Park, Cauca, Colombia.

B. l. lineola

BARRED PARAKEET

B. l. tigrinus

ANDEAN PARAKEET

RUFOUS-FRONTED PARAKEET

PLATE 110 *FORPUS* **PARROTLETS (in part)**

240 Small, stocky parrots with very short, wedge-shaped tail; sexually dimorphic, females
of most species alike, so identification often determined by attendant males, JUV like
adults. Flocks in open or semi-open country taking seeds on the ground or from
standing grasses; tame and inconspicuous when feeding on ground or in bushes,
but wary at other times; characteristic undulating, "finch-like" flight of flock in
unison; shrill chattering calls.

MEXICAN PARROTLET
Forpus cyanopygius 13cm

Identified by turquoise-blue rump, secondaries, and underwing-
coverts of ♂; all-green ♀; rolling *kreeit...kreeit* or *kree-eet...kree-eet*.
DISTRIBUTION northwestern Mexico, including Tres Marías
Islands, and only *Forpus* species in range; up to 1300m; common.
SUBSPECIES two slightly differentiated and one doubtful
subspecies. 1. *F. c. cyanopygius* lower back, rump, secondaries, and underwing-coverts turquoise-blue
(♂) or green (♀); bill horn-colored. *Range* Sinaloa and western Durango south to Colima.
2. *F. c. pallidus* paler than *cyanopygius*. *Range* southeastern Sonora and northernmost Sinaloa;
probably not separable from *cyanopygius*. 3. *F. c. insularis* upperparts darker green, underparts
glaucous-green; ♂ darker blue lower back to rump. *Range* Tres Marías Islands off coast of Nayarit.
LOCALITIES Navojoa to Alamos road, Sonora. Sewer Ponds trail, San Blas, Nayarit. Laguna la
María, 35km from Ciudad Colima, Colima.

GREEN-RUMPED PARROTLET
Forpus passerinus 12cm

Only *Forpus* parrotlet with green rump in both sexes (♂ tinged blue
in one subspecies); shrill *cheet-it...cheet-it* or *chee-sup...chee-sup*,
penetrating *tsup...tsup*. **DISTRIBUTION** northern South America; up
to 1800m; common; introduced Curaçao, Netherlands Antilles,
Tobago, Jamaica, and Barbados, and unsuccessfully to Martinique.

SUBSPECIES four discernible and one poorly differentiated subspecies. 1. *F. p. passerinus*
♂ secondary-coverts pale blue; outer secondaries and inner primary-coverts violet-blue; underwing-
coverts violet-blue tipped green; ♀ blue replaced by green. *Range* Guianas; introduced unsuccessfully
to Martinique, West Indies; introduced to Curaçao, Netherlands Antilles, and Tobago, and probably
subspecies introduced to Jamaica and Barbados. 2. *F. p. viridissimus* darker green; ♂ secondaries and
secondary-coverts darker blue, and underwing-coverts bluish green with violet-blue patch. *Range*
Trinidad, northern Venezuela, and northeastern Colombia. 3. *F. p. cyanophanes* ♂ like *viridissimus*,
but more extensively violet-blue on primary- and secondary-coverts; ♀ like *passerinus*. *Range* arid
tropical zone of northern Colombia. 4. *F. p. cyanochlorus* ♂ like *passerinus*, but lesser underwing-
coverts darker violet-blue; ♀ like *passerinus*, but more yellowish green. *Range* upper Rio Branco,
Roraima, northern Brazil. 5. *F. p. deliciosus* ♂ lower back to rump green tinged pale blue, and
primary- and secondary-coverts pale blue shaft-streaked violet-blue; ♀ like *passerinus*, but forehead
suffused yellow. *Range* lower Amazon River basin, northern Brazil. **SIMILAR SPECIES** Dusky-billed
Parrotlet *F. sclateri* (plate 112) darker green with bicolored gray/white bill; ♂ dark blue rump. Blue-
winged Parrotlet *F. xanthopterygius* (plate 111) ♀ similar; ♂ dark blue rump. Tepui Parrotlet
Nannopsittaca panychlora (plate 113) no blue on wings; black bend of wing; normally in forest canopy.
Touit parrotlets (plates 114–116) and *Brotogeris* parakeets (plates 117–119) different head and wing
markings, and normally in forest canopy. **LOCALITIES** Macuira National Park, Guajira, northern
Colombia. Finca El Siete and Sierra de San Luis National Park, Falcón, northern Venezuela.

MEXICAN PARROTLET

♂

♀

F. c. cyanopygius

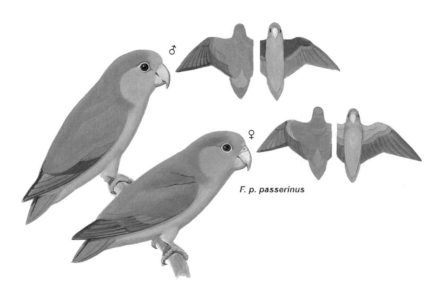

♂

♀

F. p. passerinus

GREEN-RUMPED PARROTLET

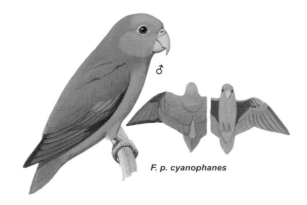

♂

F. p. cyanophanes

PLATE 111 *FORPUS* **PARROTLETS (in part)**

242

BLUE-WINGED PARROTLET
Forpus xanthopterygius 12cm

Most widespread blue-rumped *Forpus* parrotlet, and only *Forpus* species in south of range; penetrating *tseet...tseet*, twittering while feeding. **DISTRIBUTION** northwestern Colombia, and disjunctly in central South America from southern Amazonia south to northeastern Argentina, Paraguay, and eastern Bolivia; up to 500m, locally 1000m; common. **SUBSPECIES** six subspecies differentiated by intensity of coloration, and by extent of blue markings in ♂ and yellowish suffusion in ♀. 1. *F. x. xanthopterygius* ♂ primary-coverts, secondary-coverts, and underwing-coverts violet-blue; outer webs of secondaries violet-blue at bases; lower back to rump violet-blue; bill horn-colored with gray at base of upper mandible; ♀ blue replaced by green, and face greenish yellow. *Range* northeastern Argentina, in Misiones, northeastern Corrientes, eastern Chaco, and eastern Formosa, and from Paraguay north through central and mid-eastern Brazil to northern Bahia. 2. *F. x. flavissimus* ♂ ♀ paler yellowish green with yellow face. *Range* northeastern Brazil, from Maranhão, Ceará, and Paraiba south to northern Bahia. 3. *F. x. crassirostris* ♂ paler blue markings, and pale grayish-violet primary-coverts contrasting with darker violet-blue secondary-coverts; ♀ less yellowish green. *Range* northeastern Peru, eastern Ecuador, and extreme southeastern Colombia, east along both banks of Amazon River and tributaries to central Amazonas, northern Brazil. 4. *F. x. olallae* ♂ like *crassirostris*, but darker violet-blue wing-coverts, rump, and lower back; paler blue underwing-coverts; ♀ like *crassirostris*. *Range* known from only two localities, Codajas and near Itacoatiara, on north bank of Amazon River in eastern Amazonas, northern Brazil. 5. *F. x. flavescens* like *xanthopterygius*, but paler yellowish green; ♂ paler blue lower back to rump, and distinctly greenish-yellow face and underparts. *Range* central-eastern Peru south to eastern Bolivia, in Beni and Santa Cruz. 6. *F. x. spengeli* ♂ lower back to rump pale turquoise-blue; underwing-coverts turquoise intermixed violet-blue; ♀ like *xanthopterygius*, but face more yellowish. *Range* isolated in northern Colombia, from Caribbean coastal region west and south of Santa Marta mountains, Atlántico, south along lower Río Magdalena in Bolívar and Cesar. **SIMILAR SPECIES** Green-rumped Parrotlet *F. passerinus* (plate 110) ♂ ♀ green rump; ♂ less extensive, paler blue on wings. Dusky-billed Parrotlet *F. sclateri* (plate 112) darker green with bicolored gray/white bill. Amazonian Parrotlet *Nannopsittaca dachilleae* (plate 113) no blue on rump or wings; forecrown pale blue; pink bill and legs. *Touit* parrotlets (plates 114–116) and *Brotogeris* parakeets (plates 117–119) different head and wing markings, and normally in forest canopy. **LOCALITIES** Amacayacu National Park, Amazonas, southern Colombia. Cristalino Jungle Lodge, Cristalino State Park, northern Mato Grosso, Serra da Canastra National Park, Minas Gerais, and Vassununga and Fazenda Campininha State Reserves, São Paulo, Brazil. Iguazu National Park, Misiones, northeastern Argentina.

F. x. xanthopterygius

F. x. flavissimus

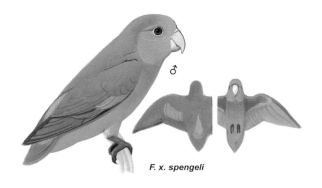

F. x. spengeli

PLATE 112 *FORPUS* **PARROTLETS (in part)**

244

DUSKY-BILLED PARROTLET
Forpus sclateri 13cm

Only *Forpus* parrotlet with bicolored gray/white bill; high-pitched *dziit* or *bzeet*, soft *jeeea-jeeea* or *weeenk-weeenk*, series of *chet* notes. **DISTRIBUTION** Guianas and Amazon River basin; up to 500m, locally 1000m; common. **SUBSPECIES** two subspecies differentiated by intensity of coloration. 1. *F. s. sclateri* dark green, brighter on forehead and cheeks; lower back to rump deep violet-blue (♂) or green (♀). *Range* from about Belém, Pará, northern Brazil, west to southeastern Colombia, thence south to eastern Peru, western Brazil, and northern Bolivia. 2. *F. s. eidos* paler green; ♂ lower back to rump paler blue-violet. *Range* French Guiana, western Guyana, eastern and southern Venezuela, and northern Brazil to eastern Colombia. **SIMILAR SPECIES** other *Forpus* species in range lack bicolored gray/white bill. *Nannopsittaca* parrotlets (plate 113) without blue on lower back to rump, and normally in forest canopy. *Touit* parrotlets (plates 114–116) and *Brotogeris* parakeets (plates 117–119) different head and wing markings, and normally in forest canopy. **LOCALITIES** La Selva Lodge and Cuyabeno Reserve, Sucumbíos, northeastern Ecuador. Amazonia National Park, Pará, and Cristalino State Park, northern Mato Grosso, northern Brazil.

SPECTACLED PARROTLET
Forpus conspicillatus 12cm

Only *Forpus* parrotlet with blue around or above eye (♂), or emerald-green encircling eye (♀); buzzing *tzit…tzit…tzit.* **DISTRIBUTION** eastern Panama and disjunctly in Colombia and western Venezuela, and only *Forpus* species in most of range; up to 1600m; common. **SUBSPECIES** three subspecies separated by intensity of coloration and extent of ocular markings. 1. *F. c. conspicillatus* ♂ cobalt-blue around eye; lower back to rump, bend of wing, and carpal edge, greater wing-coverts and underwing-coverts violet-blue; ♀ blue replaced by green, and emerald-green encircling eye. *Range* eastern Panama, west to upper Río Bayano, and northern Colombia from upper Río Sinú valley and lower Río Cauca valley east to western slopes of Cordillera Central in Boyacá and Cundinamarca. 2. *F. c. metae* ♂ blue ocular marking restricted to above and behind eye; ♀ like *conspicillatus*. *Range* eastern slopes of Cordillera Oriental in Boyacá, Cundinamarca, and Meta, central Colombia, and east through Vichada and Casanare to western Venezuela, along Río Meta in western Apure. 3. *F. c. caucae* heavier, more robust bill; ♂ greater wing-coverts and lower back to rump paler blue, less violet. *Range* western Colombia, in upper Río Cauca valley in Antioquia and Caldas, Río Dagua valley, western slopes of Cordillera Occidental in Valle, Río Patía valley in Cauca and Nariño, and coastal southwestern Nariño. **SIMILAR SPECIES** Green-rumped Parrotlet *F. passerinus* (plate 110) green rump in both sexes; ♂ without blue around eye or on bend of wing; ♀ without emerald-green encircling eye. *Touit* parrotlets (plates 114, 115) and *Brotogeris* parakeets (plates 118) different head and wing markings, and normally in forest canopy. **LOCALITIES** Paramillo National Park, Córdoba, and Sierra de la Macarena National Park, Meta, Colombia.

DUSKY-BILLED PARROTLET

♂

♀

F. s. sclateri

SPECTACLED PARROTLET

♂

♀

F. c. conspicillatus

♂

F. c. metae

PLATE 113 *FORPUS* (in part) AND *NANNOPSITTACA* PARROTLETS

246

PACIFIC PARROTLET *Forpus coelestis* 13cm

Unmistakable, one of two more highly colored *Forpus* parrotlets; ♂ greenish-gray upperparts; bright yellowish-green face with blue band behind eye; bluish-gray hindneck; lower back to rump, greater wing-coverts, and underwing-coverts violet-blue; ♀ blue replaced by green; high-pitched *tchit* or *tzit* repeated. **DISTRIBUTION** Pacific slope of Andes from Borbón district, Esmeraldas, northwestern Ecuador, south to about Trujillo, La Libertad, and middle Río Marañón valley, northwestern Peru, where possibly meets *F. xanthops*; likely to reach southernmost Nariño, extreme southwestern Colombia; mostly below 800m; common. **LOCALITIES** Cerro Blanco Reserve and Loma Alta Ecological Reserve, Guayas, Ecuador. Tumbes National Forest, Tumbes, northwestern Peru.

YELLOW-FACED PARROTLET
Forpus xanthops 15cm

Unmistakable, markedly more yellowish than *F. coelestis*, with olive-gray upperparts and blue rump in both sexes; ringing *tjeet* repeated rapidly. **DISTRIBUTION** northwestern Peru, in upper Río Marañón valley from easternmost La Libertad north to southeastern Cajamarca and southern Amazonas; mostly 800 to 1000m; locally common, but declining. **LOCALITIES** Chagual and Soquian districts, La Libertad, and Balsas district, Cajamarca.

NANNOPSITTACA PARROTLETS
Small all-green parrots with fine bill and short, squarish tail; sexes alike, JUV like adults. Poorly known; forests; arboreal; small to large flocks in swift, direct flight above canopy; well hidden amidst foliage; possibly nomadic; tinkling calls.

TEPUI PARROTLET
Nannopsittaca panychlora 14cm

Yellow around eye, black on bend of wing, and yellow carpal edge; tinkling *seize-la* or *tseez-zip*. Associated with montane forest on slopes of tepuis. **DISTRIBUTION** southeastern Venezuela, disjunctly in eastern Bolívar, southeastern Venezuela, and neighboring Kamarang River region, westernmost Guyana, with outlying Venezuelan populations in central and southern Amazonas, and north on Cerro Humo and Cerro Papelón, Sucre, and possibly in northernmost Roraima, Brazil; mostly 750 to 1850m; locally common. **SIMILAR SPECIES** *Forpus* parrotlets (plates 110, 112) ♂ with blue on rump and wings; usually on or near ground in open country; undulating flight. Lilac-tailed Parrotlet *Touit batavica* (plate 114) prominent head and wing markings, and violet tail. Golden-winged Parakeet *Brotogeris chrysoptera* (plate 119) orange primary-coverts; blue on primaries; white eye-ring. **LOCALITY** Canaima National Park, eastern Bolívar, Venezuela.

AMAZONIAN PARROTLET
Nannopsittaca dachilleae 12cm

Only parrotlet with blue forecrown; pink bill and legs; ringing *tcheereet*. Often associated with bamboo thickets. **DISTRIBUTION** eastern Peru south from Iquitos district, Loreto, neighboring westernmost Brazil and northwestern Bolivia; up to 1000m; near-threatened. **SIMILAR SPECIES** *Forpus* parrotlets (plate 112) see above. Scarlet-shouldered Parrotlet *Touit huetii* (plate 115) distinctive head markings; red underwing-coverts; multicolored tail. *Brotogeris* parakeets (plates 118, 119) larger with longer, pointed tail; blue on wings. **LOCALITY** Manú Biosphere Reserve, Madre de Dios, southeastern Peru.

PACIFIC PARROTLET

♂

♀

♂

Juv

YELLOW-FACED PARROTLET

♂

♀

TEPUI PARROTLET

AMAZONIAN PARROTLET

PLATE 114 *TOUIT* **PARROTLETS (in part)**

248 Small, stocky green parrots with short, squarish, often multicolored tail; mostly sexually dimorphic, JUV duller. Forests and secondary growth; arboreal; secretive in canopy, where detection and identification difficult; small parties or larger flocks seen in swift, direct flight above treetops; soft, high-pitched calls.

LILAC-TAILED PARROTLET
Touit batavicus 14cm

Unmistakable; distinctive coloration features yellowish-green head with dark scalloping on hindneck; upperparts black with yellow "wing-patch"; blue underwing-coverts and mauve-violet tail; sexes alike; soft, nasal *naaa-ee*, and trilling or chattering notes. **DISTRIBUTION** Trinidad, where only resident *Touit* species, Guianas, and possibly Amapá, northernmost Brazil, and disjunctly across northern Venezuela, in Bolívar, Sucre, northern Monagas, Miranda, Distrito Federal to Aragua, and southeastern Falcón, and extralimitally reaching northern Magdalena, northernmost Colombia; up to 1700m; common. **SIMILAR SPECIES** Scarlet-shouldered Parrotlet *T. huetii* (plate 115) possibly sympatric in northern Venezuela; green upperparts without yellow "wing-patch"; red underwing-coverts. *Forpus* parrotlets (plates 110, 112) all-green with blue rump; normally near or on ground in open country. Golden-winged Parakeet *Brotogeris chrysoptera* (plate 119) green upperparts and underwing-coverts; pointed all-green tail. **LOCALITIES** Asa Wright Nature Centre, Spring Hill Estate, Trinidad. Henri Pittier National Park, northern Venezuela.

SPOT-WINGED PARROTLET
Touit stictopterus 17cm

Only *Touit* parrotlet with all-green tail; ♂ upper wing-coverts dusky brown tipped dull white, and outermost primary-coverts orange; ♀ and JUV upper wing-coverts green with black bases showing through, and outermost primary-coverts green; harsh *ddreet-ddreet-ddreet*, raspy *raah-reh* or *raah-reh...reh*. **DISTRIBUTION** central Colombia, on western slopes of Cordillera Central in Cauca, western slopes of Cordillera Oriental in Cundinamarca, Sierra de la Macarena in Meta, and possibly eastern Nariño, and eastern slopes of Andes in eastern Ecuador to northern Peru; mostly 1000 to 1700m; vulnerable. **SIMILAR SPECIES** Only *Touit* parrotlet in much of range. Scarlet-shouldered Parrotlet *T. huetii* (plate 115) blue face and red underwing-coverts. Sapphire-rumped Parrotlet *T. purpuratus* (plate 116) possibly sympatric in northeastern Peru; brown band on scapulars; blue rump and red tail. Barred Parakeet *Bolborhynchus lineola* (plate 109) black barring to green plumage. Cobalt-winged Parakeet *Brotogeris cyanoptera* (plate 118) orange chin-spot; blue flight feathers; pointed tail and prominent white eye-ring. **LOCALITIES** Cordillera de los Picachos National Park, Meta-Huila, Colombia. Sangay National Park, Morona-Santiago, and Cayambe-Coca Ecological Reserve and Sierra Azul Forest Reserve, western Napo, Ecuador.

LILAC-TAILED PARROTLET

SPOT-WINGED PARROTLET

♂

♀

PLATE 115 **TOUIT** PARROTLETS (in part)

250

SCARLET-SHOULDERED PARROTLET
Touit huetii 16cm

One of three allopatric species with blue or blue-and-red on face and forewing, and prominent white cere and eye-ring; only species with red underwing-coverts and bluish-black face without red; lateral tail-feathers crimson (♂) or greenish yellow (♀) tipped black; JUV green face; disyllabic *tu-weet*, nasal *reenk* repeated. **DISTRIBUTION** scattered localities in Amazonia, but probably present in intervening areas from northern Guyana, northeastern and southern Venezuela, and southern Colombia to northernmost Bolivia, in Pando; also northeastern Brazil, south of Amazon River, between Rio Cururu and Belém, Pará, south to southernmost Pará, northernmost Goiás, and extralimitally to northwestern Mato Grosso; accidental in Trinidad and unconfirmed in French Guiana; up to 1200m; scarce and poorly known. **SIMILAR SPECIES** Lilac-tailed Parrotlet *T. batavicus* (plate 114) possibly sympatric in northern Venezuela; black upperparts with yellow "wing-patch"; no blue on face; no white eye-ring. *Forpus* (plates 110, 112) and *Nannopsittaca* (plate 113) parrotlets; no red in plumage; all-green tail; no white eye-ring. *Brotogeris* parakeets (plates 117–119) no blue on face; green or greenish-yellow underwing-coverts; pointed, all-green tail. **LOCALITIES** Sierra de la Macarena National Park, Meta, Colombia. Cristalino Jungle Lodge, Cristalino State Park, northern Brazil. La Selva Lodge, Sucumbios, Ecuador.

BLUE-FRONTED PARROTLET
Touit dilectissimus 17cm

One of two, possibly conspecific, species differentiated from *T. huetii* by yellow underwing-coverts and lateral tail-feathers yellow tipped black; this species identified by blue forecrown, red spot in front of, and behind eye, and forewing scarlet (♂) or green (♀); JUV forecrown and around eyes green; high-pitched *too-weet* repeated. **DISTRIBUTION** Cerro Jefe area, central Panama, to western Colombia, south on western slopes of Andes to El Oro, southwestern Ecuador, and northern Colombia, at northern extremity of Cordillera Oriental, Norte de Santander–Cesar border, east to Trujillo, northwestern Venezuela, mostly 500 to 2000m; uncommon and poorly known. **SIMILAR SPECIES** *Forpus* parrotlets (plates 110, 112) see above. Barred Parakeet *Bolborhynchus lineola* (plate 109) black barring to all-green plumage; wedge-shaped tail. Orange-chinned Parakeet *Brotogeris jugularis* (plate 118) green face; brown "shoulders"; pointed all-green tail. **LOCALITIES** Darién Biosphere Reserve, Darién, Panama. Los Katíos and Utría National Parks, Chocó, Los Farallones National Park, Valle, and Munchique National Park, Cauca, Colombia.

RED-FRONTED PARROTLET
Touit costaricensis 17cm

Differentiated from very similar, possibly conspecific *T. dilectissimus* by red forecrown; ♂ with red line underneath eye, and red forewing; ♀ without red underneath eye and little or no red on forewing; JUV like ♀, but forecrown red intermixed green; calls like *T. dilectissimus*. **DISTRIBUTION** Costa Rica, south from Caribbean slope of Cordillera Central at Puerto Limón and Volcán Turrialba, and western Panama, in Bocas del Toro, Chiriquí, and western Coclé; mostly 500 to 1000m, seasonally 3000m; vulnerable. **SIMILAR SPECIES** Only *Touit* species in range. Barred Parakeet *Bolborhynchus lineola* (plate 109) and Orange-chinned Parakeet *Brotogeris jugularis* (plate 118) see above. **LOCALITIES** Cerro de la Muerte, San José, and Genesis II Lodge, Cordillera Talamanca, Costa Rica. Monteverde Biological Reserve and Braulio Carillo National Park, Costa Rica. El Copé National Park, western Panama.

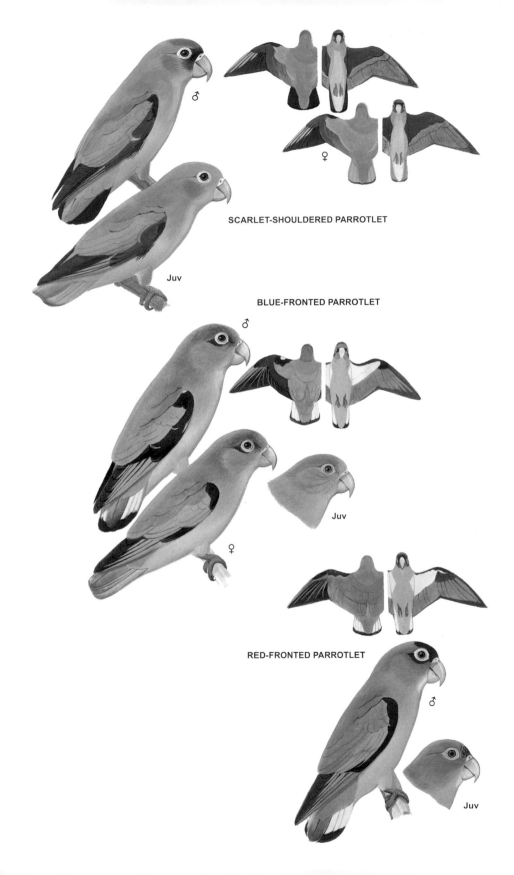

SCARLET-SHOULDERED PARROTLET

BLUE-FRONTED PARROTLET

RED-FRONTED PARROTLET

PLATE 116 *TOUIT* **PARROTLETS (in part)**

252

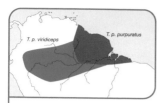

SAPPHIRE-RUMPED PARROTLET
Touit purpuratus 17cm

One of two similar, but allopatric *Touit* parrotlets with brown on back and red tail; only *Touit* parrotlet with blue rump, and differentiated from *T. melanonotus* by green, not brown, mantle; nasal *ny-aah*, trisyllabic *aa-aa-eck* and *keree-ke-ke*. **DISTRIBUTION** Amazon River basin; up to 1200m; uncommon. **SUBSPECIES** two subspecies differentiated by color of crown. 1. *T. purpuratus* crown and occiput olive-brown; scapulars and tertials dark brown forming "V"-shaped band; lateral tail-feathers violet-red tipped black (♂) or green (♀). *Range* Guianas, northern Brazil, from Amapá and northern Maranhão west to northeastern Amazonas, and southern Venezuela. 2. *T. p. viridiceps* crown and occiput green. *Range* northernmost Brazil, in northwestern Amazonas, and southern Venezuela, west of Cerro Duida in southern Amazonas, west to southeastern Colombia, eastern Ecuador, and northeastern Peru. **SIMILAR SPECIES** Lilac-tailed Parrotlet *T. batavicus* (plate 114) greenish-yellow head; upperparts black with yellow "wing-patch." Scarlet-shouldered Parrotlet *T. huetii* (plate 115) blue face; red underwing-coverts; prominent white eye-ring. *Forpus* parrotlets (plates 110–112) smaller green parrots with all-green, wedge-shaped tail; normally on or near ground in open country. *Nannopsittaca* parrotlets (plate 113) no brown scapulars, and all-green tail. Short-tailed *Brotogeris* parakeets (plates 118, 119) orange chin-spot or yellow forecrown; all-green, pointed tail. **LOCALITIES** Amacayacu National Park, Amazonas, southernmost Colombia. Cuyabeno Reserve, Sucumbíos, northeastern Ecuador. Ducke Forest Reserve, near Manaus, Amazonas, and Cristalino Jungle Lodge, Cristalino State Park, northern Mato Grosso, Brazil.

BROWN-BACKED PARROTLET
Touit melanonotus 15cm

Like *T. purpuratus*, but mantle, scapulars, and tertials dark brown; back black; green rump; rattling *tew-rew…tew-rew*. **DISTRIBUTION** southeastern Brazil, from southern Bahia south to southern São Paulo; mostly 500 to 1000m; endangered. **SIMILAR SPECIES** Golden-tailed Parrotlet *T. surdus* (see below) green back with golden-brown scapulars. Blue-winged Parrotlet *Forpus xanthopterygius* (plate 111) see above. *Brotogeris* parakeets (plate 117) green upperparts; longer all-green, pointed tail. **LOCALITIES** Itatiáia and Serra dos Orgãos National Parks, Rio de Janeiro. Serra do Mar and Intervales State Parks, São Paulo.

GOLDEN-TAILED PARROTLET
Touit surdus 16cm

Another *Touit* parrotlet with brown on back, but greenish-yellow tail; calls undescribed. **DISTRIBUTION** eastern Brazil; up to 800m; vulnerable. **SUBSPECIES** two doubtfully distinct subspecies. 1. *T. s. surdus* face golden-yellow; scapulars and tertials golden-brown forming "V"-shaped band; lateral tail-feathers greenish-yellow tipped black (♂) or green (♀). *Range* southeastern Brazil, from southern Bahia, and possibly southern Goiás, south to São Paulo. 2. *T. s. chryseura* lateral tail-feathers brownish yellow. *Range* northeastern Brazil, in Paraíba, Pernambuco, and Alagoas; extralimitally in coastal Ceará. **SIMILAR SPECIES** Brown-backed Parrotlet *T. melanonotus* (see above) dark brown back and red tail. Blue-winged Parrotlet *F. xanthopterygius* (plate 111) see above. *Brotogeris* parakeets (plate 117) see above. **LOCALITIES** Charles Darwin Ecological Refuge and Usina São José, north coast of Pernambuco. Linhares and Sooretama Reserves, Espírito Santo. Ilha do Cardoso and Intervales State Parks, São Paulo.

SAPPHIRE-RUMPED PARROTLET

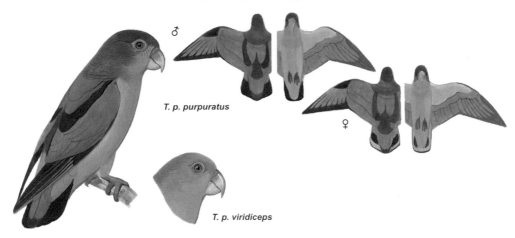

♂

T. p. purpuratus

♀

T. p. viridiceps

BROWN-BACKED PARROTLET

T. s. surdus

♂

♀

GOLDEN-TAILED PARROTLET

PLATE 117 *BROTOGERIS* PARAKEETS (in part)

254 Small green parrots with pointed, graduated tail, long in three species and shorter in others, but always longer than tails of *Forpus* or *Touit* parrotlets; proportionately long, pointed wings; sexes alike, JUV like adults. Most wooded habitats, farmlands, and some urban parklands; arboreal; highly gregarious; noisy flocks conspicuous in swift, direct flight, and large numbers at communal nighttime roosts; shrill screeching and loud squawks.

PLAIN PARAKEET *Brotogeris tirica* 23cm

All-green parakeet with long, graduated tail; bend of wing and lesser wing-coverts olive-brown; shrill, rolling screech. **DISTRIBUTION** eastern Brazil, from Alagoas and eastern Bahia south to São Paulo, and extralimitally to Santa Catarina; up to 1200m; common. **SIMILAR SPECIES** Yellow-chevroned Parakeet *B. chiriri* (see below) yellow secondary-coverts. *Touit* parrotlets (plate 116) very short, squarish tail; brown on back or scapulars. **LOCALITIES** easily seen in and around Rio de Janeiro, including Botanic Gardens, and in São Paulo city. Sete Barras State Reserve, São Paulo, and Itatiáia National Park, Rio de Janeiro.

WHITE-WINGED PARAKEET
Brotogeris versicolorus 22cm

Common cagebird; easily identified by yellow and white in wings; lores bare with scattered bluish-gray feathers; high-pitched *screek* interspersed with *weechah-weechah*. **DISTRIBUTION** northern Amazon River basin, from southeastern Colombia and northeastern Peru east to French Guiana, where no recent records, and northeastern Brazil, in Amapá, on Ilha do Mexiana at mouth of Amazon River, and Belém district, Pará; could occur in easternmost Ecuador and southernmost Suriname, but no confirmed records; up to 300m; common; feral populations in Lima district, Peru, Puerto Rico, West Indies, and with *B. chiriri* in California and southeastern Florida, U.S.A.; possibly introduced to Dominican Republic, but identification unconfirmed. **SIMILAR SPECIES** Other *Brotogeris* species (plates 118, 119) no white in wings; colored chin-spot or forecrown; shorter, pointed tail. **LOCALITIES** Amacayacu National Park and Leticia district, Amazonas, southern Colombia.

YELLOW-CHEVRONED PARAKEET
Brotogeris chiriri 22cm

Common cagebird; like *B. versicolorus*, and possibly conspecific, but no white in wing; feathered lores; shrill *chiri…chiri…ri* or *te-clee-tee*, more high-pitched than call of *B. versicolorus*. **DISTRIBUTION** interior of eastern Brazil to central Bolivia and northern Argentina; mostly below 1000m, locally 2500m in Bolivia; common; introduced to California and southeastern Florida, U.S.A., and possibly also Dominican Republic, where identification unconfirmed. **SUBSPECIES** two slightly differentiated subspecies. 1. *B. c. chiriri* feathered lores green; secondary-coverts yellow; primaries green tinged blue. *Range* interior of eastern and southern Brazil, from Ceará, Maranhão and southern Pará south to Rio de Janeiro, western São Paulo and Mato Grosso, to eastern Bolivia, south to Santa Cruz, and through Paraguay to northern Argentina, in Formosa, Chaco, Misiones, and northern Corrientes. 2. *B. c. behni* darker, less yellowish-green, and larger. *Range* central Bolivia to northwestern Argentina, in Salta. **SIMILAR SPECIES** Plain Parakeet *B. tirica* (see above) no yellow in wings. Other *Brotogeris* species (plates 118, 119) no yellow in wings; colored chin-spot or forecrown. **LOCALITIES** Serra da Canastra National Park, Minas Gerais, and Caiman Lodge Wildlife Refuge, Mato Grosso do Sul, Brazil.

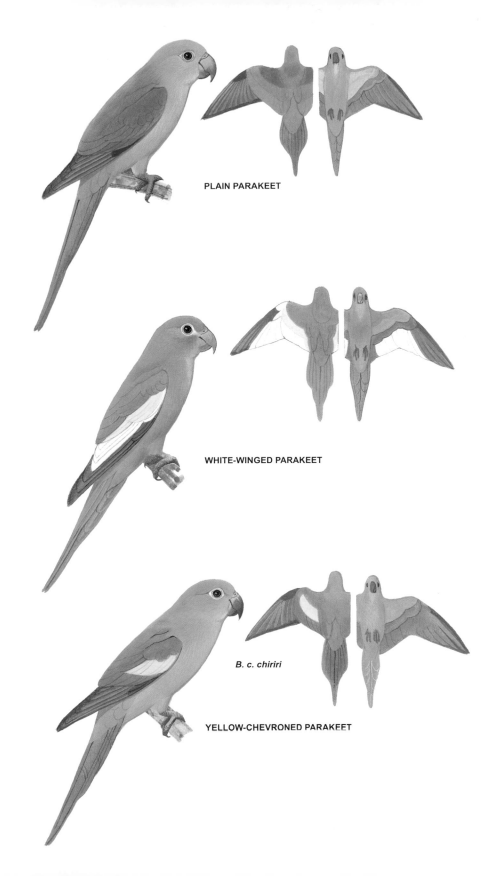

PLAIN PARAKEET

WHITE-WINGED PARAKEET

B. c. chiriri

YELLOW-CHEVRONED PARAKEET

PLATE 118 *BROTOGERIS* **PARAKEETS (in part)**

256

GRAY-CHEEKED PARAKEET
Brotogeris pyrrhoptera 20cm

Identified by distinctive coloration featuring greenish-blue crown, buff-gray cheeks, and orange underwing-coverts; trilling *tsleet-tsleet-tsleet*. **DISTRIBUTION** west of Andes in southwestern Ecuador, north to central Manabí, and northwestern Peru, south to northern Píura; mostly below 1000m; endangered. **SIMILAR SPECIES** Only *Brotogeris* parakeet in range. Pacific Parrotlet *Forpus coelestis* (plate 113) yellowish-green crown and cheeks; underwing-coverts blue (♂) or green (♀); rump blue (♂); normally on or near ground in open country. **LOCALITIES** Cerro Blanco Reserve, Guayas, Ecuador. Tumbes National Forest Reserve, Tumbes, Peru.

ORANGE-CHINNED PARAKEET
Brotogeris jugularis 18cm

Common cagebird; one of three short-tailed species with orange chin-spot and prominent white eye-ring; identified by olive-brown upper wing-coverts forming conspicuous "shoulder-patch"; harsh *ack-ack-ack*, scratchy *ra-aa-aa-aa*, musical *kweek-kweek…kweek-kee…roo-kree-roo*, sharp *ki-ki* or *chee…chee-chit*. **DISTRIBUTION** southwestern Mexico south to central Colombia and northern Venezuela, but absent from Belize; mostly below 1000m; common. **SUBSPECIES** two slightly differentiated subspecies. 1. *B. j. jugularis* mantle washed olive; lower back to rump and lower underparts tinged blue. *Range* southwestern Mexico, in eastern Oaxaca, south through Central America, mainly on Pacific slope, to central Colombia, in middle Río Magdalena valley, and northwestern Venezuela, in Maracaibo basin and western Andes in Táchira and Mérida. 2. *B. j. exsul* thighs and undertail-coverts bright green not tinged blue; paler, less extensive orange chin-spot; more pronounced olive suffusion on mantle; darker brown "shoulder-patch." *Range* northeastern Colombia and northern Venezuela from Trujillo and Lara east to Guarico. **SIMILAR SPECIES** Only *Brotogeris* species in range. Barred Parakeet *Bolborhynchus lineola* (plate 109) black barring to green plumage; no brown "shoulder-patch," orange chin-spot, or white eye-ring. **LOCALITIES** Bosque del Río Tigre Sanctuary, Dos Brazos, and Lapa Ríos Nature Reserve, Golfo Dulce, Costa Rica. Tayrona National Park, Magdalena, and Catatumbo-Barí National Park, Norte de Santander, Colombia.

COBALT-WINGED PARAKEET
Brotogeris cyanoptera 18cm

Another short-tailed species with orange chin-spot and white eye-ring, but no brown "shoulder-patch"; blue in flight feathers; clear *splink…splink*, ringing *jeet…jeet*. **DISTRIBUTION** western Amazonia; up to 1000m; common. **SUBSPECIES** "green-shouldered" and "yellow-shouldered" populations, with slight differentiation in latter. 1. *B. c. cyanoptera* bend of wing and carpal edge green. *Range* southern Venezuela, northwestern Brazil, and southeastern Colombia to eastern Ecuador, eastern Peru, except upper Río Huallaga valley, and northernmost Bolivia. 2. *B. c. gustavi* bend of wing and carpal edge yellow. *Range* upper Río Huallaga valley, northern Peru. 3. *B. c. beniensis* like *gustavi*, but paler green; forehead tinged yellow; crown suffused blue. *Range* northern Bolivia, in Beni and Cochabamba. **SIMILAR SPECIES** White-winged Parakeet *B. versicolorus* and Yellow-chevroned Parakeet *B. chiriri* (plate 117) yellow or yellow-and-white in wing; no orange chin-spot. Golden-winged Parakeet *B. chrysoptera* (plate 119) darker green; orange or yellow primary-coverts. Tui Parakeet *B. sanctithomae* (plate 119) yellow forecrown; no orange chin-spot. **LOCALITIES** Tinigua National Park, Meta, and Amacayacu National Park, Amazonas, Colombia. Río Bombuscaro, Podocarpus National Park, Zamora-Chinchipe, Ecuador. Manú Biosphere Reserve, Madre de Dios, Peru.

GRAY-CHEEKED PARAKEET

B. j. jugularis

ORANGE-CHINNED PARAKEET

COBALT-WINGED PARAKEET

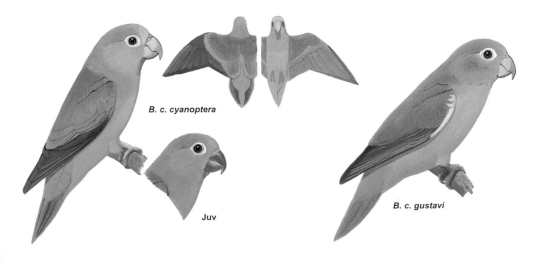

B. c. cyanoptera

Juv

B. c. gustavi

PLATE 119 *BROTOGERIS* PARAKEETS (in part)

258

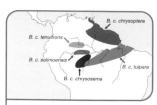

GOLDEN-WINGED PARAKEET
Brotogeris chrysoptera 16cm

A third short-tailed species with orange chin-spot and white eye-ring; identified by orange to dark brown frontal band and yellow or orange primary-coverts; harsh *churr-churr-churr*, high-pitched *chil...chil...chil* or *chit-chit*. **DISTRIBUTION** eastern Amazonia; up to 1200m; common. **SUBSPECIES** five subspecies differentiated mainly by head and wing markings. 1. *B. c. chrysoptera* blackish-brown frontal band; orange-brown chin-spot; primary-coverts orange. *Range* Guianas, northeastern Venezuela, and north of Amazon River in northernmost Brazil, from Roraima east to Amapá. 2. *B. c. tuipara* narrow frontal band and chin-spot orange. *Range* south of Amazon River in northern Brazil, from Pará to northeastern Maranhão. 3. *B. c. chrysosema* forehead yellowish orange; orange chin-spot; primary-coverts yellow. *Range* Rio Madeira and tributaries in Amazonas and northern Mato Grosso, northern Brazil. 4. *B. c. solimoensis* like *chrysoptera*, but paler reddish-brown frontal band and yellowish-brown chin-spot. *Range* upper Amazon River in Codajas and Manaus districts, Amazonas, northern Brazil. 5. *B. c. tenuifrons* like *tuipara*, but little or no orange frontal band. *Range* upper Rio Negro at Santa Isabel and confluence of Rio Cauaburi, Amazonas, northern Brazil. **SIMILAR SPECIES** White-winged Parakeet *B. versicolorus* (plate 117) white and yellow in wings; no frontal band or chin-spot; longer tail. Cobalt-winged Parakeet *B. cyanoptera* (plate 118) paler green; primary-coverts blue; no frontal band. Tui Parakeet *B. sanctithomae* (see below) yellow forecrown; no chin-spot. **LOCALITIES** Amazonia National Park, Pará, Ducke Forest Reserve, near Manaus, eastern Amazonas, and Cristalino Jungle Lodge, Cristalino State Park, northern Mato Grosso, northern Brazil.

TUI PARAKEET *Brotogeris sanctithomae* 17cm

Short-tailed species without chin-spot, and only *Brotogeris* parakeet with yellow forecrown; high-pitched *screek* repeated rapidly. **DISTRIBUTION** Amazon River basin, west to eastern Peru and northernmost Bolivia; up to 300m; common. **SUBSPECIES** two well-marked, possibly isolated subspecies. 1. *B. s. sanctithomae* forehead, lores, and forecrown yellow. *Range* upper Amazon River at confluence of Rio Madeira, eastern Amazonas, northern Brazil, west to extreme southeastern Colombia, northeastern Ecuador, and northeastern Peru, and south along Rio Madeira and tributaries to northernmost Bolivia, in Pando and northern Beni, and southeastern Peru. 2. *B. s. takatsukasae* yellow stripe behind eye to ear-coverts. *Range* northern Brazil, along lower Amazon River from about confluence of Rio Madeira east on north bank to Amapá, and on south bank to confluence of Rio Curuá or possibly Belém district, eastern Pará. **SIMILAR SPECIES** Other *Brotogeris* species (plates 117, 118) without yellow forecrown. Amazonian Parrotlet *Nannopsittaca dachilleae* (plate 113) no yellow forecrown; pale blue crown; smaller. **LOCALITIES** Amacayacu National Park, Amazonas, southern Colombia. La Selva Lodge, Sucumbios, northeastern Ecuador. Manú Biosphere Reserve, Madre de Dios, eastern Peru.

GOLDEN-WINGED PARAKEET

B. c. chrysoptera

B. c. tuipara

B. c. chrysosema

TUI PARAKEET

B. s. sanctithomae

B. s. takatsukasae

PLATE 120 *PIONITES* **PARROTS**

260 Midsized, stocky parrots with short, squarish tail and fairly short, rounded wings; distinctive coloration features white and yellow underparts; sexes alike, JUV duller than adults. Forests, forest edges, tall secondary growth; arboreal; pairs and small flocks; noisy and conspicuous; communal roosting, often in tree hollows; swift, direct flight with audible "whirring" of wingbeats.

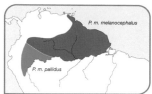

BLACK-CAPPED PARROT
Pionites melanocephalus 23cm

Unmistakable; distinctive white underparts and black crown; squealing *cleeeooo-cleeeooo* or *heeyah…heeyah*, screeching *wheech-wheech-wheech*, shrill *wey-ak*, loud *kleek*. **DISTRIBUTION** northern Amazonia, north of Amazon River; up to 1000m; common. **SUBSPECIES** two subspecies differentiated by intensity of plumage coloration. 1. *P. m. melanocephalus* cheeks, throat, and lower underparts orange-yellow; black bill. *Range* Amapá, northern Brazil, and Guianas west to eastern Colombia, in eastern Vichada and northeastern Guainía. 2. *P. m. pallidus* cheeks, throat, and lower underparts clear yellow. *Range* southeastern Colombia, south from western Meta and Vaupes, to eastern Ecuador and northeastern Peru, west of Río Ucayali. **LOCALITIES** La Selva Lodge, Sucumbios, northeastern Ecuador. Tinigua National Park, Meta, and Amacayacu National Park, Amazonas, Colombia.

WHITE-BELLIED PARROT
Pionites leucogaster 23cm

Unmistakable; only white-bellied parrot in range, and differentiated from allopatric *P. melanocephalus* by orange crown and pale bill; screeching calls similar to *P. melanocephalus*. **DISTRIBUTION** southern Amazonia, south of Amazon River; up to 800m; common. **SUBSPECIES** three well-marked subspecies, but with much intergradation. 1. *P. l. leucogaster* flanks and thighs green; undertail-coverts yellow; tail green. *Range* northern Brazil, from northwestern Maranhão and eastern Pará to northwestern Mato Grosso and west to lower Rio Madeira, northeastern Amazonas. 2. *P. l. xanthurus* thighs, flanks, undertail-coverts, and entire tail yellow. *Range* northern Brazil, from Rio Madeira catchment in Amazonas and northern Rondônia west to Rio Juruá, western Amazonas. 3. *P. l. xanthomerius* thighs and flanks yellow, but tail green. *Range* western Amazonas, northern Brazil, and eastern Peru, east of Río Ucayali, to northern Bolivia. **LOCALITIES** Amazonia National Park, Pará, and Cristalino Jungle Lodge, Cristalino State Park, northern Mato Grosso, northern Brazil. Manú Biosphere Reserve, Madre de Dios, southeastern Peru. Los Ferros Lodge, Noel Kempff Mercado National Park, Santa Cruz, Bolivia.

BLACK-CAPPED PARROT

P. m. melanocephalus

Juv

P. m. pallidus

WHITE-BELLIED PARROT

P. l. leucogaster

Juv

P. l. xanthurus

P. l. xanthomerius

PLATE 121 *GYPOPSITTA* PARROTS (in part)

262 Midsized, stocky parrots with short, rounded tail and broad, pointed wings; bare eye-ring, and two species bare-headed; sexes alike, JUV duller than adults. Forests, tall secondary growth, plantations; pairs or small flocks, larger flocks at roosts; not noisy, and inconspicuous in treetops; swift flight with characteristic side-to-side rolling.

BROWN-HOODED PARROT
Gypopsitta haematotis 21cm

Only *Gypopsitta* species in much of range; distinctive brownish face with contrasting white lores; red axillaries displayed in flight; *pileek-pileek* alternating with harsh *zapp-zapp*, warbling *kree-ee...tee...yer*. **DISTRIBUTION** southernmost Mexico to northwestern Colombia; up to 1200m; common. **SUBSPECIES** two well-marked subspecies.

1. *G. h. haematotis* foreneck and breast dull brownish olive. *Range* southernmost Mexico, in Oaxaca and southern Veracruz, south mainly along Caribbean slope to western Panama.
2. *G. h. coccinicollaris* foreneck and upper breast marked red, often forming collar in ♂. *Range* eastern Panama to northwestern Colombia, in northernmost Chocó and northern Bolívar. **LOCALITIES** Montes Azules Biosphere Reserve, Chiapas, Mexico. Tikal National Park, El Petén, Guatemala. Lamanai Field Research Center, Belize. Soberanía National Park, Panama. Los Katíos National Park, Chocó, Colombia.

ROSE-FACED PARROT
Gypopsitta pulchra 23cm

Identified by rose-pink face; crown to nape grayish brown; neck and breast dull yellow-brown; shrieking *skreek-skreek*. **DISTRIBUTION** west of Andes in Colombia, south from Chocó and Antioquia, and western Ecuador, south to El Oro; up to 1600m; uncommon. **LOCALITIES** Farallones de Cali National Park, Valle, and Munchique National Park, Cauca, Colombia. Cotacachi-Cayapas Ecological Reserve, Esmeraldas, and Buenaventura forest, near Piñas, El Oro, Ecuador.

SAFFRON-HEADED PARROT
Gypopsitta pyrilia 24cm

Unmistakable; only short-tailed green parrot with all-yellow head (green in JUV) and red underwing-coverts; upper breast olive; bend of wing and carpal edge red; scraping *che-week*, high-pitched *keek*. **DISTRIBUTION** Darién, easternmost Panama, and northwestern Colombia, south to middle Río Magdalena valley and Boyaca, to northwestern Venezuela, east to northwestern Barinas and southeastern Lara; sight records from northwestern Ecuador probably erroneous; mostly 150 to 1700m; uncommon. **LOCALITIES** Darién Biosphere Reserve, Darién, eastern Panama. Utría National Park, Chocó, and Paramillo National Park, Córdoba, Colombia.

CAICA PARROT *Gypopsitta caica* 23cm

Only *Gypopsitta* parrot with all-black head (green in JUV); neck yellowish brown with fine blackish scalloping on hindneck; green underwing-coverts; nasal *queek* or *skrek*, nasal *kunk* or *aank* when perched, low-pitched *wee-uck* or *who-cha*. **DISTRIBUTION** eastern Amazonia, north of Amazon River, in Guianas, northern Brazil, between Amapá and western Roraima, and eastern Venezuela, in eastern Bolívar. **SIMILAR SPECIES** Only *Gypopsitta* species in much of range. Orange-cheeked Parrot *G. barrabandi* (plate 122) black head with contrasting orange cheeks; red underwing-coverts. **LOCALITIES** Caura Forest Reserve, Bolívar, Venezuela. Ducke Forest Reserve, near Manaus, Amazonas, Brazil.

G. h. haematotis

G. h. coccinicollaris

BROWN-HOODED PARROT

Juv

ROSE-FACED PARROT

Juv

SAFFRON-HEADED PARROT

Juv

CAICA PARROT

Juv

PLATE 122 *GYPOPSITTA* PARROTS (in part)

264

ORANGE-CHEEKED PARROT
Gypopsitta barrabandi 25cm

Identified by black head with contrasting yellow or orange cheeks; red underwing-coverts displayed in flight; reedy *chew-it* or *hoy-et*, guttural *kuk* or *kek*. **DISTRIBUTION** western Amazonia and upper Río Orinoco drainage; up to 400m; common. **SUBSPECIES** two subspecies with ranges separated by Amazon River. 1. *G. b. barrabandi* cheeks, bend of wing to lesser wing-coverts, and thighs orange-yellow; JUV head and foreneck greenish brown. *Range* north of Amazon River from eastern Amazonas, northern Brazil, and southern Venezuela, in Amazonas and northwestern Bolívar, to southeastern Colombia, eastern Ecuador, and northeastern Peru, west of Río Ucayali. 2. *G. b. aurantiigena* cheeks, bend of wing to lesser wing-coverts, and thighs deep orange. *Range* south of Amazon River from northern Brazil to eastern Peru, east of Río Ucayali, and northern Bolivia, south to Beni and La Paz. **SIMILAR SPECIES** Caica Parrot *G. caica* (plate 121) black head, but no orange-yellow cheeks; green underwing-coverts. Vulturine Parrot *G. vulturina* and Orange-headed Parrot *G. aurantiocephala* (see below) black bare face or orange bare head without contrasting orange cheeks. **LOCALITIES** Amazonia National Park, Pará, and Cristalino Lodge, Cristalino State Park, northern Mato Grosso, northern Brazil. Tinigua National Park, Meta, and Amacayacu National Park, Amazonas, Colombia. La Selva Lodge, Sucumbios, and Jatun Sacha Biological Station, Pichincha, Ecuador. Manú Biosphere Reserve, Madre de Dios, southeastern Peru.

VULTURINE PARROT *Gypopsitta vulturina* 23cm

One of two bare-headed *Gypopsitta* parrots (JUV with feathered head); identified by black bare-skinned crown and face with prominent black and yellow collars on feathered occiput to nape; red underwing-coverts displayed in flight; JUV bare lores to around eyes greenish buff, and remainder of head feathered green; distinctive *fee-chu...fee-chu* cry, warbling *iz-teret...tre-trayeh*. **DISTRIBUTION** eastern Amazonia, south of Amazon River, in northeastern Brazil from east bank of lower Rio Madeira, eastern Amazonas, east to Rio Gurupí region, along Pará–Maranhão border, and south to Serra do Cachimbo in southern Pará; up to 400m; uncommon and poorly known. **SIMILAR SPECIES** Orange-headed Parrot *G. aurantiocephala* (see below) uniformly orange bare head without yellow or black nuchal collars. Orange-cheeked Parrot *G. barrabandi* (see above) black feathered head with contrasting orange cheeks; prominent white eye-ring. **LOCALITY** Amazonia National Park, Pará, northern Brazil.

ORANGE-HEADED PARROT
Gypopsitta aurantiocephala 23cm

Another bare-headed *Gypopsitta* species (JUV with feathered head); identified by orange bare skin extending to neck without feathered yellow or black nuchal collars; red underwing-coverts displayed in flight; JUV bare lores to around eyes orange, and remainder of head feathered dull green; calls undescribed. **DISTRIBUTION** known from only few localities along lower Rio Madeira and upper Rio Tapajós in eastern Amazonas and western Pará, northern Brazil; up to 300m; locally common, but poorly known. **SIMILAR SPECIES** Vulturine Parrot *G. vulturina* (see above) black bare crown and face with yellow and black collars on feathered occiput to nape. Orange-cheeked Parrot *G. barrabandi* (see above) black feathered head with contrasting orange cheeks; prominent white eye-ring. **LOCALITY** Thaimaçu Lodge, Pesca Esportiva State Reserve, Pará, northern Brazil.

ORANGE-CHEEKED PARROT

G. b. barrabandi

Juv

G. b. aurantiigena

VULTURINE PARROT

Juv

ORANGE-HEADED PARROT

PLATE 123 *PIONOPSITTA* AND *TRICLARIA* PARROTS

266

PILEATED PARROT *Pionopsitta pileata* 22cm

Midsized, stocky green parrot with short, rounded tail and broad, pointed wings; sexually dimorphic in both adults and juveniles— AD ♂ crown and lores to around eyes red; ear-coverts brownish purple; JUV ♂ red frontal band bordered behind by orange-yellow patch; AD & JUV ♀ crown and lores to around eyes green, and forecrown tinged blue; high-pitched shrieks, trisyllabic *ch-ch-chee* with last note higher pitched. Forests, including *Araucaria*-dominated stands, secondary growth, remnant woodlots in cleared lands, visits orchards; arboreal; pairs or small flocks, sometimes all-male groups when breeding; in Paraguay attracted to *Euterpe* fruits; swift, direct flight high above treetops, diving down steeply into canopy. **DISTRIBUTION** southeastern Brazil from southern Bahia south to Rio Grande do Sul, eastern Paraguay, and northeastern Argentina in Misiones, Corrientes, and eastern Chaco; up to 1500m; near-threatened, CITES I. **SIMILAR SPECIES** Purple-bellied Parrot *Triclaria malachitacea* (see below) all-green ♀ like *P. pileata* ♀, but with longer tail and larger, pale bill; normally in lower stages of forest. **LOCALITIES** Estancia Itabó Nature Reserve, Canindeyú, eastern Paraguay. Intervales State Park and Carlos Botelho Reserve, São Paulo, Brazil. Iguazu National Park, Misiones, northeastern Argentina.

PURPLE-BELLIED PARROT
Triclaria malachitacea 28cm

Midsized green parrot with fairly long, rounded tail and stout, pale bill; sexually dimorphic in both adults and juveniles—AD♂ center of abdomen and lower breast purple, and undertail and underwings dull blue; JUV♂ lower breast green, and little purple on center of abdomen; AD & JUV♀ underparts uniformly green. Favors wet lower montane forest, less commonly in second-growth woodland, plantations, or cultivation, occasional in urban parks or gardens; arboreal; pairs or small groups in mid to lower stages of forest interior; quiet and unobtrusive, so easily overlooked until flushed; buoyant flight with shallow, steady wingbeats; distinctive trilling *soo-see-soo-soo-see-soo*. **DISTRIBUTION** coastal southern Brazil, from southern Bahia and eastern Minas Gerais south to Rio Grande do Sul, and extreme northeastern Argentina, in northernmost Misiones; mostly 300 to 700m; near-threatened. **SIMILAR SPECIES** Pileated Parrot *Pionopsitta pileata* (see above) short, rounded tail; grayish bill; ♂ red forecrown, but no purple on underparts; normally fast-flying pairs or flocks in or above forest canopy. **LOCALITIES** Intervales State Park, São Paulo, and Monte Alverne district, Rio Grande do Sul, southeastern Brazil.

PILEATED PARROT

PURPLE-BELLIED PARROT

PLATE 124 *HAPALOPSITTACA* **PARROTS (in part)**

268 Midsized, stocky green parrots with short, rounded tail and broad, pointed wings; sexes alike, JUV duller than adults. Poorly known; wet mountain forests, occasional in subtropical forests and tall secondary growth; arboreal; pairs or small flocks in upper stages and canopy where difficult to detect amidst foliage; some seasonal altitudinal movements; swift flight with rapid, deep wingbeats.

RUSTY-FACED PARROT
Hapalopsittaca amazonina 26cm

One of three similar, allopatric species forming single superspecies, or sometimes considered conspecific; identified by distinctive coloration featuring rusty-red face, red on bend of wing to lesser wing-coverts, and blue-tipped red tail; loud, metallic *jiink* or *jeenk*, disyllabic *eea-reek*. **DISTRIBUTION** northwestern Venezuela, western Colombia, and possibly northernmost Ecuador; mostly 2200 to 3500m; endangered. **SUBSPECIES** two well-marked and one less-discernible subspecies occurring in separate Andean cordilleras. 1. *H. a. amazonina* forecrown, chin, and forecheeks dark red; lores and around eyes yellow; throat and breast olive; underwing-coverts red. *Range* extreme northwestern Venezuela, in southern Táchira, and western slopes of Cordillera Central in Norte de Santander to Cundinamarca, northwestern Colombia. 2. *H. a. theresae* forecrown, chin and forecheeks darker rufous-red; throat and upper breast darker olive-brown. *Range* northwestern Venezuela, from northern Táchira north to central Trujillo. 3. *H. a. velezi* forecrown, chin, and forecheeks paler orange-red; nape to neck and upper breast golden-olive. *Range* western Colombia, along western slopes of Cordillera Central in Caldas to Tolima and possibly south along eastern slopes to head of Río Magdalena valley, Huila; possibly also in eastern Carchi, northernmost Ecuador. **SIMILAR SPECIES** *Pionus* parrots (plates 126–128) different head patterns; red undertail-coverts, but no red "shoulders" or underwings; different flight pattern. **LOCALITIES** El Tamá National Park, Táchira, and Guaramacal National Park, Trujillo, Venezuela. Cueva de los Guácharos National Park and Finca Meremberg Reserve, Huila, Chingaza National Park, Cundinamarca, and Ucumarí Regional Park, Risaralda, Colombia.

INDIGO-WINGED PARROT
Hapalopsittaca fuertesi 23cm

Like *H. amazonina* but with narrow red frontal band; crown blue; chin and forecheeks green; dull red suffusion on abdomen; bend of wing and lesser wing-coverts dark crimson; calls undescribed. **DISTRIBUTION** known only from below Nevado de Santa Isabel, in Cordillera Central, Risaraldo–Quindío border region, western Colombia, where rediscovered in 2002; possible sighting on eastern slope of Cordillera Central, near Ibague, Tolima; mostly 2600 to 3800m; critically endangered. **SIMILAR SPECIES** recorded within 25km of range of *H. amazonina*, which has red face and golden-olive neck to upper breast. *Pionus* parrots (plates 126, 127) see above. **LOCALITIES** Alto Quindío Acaime Natural Reserve and Cañón del Quindío Natural Reserve, Quindío, Colombia.

RED-FACED PARROT
Hapalopsittaca pyrrhops 23cm

Identified by blue-tipped green tail and dark red face, including lores; crown green streaked blue; loud *chek-chek…chek-chek*, high-pitched *eek…eek…eek*. **DISTRIBUTION** interandean slopes in southern Cañar and western Morona-Santiago, southern Ecuador, south to east slope in northern Peru, north and west of Río Marañón; mostly 2400 to 3500m; vulnerable. **SIMILAR SPECIES** *Pionus* parrots (plates 126, 127) see above. **LOCALITIES** Sangay National Park, Morona-Santiago, and Podocarpus National Park, Loja, Ecuador. Cerro Chinguela, Píura, northern Peru.

RUSTY-FACED PARROT

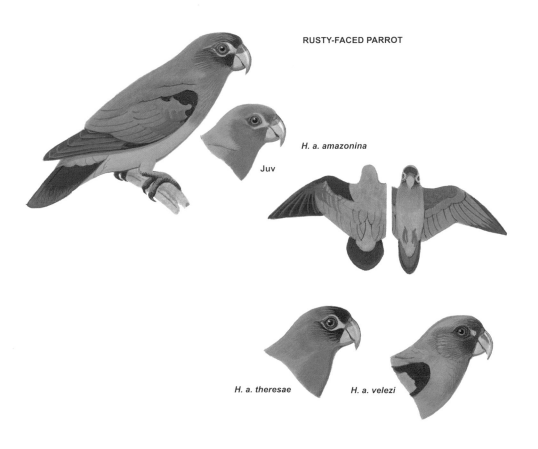

H. a. amazonina

Juv

H. a. theresae

H. a. velezi

INDIGO-WINGED PARROT

RED-FACED PARROT

PLATE 125 *HAPALOPSITTACA* **(in part) AND AMAZONIAN PARROTS**

270

BLACK-WINGED PARROT
Hapalopsittaca melanotis 24cm

Black upper wing-coverts form conspicuous "wing-patch"; rapid *chuit-chuit-chuit* rising in pitch, squeaky *ka-reet*, soft *churt*.
DISTRIBUTION eastern slopes of Andes in central Peru and central-western Bolivia; mostly 1500 to 3500m; uncommon. **SUBSPECIES** two apparently isolated subspecies differentiated by color of ear-coverts. 1. *H. m. melanotis* lores and narrow frontal band blue; crown, neck, and throat suffused blue and forming broad collar; ear-coverts black. *Range* eastern slopes of Andes in La Paz and Cochabamba, Bolivia. 2. *H. m. peruviana* ear-coverts orange-brown; narrower blue collar. *Range* eastern slopes of Andes from Huánuco to Cuzco, Peru. **SIMILAR SPECIES** *Pionus* parrots (plates 126, 127) no black on wings; red undertail-coverts; different flight pattern. **LOCALITIES** Huailaspampa forest, in Cordillera Carpish, north of Huánuco city, and Yanachaga-Chemillén National Park, in Cordillera Yanachaga, Pasco, central Peru.

SHORT-TAILED PARROT
Graydidascalus brachyurus 24cm

Midsized, green parrot with very short tail and large bill giving "top-heavy" appearance; dark red on forewing and at base of tail; sexes alike, JUV without red at base of tail; ringing *zhree-ree-ree* in flight, repeated *zee-craak*, hornlike *fuu-uudle…fuu-uudle* while feeding.
Forests along waterways and tall, swampy secondary growth, seldom far from water; arboreal; normally small flocks; gregarious, very noisy, and highly conspicuous; fast flight with rapid, deep wingbeats, and rolling or turning in air. **DISTRIBUTION** southeastern Colombia, northeastern Ecuador, northeastern Peru, and east in northern Brazil along Amazon River and tributaries to river mouth, and north through coastal Amapá to French Guiana. **SIMILAR SPECIES** *Pionus* parrots (plates 126,128) distinctive head patterns; red undertail-coverts. *Amazona* parrots (plates 132, 133, 136–138, 141) larger with longer tail; often with red wing-speculum; shallower wingbeats in flight. **LOCALITIES** La Paya National Park, Putumayo, and Amacayacu National Park, Amazonas, Colombia. Cuyabeno Wildlife Reserve, Sucumbíos, northeastern Ecuador.

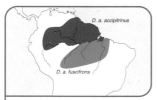

HAWK-HEADED PARROT
Deroptyus accipitrinus 35cm

Unmistakable; midsized green parrot with long, rounded tail; elongated feathers on nape and hindneck forming erectile ruff; sexes alike, JUV like adults; in flight several loud *chack* notes followed by high-pitched *tak-heeya-heeya* or *tak…tak…heeya-heeya*, drawn-out *yaag*, honking *naaaaa-unk*, and shrill *slit*. Mainly in interior of terra firme forest in lowlands and foothills; pairs or small groups; arboreal; noisy; undulating flight.
DISTRIBUTION Amazon River basin; up to 500m; uncommon. **SUBSPECIES** two subspecies with ranges separated by the Amazon River. 1. *D. a. accipitrinus* crown buff-white; occiput and sides of head brown shaft-streaked buff-white; long feathers of hindneck dark red broadly edged blue; breast and abdomen dark red strongly scalloped blue; concealed maroon bases of tail-feathers; gray eye-ring; JUV white restricted to forehead, and grayish-white eye-ring. *Range* north of Amazon River, from northeastern Brazil and Guianas to eastern Venezuela, southeastern Colombia, southeastern Ecuador, and northeastern Peru. 2. *D. a. fuscifrons* crown dusky brown faintly shaft-streaked buff-white; no maroon at base of tail. *Range* south of Amazon River, in northern Brazil and possibly neighboring northernmost Bolivia. **LOCALITIES** Amazonia National Park, Pará, and Cristalino Jungle Lodge, Cristalino State Park, northern Mato Grosso, Brazil.

BLACK-WINGED PARROT

H. m. melanotis

H. m. peruviana

Juv

SHORT-TAILED PARROT

HAWK-HEADED PARROT

D. a. accipitrinus

Juv

HAWK-HEADED PARROT

D. a. fuscifrons

PLATE 126 *PIONUS* **PARROTS (in part)**

272 Midsized, stocky, and mostly green parrots with short, squarish tail; all species with red undertail-coverts, differentiating them from larger *Amazona* parrots; sexes alike, JUV duller than adults and similar among green species so identification determined by attendant adults. Most wooded habitats; arboreal; normally small or large flocks, with very large numbers at communal nighttime roosts; gregarious, noisy, and highly conspicuous; distinctive flight pattern with rapid, deep wingbeats below body level.

WHITE-CROWNED PARROT
Pionus senilis 24cm

Only *Pionus* species in most of range; identified by white crown and prominent brownish-pink eye-ring; white patch on chin and center of throat; cheeks and breast green suffused dark blue; upper wing-coverts golden-brown edged paler; JUV forehead and forecrown buff-white tinged green and no blue on green cheeks or breast; screeching *kreeek…kreeek…kreeek* or *kree-ah…kree-ah…kree-ah*. **DISTRIBUTION** Central America, mainly Caribbean slope from southern Tamaulipas and eastern San Luis Potosí, southeastern Mexico, to central Costa Rica, and both slopes from central Costa Rica to western Panama, in western Chiriquí and western Bocas del Toro; up to 2300m; common. **SIMILAR SPECIES** Brown-hooded Parrot *Gypopsitta haematotis* (plate 121) white lores, but crown not white; red axillaries displayed in flight, but undertail-coverts not red. **LOCALITIES** Montes Azules Ecological Reserve, Chiapas, southern Mexico. Tikal National Park, El Petén, Guatemala. Bosque del Río Tigre Sanctuary and Lodge, Dos Brazos, and Lapa Rios Nature Reserve, Golfo Dulce, Costa Rica.

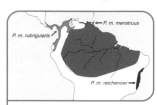

BLUE-HEADED PARROT
Pionus menstruus 28cm

Unmistakable; only short-tailed, green parrot with blue head and bicolored red/gray bill; harsh *kee-wenk…kee-wenk…kee-wenk*, shrill *krit-krit* or *chitty-wit-wit*, more liquid *chil-chil*. **DISTRIBUTION** southern Costa Rica to western Colombia and western Ecuador, throughout Amazonia, including Trinidad, and eastern Brazil; mostly below 1100m; common, and one of most ubiquitous of neotropical parrots. **SUBSPECIES** three subspecies differentiated by intensity of coloration. 1. *P. m. menstruus* head and neck blue with red marking on throat; undertail-coverts red tipped bluish green; lesser wing-coverts olive-brown; JUV head and neck green, and undertail-coverts green marked rose-red. *Range* Trinidad, Guianas, Venezuela, mostly south of Río Orinoco and in or near northern cordilleras, eastern Colombia, and throughout Amazonia, east to Piauí, northern Brazil, and south to central Bolivia, in Santa Cruz. 2. *P. m. reichenowi* head, neck, and throat darker blue, extending to upper breast; lower underparts suffused blue; undertail-coverts red tipped blue. *Range* eastern Brazil, from Alagoas south to Espírito Santo, or possibly Rio de Janeiro. 3. *P. m. rubrigularis* head and neck duller blue; more extensively red on throat. *Range* southern Costa Rica, mainly on Caribbean slope, Panama and western Colombia south to Manabí, western Ecuador. **SIMILAR SPECIES** Red-billed Parrot *P. sordidus* (plate 127) green-headed JUV probably indistinguishable in field from JUV *P. menstruus*. **LOCALITIES** Soberanía National Park in former Canal Zone, and Darién Biosphere Reserve, Panama. La Paya National Park, Putumayo, and Amacayacu National Park, Amazonas, Colombia. La Selva Lodge, Sucumbíos, and Jatun Sacha Biological Station, Pichincha, Ecuador. Sooretama Biological Reserve and Linhares Reserve, Espírito Santo, and Cristalino Jungle Lodge, Cristalino State Park, Mato Grosso, Brazil.

WHITE-CROWNED PARROT

Juv

BLUE-HEADED PARROT

P. m. menstruus

Juv

P. m. reichenowi

PLATE 127 *PIONUS* PARROTS (in part)

274

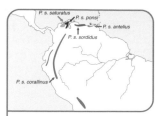

RED-BILLED PARROT *Pionus sordidus* 28cm

Only green-headed *Pionus* species with all-red bill; largely replaces *P. menstruus* at higher elevations; harsh *kee-aank…kee-aank* more high-pitched than *P. menstruus*. **DISTRIBUTION** disjunctly in northern Venezuela, western Colombia, Ecuador to northern Peru, and northern Bolivia; mostly 350 to 2000m; uncommon to fairly common. **SUBSPECIES** five isolated subspecies differentiated mainly by intensity of coloration. 1. *P. s. sordidus* head olive-green, feathers edged blue on crown and occiput; chin and band across throat blue; JUV head and throat pale green, and undertail-coverts green marked red. *Range* highlands of northwestern Venezuela. 2. *P. s. antelius* paler yellowish green; little or no blue on throat. *Range* highlands of northeastern Venezuela. 3. *P. s. ponsi* darker green; no blue band on throat. *Range* extreme northwestern Venezuela and neighboring northern Colombia. 4. *P. s. saturatus* like *ponsi*, but paler yellowish green on lesser wing-coverts, upper tail-coverts and lower underparts. *Range* Sierra Nevada de Santa Marta district, northern Colombia. 5. *P. s. corallinus* feathers of head edged blue; chin and broad band across foreneck purple-blue. *Range* eastern slopes of Cordillera Oriental in Colombia south to Andean slopes in Ecuador and northern Peru, and eastern Andean slopes in western Bolivia. **SIMILAR SPECIES** Blue-headed Parrot *P. menstruus* (plate 126) green-headed JUV probably indistinguishable in field from JUV *P. sordidus*. **LOCALITIES** Henri Pittier National Park, northern Aragua, Venezuela. Tayrona National Park, Magdalena, Catatumbo-Barí National Park, Norte de Santander, and Cueva de los Guácharos National Park, Huíla, Colombia. Mindo Nambillo Protection Forest, Mindo, Pichincha, Ecuador.

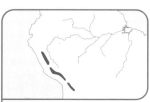

PLUM-HEADED PARROT
Pionus tumultuosus 29cm

Only red-headed *Pionus* parrot; face speckled white and sides of neck to upper breast maroon-purple; bill olive-yellow; JUV green with some dark red showing on crown and throat, and faint gray-white speckling on face; smooth *reenk* or *careenk*, harsher *kiaank*, nasal *ra-aaa*. **DISTRIBUTION** mountains of Peru, south from Carpish region, Huánuco, to northern Bolivia, in La Paz, Cochabamba, and Santa Cruz; mostly 1400 to 3300m; uncommon. **LOCALITIES** Río Abiseo and Yanachaga-Chemillén National Parks, Peru. Zongo valley in La Paz, and in Cochabamba along main road through yungas to Villa Tunari, Bolivia.

WHITE-CAPPED PARROT
Pionus seniloides 30cm

Like allopatric, but possibly conspecific *P. tumultuosus*, but with white forecrown; remainder of head pink-gray with white speckling on face; breast reddish mauve, becoming brownish pink on abdomen; bill pale olive-yellow; JUV green with buff-white speckling on face; calls like *P. tumultuosus*. **DISTRIBUTION** mountains of westernmost Venezuela, from Táchira and Mérida to Trujillo–Lara border, through western Colombia, though in Cordillera Occidental recorded only locally on eastern slope in Río Cauca valley, southwestern Antioquia, and south on both slopes of Andes in Ecuador to northwestern Peru, in Cajamarca and La Libertad; mostly 1500 to 3200m; uncommon. **LOCALITIES** Guaramacal and Dinira National Parks, Trujillo, northern Venezuela. Guanentá-Alto Río Fonce Fauna and Flora Sanctuary, Santander, and Los Nevados National Park, Colombia. Podocarpus National Park, Zamora-Chinchipe, and San Isidro Lodge, near Cosanga, Napo, Ecuador.

RED-BILLED PARROT

P. s. sordidus

P. s. corallinus

P. s. saturatus

PLUM-HEADED PARROT

Juv

WHITE-CAPPED PARROT

Juv

PLATE 128 *PIONUS* PARROTS (in part)

276

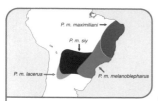

SCALY-HEADED PARROT
Pionus maximiliani 29cm

Only green-headed *Pionus* parrot with bicolored gray/yellow bill and prominent white eye-ring; resonant *choik-choik...choik-choik*, harsh squawks with low querulous clucks. **DISTRIBUTION** northeastern Brazil to northern Argentina and central Bolivia; up to 2000m; common. **SUBSPECIES** four subspecies differentiated mainly by intensity of blue on foreneck; subspecific ranges in south poorly documented. 1. *P. m. maximiliani* throat and interrupted band across foreneck dull blue; JUV paler green with less dark edging to feathers, and little blue on foreneck. *Range* northeastern Brazil, from Ceará, Piauí, and southern Maranhão south to Espírito Santo, central Minas Gerais and southern Goiás. 2. *P. m. melanoblepharus* throat and foreneck darker blue. *Range* central Brazil, from southernmost Goiás and southern Minas Gerais, south to eastern Paraguay and northeastern Argentina, in Misiones and Corrientes. 3. *P. m. siy* like *melanoblepharus*, but throat and foreneck bluish purple; reddish bases to some feathers on breast. *Range* Mato Grosso, Brazil, to central Bolivia, Paraguay, except the east, and northern Argentina, in Formosa and Chaco. 4. *P. m. lacerus* like *siy*, but more extensive deeper blue on foreneck. *Range* northwestern Argentina, in Catamarca and Tucumán to southern Salta and western Chaco. **SIMILAR SPECIES** Blue-headed Parrot *P. menstruus* (plate 126) adults well differentiated, but green-headed JUV much alike. **LOCALITIES** Morro do Diabo State Reserve and Intervales State Park, São Paulo, and Caiman Lodge Wildlife Refuge, Mato Grosso do Sul, Brazil. Noel Kempff Mercado National Park, Santa Cruz, Bolivia.

BRONZE-WINGED PARROT
Pionus chalcopterus 29cm

Unmistakable; only short-tailed dark blue parrot with bronze-brown upper wing-coverts; white chin and pink foreneck; yellow bill and pink eye-ring; JUV dark bluish green with paler brown upper wing-coverts; screeching *chee-ee...chee-ee...chi-ri-ree*. **DISTRIBUTION** extreme northwestern Venezuela, in Sierra de Perijá, Zulia, and mountains of eastern Táchira and western Mérida, and northern Colombia, in Sierra de Perijá, Guajira, and patchily through Andean cordilleras, on western slopes of Cordillera Occidental south from Río Atrato, to both slopes of Andes in western Ecuador, and to extreme northwestern Peru, in Tumbes and Piura; mostly 900 to 2800m; uncommon. **LOCALITIES** Cueva de los Guácharos National Park, Huila, Los Nevados National Park, Tolima, and La Planada Reserve, Nariño, Colombia. Loma Alta Ecological Reserve, Guayas, and Machalilla National Park, Manabí, Ecuador. Tumbes National Forest, Tumbes, northwestern Peru.

DUSKY PARROT *Pionus fuscus* 26cm

Unmistakable; only short-tailed dusky brown parrot with red undertail-coverts; throat and sides of neck streaked white; JUV upper wing-coverts and secondaries tinged green; nasal *tell-it...tell-it* or *feel-it...feel-it*. **DISTRIBUTION** eastern Venezuela, along lower Río Caura and Sierra de Imataca to upper Río Cuyuní and Sierra de Lema, northern Bolívar, and Guianas to northeastern Brazil, north of Amazon River from Amapá inland to Rio Negro, and south of Amazon River from northwestern Maranhão inland to lower Rio Madeira; isolated population on western slopes of Sierra de Perijá, Guajira, northernmost Colombia; up to 1200m; uncommon. **LOCALITIES** Mabaruma district, Barima-Waini, northern Guyana. Caura Forest Reserve, Bolívar, Venezuela. Ducke Forest Reserve, near Manaus, Amazonas, northern Brazil.

SCALY-HEADED PARROT

P. m. maximiliani

P. m. siy

BRONZE-WINGED PARROT

Juv

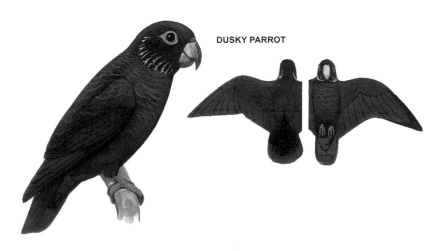

DUSKY PARROT

PLATE 129 *AMAZONA* PARROTS (in part)

278 Midsized to large, stocky parrots with short, slightly rounded tail and broad, rounded wings; dark edging to feathers produces barred appearance in most species; little or no sexual dimorphism, JUV duller than adults. Most wooded habitats; arboreal; pairs or small to large flocks, and large numbers at communal nighttime roosts; noisy and conspicuous; distinctive fast flight with rapid, shallow wingbeats below body level; most widespread and familiar of neotropical parrots.

BLACK-BILLED AMAZON
Amazona agilis 25cm
Smaller of two *Amazona* parrots in Jamaica; identified by all-green coloration with dark bill; primary-coverts red (♂) or green (♀ & JUV); bugling *tuh-tuk*, growling *rrak* or *muh-weep*, more high-pitched than calls of *A. collaria*. **DISTRIBUTION** Jamaica, West Indies; mostly 300 to 1200m; vulnerable. **SIMILAR SPECIES** Yellow-billed Amazon *A. collaria* (see below) prominent blue crown and pink throat; yellow bill. **LOCALITIES** Mount Diablo and forests of Cockpit Country, Jamaica.

YELLOW-BILLED AMAZON
Amazona collaria 28cm
Larger of two *Amazona* parrots in Jamaica; identified by blue crown, pink throat, and yellow bill; no red wing-speculum in ♂; bugling *tuk-tuk-tuk-taaah* more low-pitched than call of *A. agilis* with prolonged last syllable, also high-pitched *tah-tah-eeeeep*. **DISTRIBUTION** Jamaica, West Indies; up to 1200m; near-threatened. **SIMILAR SPECIES** Black-billed Amazon *A. agilis* (see above) all-green coloration with dark bill; red wing-speculum in ♂. **LOCALITIES** Hope Gardens, Kingston, and Mount Diablo, Jamaica.

HISPANIOLAN AMAZON
Amazona ventralis 28cm
Identified by prominent white forecrown and lores, with rose-red spot on chin, but no pink on throat; variable maroon patch on abdomen; no red wing-speculum; pale bill; loud screeching in flight, soft growl or chatter. **DISTRIBUTION** endemic to Hispaniola (Haiti and Dominican Republic), Greater Antilles, where only *Amazona* parrot; up to 1500m; vulnerable; introduced to St. Croix and St. Thomas, Virgin Islands, and Puerto Rico. **SIMILAR SPECIES** only on Puerto Rico, where occurs with other *Amazona* species (plates 132, 135, 138), no white forecrown and no maroon abdominal patch; red or orange wing-speculum. **LOCALITIES** Sierra de Baoruco Park and Sierra de Naiba, and Del Este National Park, Dominican Republic. Los Haitises Reserve, Haiti. Isla Grande Naval Reserve, Puerto Rico (feral population).

BLACK-BILLED AMAZON

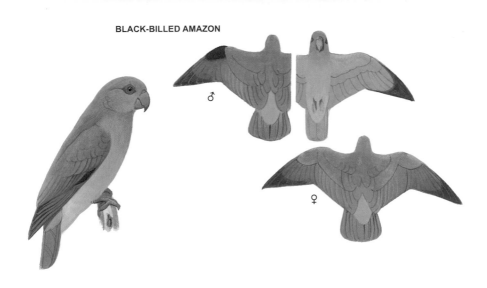

♂

♀

YELLOW-BILLED AMAZON

HISPANIOLAN AMAZON

PLATE 130 *AMAZONA* **PARROTS (in part)**

280

PUERTO RICAN AMAZON
Amazona vittata 29cm

All-green *Amazona* parrot with red frontal band and prominent white eye-ring; no red wing-speculum; strong black edging to feathers of head, neck, and back; bugling *kar…kar*, loud squawks and chuckling. **DISTRIBUTION** endemic to Puerto Rico, and formerly nearby Culebra Island, Greater Antilles, West Indies; mostly 200 to 600m; critically endangered, CITES I. **SIMILAR SPECIES** Other *Amazona* parrots introduced to Puerto Rico. Hispaniolan Amazon *A. ventralis* (plate 129) white forecrown and maroon abdominal patch. Other *Amazona* parrots (plates 133, 135, 138) different head markings; red or orange wing-speculum. **LOCALITY** Luquillo National Forest Reserve, Puerto Rico.

CUBAN AMAZON *Amazona leucocephala* 32cm

Unmistakable, only *Amazona* parrot in range; identified by white forecrown or face and rose-red throat to foreneck; scolding *yaaart…yaaart*, disyllabic cries like braying of donkey. **DISTRIBUTION** Cayman Islands, Bahamas, and Cuba and nearby Isla de Pinos (Isla de la Juventud); up to 1000m; near-threatened, CITES I. **SUBSPECIES** three well-marked, insular subspecies, and two poorly differentiated subspecies. 1. *A. l. leucocephala* forecrown and around eyes white; lores and cheeks to foreneck rose-red; maroon abdominal patch. *Range* eastern and central Cuba, west to Villa Clara province.
2. *A. l. palmarum* darker green with more extensive purplish-red abdominal patch. *Range* Isla de Pinos and western Cuba, east to Villa Clara province; doubtfully distinct from *leucocephala*.
3. *A. l. caymanensis* white of forecrown less extensive posteriorly; green on sides of neck extending forward to separate rose-red of throat from that of cheeks. *Range* Grand Cayman Island.
4. *A. l. hesterna* cheeks and throat deeper red; smaller size. *Range* Cayman Brac, and formerly Little Cayman Island, where recent sightings probably of visiting birds from Cayman Brac.
5. *A. l. bahamensis* white of crown extending to upper cheeks, lores, and below eyes; maroon abdominal patch small or absent. *Range* Great Inagua and Abaco, Bahamas; formerly on other islands in Bahamas. **LOCALITIES** Forests of Zapata and Guanahacabibes Peninsulas, Cuba. Abaco National Park, south Abaco, and Great Inagua National Park, Great Inagua Island, Bahamas. Forest Glen and environs of George Town, Grand Cayman Island, and Amazon Parrot Reserve, Cayman Brac Island.

PUERTO RICAN AMAZON

CUBAN AMAZON

A. l. leucocephala

A. l. hesterna

A. l. bahamensis

PLATE 131 *AMAZONA* **PARROTS (in part)**

282

WHITE-FRONTED AMAZON
Amazona albifrons 26cm

One of two small amazons with white forecrown and red on face; shrill *ca-ca-ca-ca* or *kyi…kyeh-kyeh-kyeh*, yapping *kyak-yak-yak-yak* or *rek…rek-rek-rek*, harsh *screet* or *scree-at*. Wide habitat tolerance, but where sympatric with *A. xantholora* favors more humid, closed vegetation. **DISTRIBUTION** western Mexico to northwestern Costa Rica; up to 1850m; common. **SUBSPECIES** two discernible and one doubtfully differentiated subspecies. 1. *A. a. albifrons* lores to around eyes red; forecrown white; hindcrown blue; alula and primary-coverts red (♂) or green (♀ & JUV); bill yellow. *Range* Pacific slope of southwestern Mexico, south from Nayarit, to southwestern Guatemala. 2. *A. a. saltuensis* green upperparts suffused blue; blue of hindcrown extending to nape. *Range* northwestern Mexico, in southern Sonora, Sinaloa, and western Durango. 3. *A. a. nana* like *albifrons*, but smaller. *Range* southern Mexico, in southeastern Veracruz, south to northwestern Costa Rica. **SIMILAR SPECIES** Yellow-lored Amazon *A. xantholora* (see below) difficult to distinguish in field; yellow lores and black (♂) or dark gray (♀) ear-coverts; ♀ without white forecrown. White-crowned Parrot *Pionus senilis* (plate 126) blue cheeks and upper breast; no red on face, but red undertail-coverts. **LOCALITIES** Alamos district, Sonora, and San Blas district, Nayarit, Mexico. Tikal National Park, El Petén, Guatemala. Shipstern Nature Reserve, Lamanai Outpost Lodge, and Crooked Tree Wildlife Sanctuary, Belize. Barra de Santiago, 20km south of El Impossible National Park, and Hotel de Montaña, Perkin Lenca, Perquín, El Salvador. Chocoyero-El Brujo Nature Reserve, Ticuantepe, Managua, and Volcan Mombacho Nature Reserve, Nicaragua.

YELLOW-LORED AMAZON
Amazona xantholora 26cm

A small amazon with forecrown white (♂) or green suffused blue (♀ & JUV); ear-coverts black (♂) or dark gray (♀); yellow lores, duller in ♀; red around eye (♂) or line beneath eye (♀); rolling *reeeah-ah* or *kyeh-kyeh…keee-i-irr*, screeching *ree-o-rak…zeek…ree-o-rah*, barking *rek-rek-rek-rek…rek…rek-rek* or *rek-rek…rek-rek…rrehr*. On Yucátan Peninsula less common in light rainforest, the preferred habitat of *A. albifrons*. **DISTRIBUTION** Yucátan Peninsula and offshore Isla Cozumel, extreme southeastern Mexico, and northern Belize; mainly below 300m; common. **SIMILAR SPECIES** White-fronted Amazon *A. albifrons* (see above) difficult to distinguish in field; red, not yellow lores and green ear-coverts; white forecrown in both sexes. White-crowned Parrot *Pionus senilis* (plate 126) see above. **LOCALITIES** Ría Lagartos Natural Park, Yucátan, and Isla Cozumel (where *A. albifrons* absent), Quintana Roo, Mexico. Shipstern Nature Reserve, Lamanai Outpost Lodge, and Crooked Tree Wildlife Sanctuary, Belize.

WHITE-FRONTED AMAZON

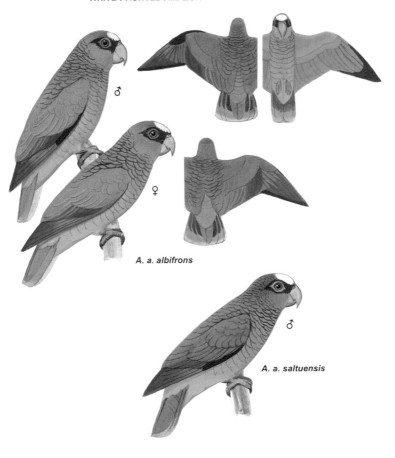

A. a. albifrons

♂

A. a. saltuensis

YELLOW-LORED AMAZON

PLATE 132 *AMAZONA* **PARROTS (in part)**

284

GREEN-CHEEKED AMAZON
Amazona viridigenalis 33cm

One of two allopatric amazons with red forecrown, blue from above eyes to sides of neck, and green cheeks; red wing-speculum; bill yellow; JUV with red only on forehead; rolling *rreeoo* or *keer-yoo...keer-yoo*, barking *rreh-rreh-rreh* or *rrak-rrak-rrak* often combined in flight to *clee-oo...clee-oo...ahk-ahk-ahk*, quieter *rreeah...rreeah* and *clee-ik*. **DISTRIBUTION** eastern Mexico, from Tamaulipas, eastern Nuevo León and eastern San Luis Potosí to northern Veracruz; up to 1000m; endangered, CITES I; feral populations in Los Angeles urban areas, southern California, southeastern Florida, and on Oahu, Hawaii, U.S.A., and Puerto Rico, West Indies; status in southern Texas not resolved. **SIMILAR SPECIES** Red-lored Amazon *A. autumnalis* (plate 133) red only on forehead; paler blue crown; yellow lores to upper cheeks; bicolored gray/yellow bill. **LOCALITIES** Gómez Farías district, Tamaulipas, and El Naranjo district, San Luis Potosí, Mexico. Pasadena area and Los Angeles Arboretum, Los Angeles, California, U.S.A. (feral population).

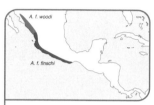

LILAC-CROWNED AMAZON
Amazona finschi 33cm

Plumage coloration like allopatric *A. viridigenalis*; shrill *krih-krih* or *kreeih-kreeih*, rolling *krreeeih* or deeper *kyah-ha*, raven-like *krra...krra*. **DISTRIBUTION** western Mexico, from extreme southeastern Sonora and southwestern Chihuahua south to Oaxaca; up to 2000m; locally common; feral population in Los Angeles and San Diego areas, southern California, U.S.A. **SUBSPECIES** two poorly differentiated subspecies. 1. *A. f. finschi* forecrown and lores maroon; hindcrown to sides of neck blue-mauve; red wing-speculum; pale bill. *Range* central-western to southwestern Mexico. 2. *A. f. woodi* narrower, duller maroon frontal band. *Range* northwestern Mexico. **SIMILAR SPECIES** White-fronted Amazon *A. albifrons* (plate 131) white forecrown and red face. **LOCALITIES** Barranca Rancho Liebre, near El Palmito, Sinaloa, San Blas district, Nayarit; and Barranca el Choncho, near Barra de Navidad, Jalisco, Mexico.

GREEN-CHEEKED AMAZON

Juv

LILAC-CROWNED AMAZON

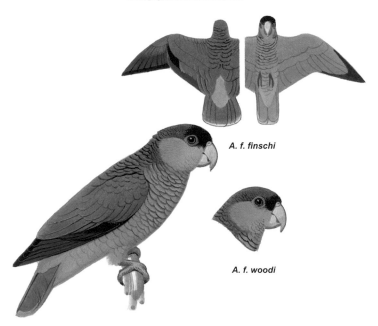

A. f. finschi

A. f. woodi

PLATE 133 *AMAZONA* PARROTS (in part)

286

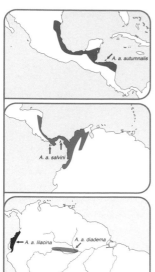

RED-LORED AMAZON
Amazona autumnalis 34cm

Polytypic amazon with red forehead, blue crown, yellow to yellowish-green cheeks, and red wing-speculum; discordant *kiak…kiak…kiak* or *yoik…yoik…yoik*, repeated *ack-ack* or *chek-chek*. **DISTRIBUTION** Central America to northwestern South America and western Amazonia; mostly below 800m; locally common. **SUBSPECIES** four well-differentiated subspecies. 1. A. *a. autumnalis* upper cheeks to ear-coverts yellow; white eye-ring; bicolored gray/ yellowish bill. *Range* Caribbean slope from Tamaulipas, Mexico, south to northern Nicaragua, and on Bay Islands, Honduras. 2. A. *a. salvini* upper cheeks to ear-coverts yellowish green; pale yellow eye-ring. *Range* southeastern Nicaragua and eastern and southwestern Costa Rica south to western Colombia and extreme northwestern Venezuela. 3. A. *a. lilacina* red extending as line above eye; crown lilac; all-black bill. *Range* southwestern Colombia, in Nariño, and western Ecuador, south to El Oro. 4. A. *a. diadema* feathered cere and forehead crimson; lores dark purple; bicolored bill. *Range* western Amazonia, between lower Rio Negro and upper Amazon River, Amazonas, northern Brazil. **SIMILAR SPECIES** Green-cheeked Amazon A. *viridigenalis* (plate 132) red extending to crown; no yellow on face; yellow bill. **LOCALITIES** Montes Azules Biosphere Reserve, Chiapas, Mexico. Tikal National Park, El Petén, Guatemala. La Selva Biological Reserve, Costa Rica. Utría and Los Katíos National Parks, Chocó, Colombia. Steve's Lodge, Río Cayapas, Esmeraldas, Ecuador. Ducke Forest Reserve, near Manaus, Amazonas, Brazil.

RED-LORED AMAZON

A. a. autumnalis

Juv

A. a. salvini

A. a. lilacina

A. a. diadema

PLATE 134 *AMAZONA* **PARROTS (in part)**

288

YELLOW-CROWNED AMAZON *Amazona ochrocephala* 31–38cm

Midsized to large, strongly polytypic and locally polymorphic amazon with variable yellow head markings, red wing-speculum, and red at base of tail. **DISTRIBUTION** northern Mexico to Amazonia, including Trinidad where possibly introduced; up to 750m; uncommon or scarce to locally common; CITES I; feral populations in Los Angeles urban area, southern California, U.S.A., and on Puerto Rico and Grand Cayman Island, West Indies. **SUBSPECIES** (in part; see plates 135, 136) subspecies categorized in three groups, which often are treated as separate species—A: yellow-crowned or "*ochrocephala*" group (1–4), B: yellow-naped or "*auropalliata*" group (5–7), C: yellow-headed or "*oratrix*" group (8–11).

A: YELLOW-CROWNED OR "OCHROCEPHALA"

GROUP—yellow on crown or face and bicolored gray/orange bill; distinctive mellow, rolling *bow-wow*. 1. *A. o. ochrocephala* (Yellow-crowned Amazon 35cm) forehead and lores yellow variably marked green; forecrown yellow; bend of wing red; carpal edge yellowish green; eye-ring white. *Range* Trinidad, where possibly introduced, Guianas, and northeastern Brazil, west from northern Pará along both sides of lower Amazon River to Venezuela, except arid northeast and northwest, and eastern Colombia, east of Cordillera Oriental in Norte de Santander south to Amazonas and western Caquetá. 2. *A. o. xantholaema* yellow extending to nape, ear-coverts, and upper cheeks; green frontal band; thighs yellow; larger. *Range* Ilha do Marajó, at mouth of Amazon River, northeastern Brazil. 3. *A. o. nattereri* like *ochrocephala*, but darker green; cheeks, ear-coverts, and foreneck suffused blue. *Range* southern Colombia, in Putumayo and western Caquetá, south through eastern Ecuador and eastern Peru to Santa Cruz, eastern Bolivia, and Acre to northwestern Mato Grosso, Brazil. 4. *A. o. panamensis* 31cm forehead yellow; hindcrown bluish green; little red on bend of wing; thighs green; smaller. *Range* Panama, east from central Chiriquí and western Bocas del Toro, including Archipíelago de las Perlas, and northwestern Colombia, from Río Atrato valley, Chocó, to north of Andes, between foothills of Sierra Nevada de Santa Marta and Sierra de Perijá, and south in Río Magdalena valley to southern

Huila. **SIMILAR SPECIES** Mealy Amazon *A. farinosa* (plate 137) little or no yellow on crown; glaucous suffusion on upperparts; red carpal edge, but no red on bend of wing; little or no red at base of tail; pale bill and very prominent white eye-ring; different calls. Orange-winged Amazon *A. amazonica* (plate 139) blue lores and superciliary band; no red on bend of wing; orange wing-speculum; pale bill. **LOCALITIES** Utría National Park, Chocó, Tinigua National Park, Meta, and Amacayacu National Park, Amazonas, Colombia. La Selva Lodge, Sucumbíos, northeastern Ecuador. Santos Luzardo (Cinaruco-Capanaparo) National Park, Apure, and Caura Forest Reserve, Bolívar, Venezuela. Cristalino Jungle Lodge, Cristalino State Park, northern Mato Grosso, Brazil.

YELLOW-CROWNED AMAZON

A. o. ochrocephala

Juv

A. o. xantholaema

A. o. nattereri

A. o. panamensis

PLATE 135 *AMAZONA* PARROTS (in part)

290

YELLOW-CROWNED AMAZON *Amazona ochrocephala* (in part)

SUBSPECIES (in part; see plates 134, 136) subspecies categorized in three groups, which often are treated as separate species—A: yellow-crowned or "*ochrocephala*" group (1–4), B: yellow-naped or "*auropalliata*" group (5–7), C: yellow-headed or "*oratrix*" group (8–11).

B: YELLOW-NAPED OR "*AUROPALLIATA*" GROUP— yellow on nape to hindneck; bicolored dark gray/pale gray bill; raucous *ke-chow…chow* or *ke-chep…chep*, mellow *churr…uhrr-rr* or low-pitched *hrrah…h-rrah*, also gruff *rrrow-ow* or *grrr-ow*, loud *kwok* notes. 5. A. o. *auropalliata* (Yellow-naped Amazon 35cm) variable yellow patch on lower nape to hindneck, lacking in JUV; no red on bend of wing, *Range* Pacific slope of Central America, from eastern Oaxaca, southern Mexico, to northwestern Costa Rica, south to about Tárcoles district. 6. A. o. *parvipes* like *auropalliata*, but yellow more extensive on nape, and sometimes yellow markings on forehead; red on bend of wing. *Range* Caribbean slope in La Mosquitia of easternmost Honduras and neighboring northeastern Nicaragua. 7. A. o. *caribaea* like *parvipes*, but underparts more olive-green; paler grayish-horn lower mandible. *Range* confined to Roatán, Barbareta, and Guanaja, in Bay Islands, Honduras; doubtfully distinct from *parvipes*. **SIMILAR SPECIES** Red-lored Amazon A. *autumnalis* (plate 133) red lores and forecrown; blue crown; no red on bend of wing. Mealy Amazon A. *farinosa* (plate 137) no yellow on nape to hindneck; glaucous suffusion on upperparts; no red on bend of wing; very prominent white eye-ring; different calls. **LOCALITIES** Río Platano National Park, La Mosquitia, and Guanaja, Bay Islands, Honduras.

A. o. auropalliata

Juv

A. o. parvipes

PLATE 136 *AMAZONA* PARROTS (in part)

292

YELLOW-CROWNED AMAZON *Amazona ochrocephala* (in part)

SUBSPECIES (in part; see plates 134, 135) subspecies categorized in three groups, which often are treated as separate species—A: yellow-crowned or "*ochrocephala*" group (1–4), B: yellow-naped or "*auropalliata*" group (5–7), C: yellow-headed or "*oratrix*" group (8–11).

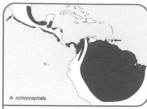

A. ochrocephala

A. o. tresmariae

A. o. oratrix

A. o. belizensis

A. o. hondurensis

C: YELLOW-HEADED OR "*ORATRIX*" GROUP—yellow crown, yellow face, or entirely yellow head; pale yellowish-horn bill, and white eye-ring; screaming *kyaa-aa-ah* or *krra-aah-aa-ow*, rolling *ahrrr* or *ahrhrrr*, also *whoh-oh-ohr* and *rolling rrohrr*. 8. A. o. oratrix (Yellow-headed Amazon 38cm) head and neck yellow; thighs yellow; bend of wing and lesser wing-coverts orange-red intermixed yellow; carpal edge yellow often intermixed orange-red; red wing-speculum; JUV yellow only on crown to lores. *Range* Pacific slope of central Mexico, from Jalisco to central Oaxaca (recent records mostly from southern Jalisco and Michoacán) and Caribbean slope, from eastern Nuevo León and Tamaulipas to Tabasco and northern Chiapas; feral populations in Los Angeles urban area, southern California, U.S.A. 9. A. o. tresmariae like *oratrix*, but paler yellow of head extending to foreneck and upper breast; underparts suffused blue. *Range* Islas Marías, off coast of Nayarit, western Mexico.

10. A. o. belizensis (two populations, one dimorphic) yellow restricted to lores, forehead, crown, and around eyes to ear-coverts and upper cheeks, together with (yellow-naped morph) or without (yellow-faced morph) yellow patch on nape to hindneck; little or no yellow on carpal edge. *Range* central Belize and El Petén, northern Guatemala (yellow-faced morph only), and disjunctly northeastern Guatemala to extreme northwestern Honduras (both morphs). 11. A. o. hondurensis (variable plumage) differs from *belizensis* by having yellow restricted to forehead and crown, thus approaching *ochrocephala*, but with or without yellow patch on nape to hindneck; carpal edge green. *Range* Valle de Sula, northwestern Honduras. **SIMILAR SPECIES** yellow-headed adults readily identified, but yellow-crowned juveniles can be misidentified. **LOCALITIES** La Pesca district, Tamaulipas, northeastern Mexico. Crooked Tree Wildlife Sanctuary, Belize. Punta Sal National Park, Valle de Sula, Honduras.

YELLOW-CROWNED AMAZON

A. o. oratrix

Juv

A. o. tresmariae

yellow-faced morph

yellow-naped morph

A. o. belizensis

A. o. hondurensis

PLATE 137 *AMAZONA* PARROTS (in part)

294

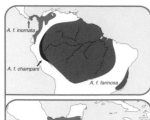

MEALY AMAZON *Amazona farinosa* 38cm

Large amazon with glaucous suffusion on upperparts giving grayish appearance, and prominent white eye-ring; loud *catch-it...catch-it*, trisyllabic *taa-kaa-ee...taa-kaa-ee*, and bell-like *kwok-kwok-kwok*. **DISTRIBUTION** southern Mexico south to central Bolivia and central-eastern Brazil; up to 1100m; locally common. **SUBSPECIES** five subspecies, but much individual variation. 1. A. *f. farinosa* variable yellow patch on crown; red wing-speculum and red carpal edge; little or no red at base of tail; pale bill. *Range* Guianas, northern Brazil, and southeastern Venezuela, in southern Delta Amacuro, eastern Bolívar, and Amazonas, to southeastern Colombia, and south to central Bolivia and northern Mato Grosso, Brazil; isolated population in coastal Brazil, from southern Bahia to northern São Paulo. 2. A. *f. chapmani* larger than *farinosa*. *Range* eastern slopes of Andes, from southern Colombia, in Meta, Caquetá, and Putumayo, to eastern Ecuador, eastern Peru, and northern Bolivia. 3. A. *f. inornata* less glaucous suffusion on upperparts; crown green with few yellow feathers. *Range* Veraguas and Isla Coiba, western Panama, to northern Colombia, north of Andes, excluding north coast and Sierra Nevada de Santa Marta, south in Río Magdalena valley to Cundinamarca, and east to northwestern Venezuela, and Pacific lowlands in western Colombia to western Ecuador. 4. A. *f. virenticeps* more yellowish green; forehead and lores tinged blue. *Range* westernmost Panama, in western Chiriquí and western Bocas del Toro, north through Costa Rica and Nicaragua to Valle de Sula, northwestern Honduras. 5. A. *f. guatemalae* like *virenticeps*, but forehead and crown blue; bill dark gray. *Range* Caribbean slope from northwestern Honduras north to Oaxaca and southern Veracruz, southern Mexico. **SIMILAR SPECIES** White-chinned Amazon A. *kawalli* (see below) white bare skin at base of bill; red at base of tail; no glaucous suffusion on upperparts; gray eye-ring; different calls; replaces A. *farinosa* in várzea forest. Scaly-naped Amazon A. *mercenaria* (plate 138) no glaucous suffusion on upperparts; tail banded red and purple-blue; normally at higher elevations. **LOCALITIES** Monte Azules Biosphere Reserve, Chiapas, Mexico. Tikal National Park, El Petén, Guatemala. Cristalino Lodge, Cristalino State Park, northern Mato Grosso, and Ducke Forest Reserve, near Manaus, Amazonas, Brazil. Manu Biosphere Reserve, Madre de Dios, eastern Peru.

WHITE-CHINNED AMAZON
Amazona kawalli 38cm

Large amazon with white bare skin at base of bill; red wing-speculum and carpal edge; red at base of tail; gray bill and eye-ring; distinctive whistling *wee-ou*. Closely associated with várzea forest. **DISTRIBUTION** central Amazon River basin, northern Brazil, where recorded in Amazonas, from upper Rio Jurua below Eirunepé, near mouth of Rio Tefé, and at confluence of Rios Roosevelt and Aripuana, in northernmost Mato Grosso in headwaters of Rio Tapajós and tributaries and along lower reaches of Rios Teles Pires and Juruena, and in Pará at Itaituba; below 100m; locally common. **SIMILAR SPECIES** Mealy Amazon A. *farinosa* (see above) no white bare skin at base of pale bill; glaucous suffusion on upperparts; little or no red at base of tail; replaces A. *kawalli* in drier habitats. **LOCALITY** Apíacás Ecological Station, northernmost Mato Grosso.

MEALY AMAZON

A. f. farinosa

A. f. inornata

A. f. guatemalae

A. f. virenticeps

MEALY AMAZON

WHITE-CHINNED AMAZON

PLATE 138 *AMAZONA* **PARROTS (in part)**

296

SCALY-NAPED AMAZON
Amazona mercenaria 34cm

Midsized amazon with lateral tail-feathers subterminally banded red and purple-blue; distinctive *ka-lee…ka-lee…ka-lee* repeated in long sequences. Closely associated with wet montane forest, where usually only *Amazona* species. **DISTRIBUTION** highlands of northwestern Venezuela and western Colombia to northern Bolivia; mostly 1500 to 3000m; fairly common. **SUBSPECIES** two subspecies separated by differences in wing-speculum. 1. *A. m. mercenaria* red wing-speculum; carpal edge yellow intermixed orange-red; gray bill. *Range* mountains of northern Bolivia, in La Paz, Cochabamba, and western Santa Cruz, north to northern Peru and western Ecuador. 2. *A. m. canipalliata* red wing-speculum replaced by concealed maroon markings at bases of s1 to s3. *Range* both slopes of Andes in western Ecuador to western Colombia, north to Cundinamarca, though probably continuing northward into populations in Sierra Nevada de Santa Marta, Sierra de Perijá, and mountains of northwestern Venezuela. **SIMILAR SPECIES** Mealy Amazon *A. farinosa* (plate 137) glaucous suffusion on upperparts; no tail-bands; usually at lower altitudes. **LOCALITIES** Yacambú National Park, Lara, and Guaramacal National Park, Trujillo, Venezuela. Guanentá-Alto Río Fonce Fauna and Flora Sanctuary, Santander, and Cueva de los Guacharos National Park, Huíla, Colombia. Podocarpus National Park, Zamora-Chinchipe, Ecuador. Machu Picchu Hotel (Hotel Ruinas), Cuzco, Peru.

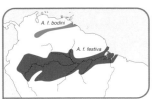

FESTIVE AMAZON *Amazona festiva* 34cm

Unmistakable; only amazon with red lower back to rump (lacking in JUV); gruff notes ending with ringing *waghh-t*. Favors várzea forest, and rarely far from water. **DISTRIBUTION** Amazonia, throughout drainages of Río Orinoco and Amazon River; up to 500m; common. **SUBSPECIES** two well-marked, isolated subspecies. 1. *A. f. festiva* lores and frontal band dark red (absent in JUV); blue above and behind eyes, and blue chin; outer webs of primaries blue. *Range* Amazon River drainage from mouth at Ilha do Mexiana, northeastern Brazil, west to southeastern Colombia, in Vaupés and Amazonas, eastern Ecuador, mainly along Ríos Napo and Aguarico and tributaries in Sucumbíos and eastern Napo, and northeastern Peru. 2. *A. f. bodini* red forehead and forecrown; forecheeks suffused blue; outer webs of primaries green. *Range* lower Ríos Meta and Casanare, eastern Colombia, and central Venezuela, from Río Meta in southern Apure north to Río Capanaparo and east along Río Orinoco from mouth of Río Meta to lower Río Caura, ranging sporadically into northwestern Guyana. **LOCALITIES** Santos Luzardo (Cinaruco-Capanaparo) National Park, Apure, Venezuela. Amacayacu National Park, Amazonas, Colombia. Imuya Camp, Cuyabeno Wildlife Reserve, Sucumbíos, Ecuador.

SCALY-NAPED AMAZON

A. m. mercenaria

A. m. canipalliata

FESTIVE AMAZON

A. f. festiva

Juv

A. f. bodini

PLATE 139 *AMAZONA* PARROTS (in part)

298

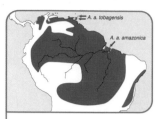

ORANGE-WINGED AMAZON
Amazona amazonica 31cm

Midsized amazon with yellow face and orange wing-speculum; distinctive shrill *kee-wik…kee-wik* or *kik-kik…kik-kik.* **DISTRIBUTION** Colombia, Venezuela, and Guianas south to southern and eastern Brazil; also Trinidad and Tobago; mostly below 600m; common; introduced to Puerto Rico and Martinique, West Indies.

SUBSPECIES two doubtfully distinct subspecies. 1. *A. a. amazonica* lores and superciliary band violet-blue; orange bases to s1 to s3. *Range* northern and eastern Colombia to northern Venezuela, including Isla Margarita, and south of Río Orinoco throughout Amazonas and Bolívar, and the Guianas and Amazon River basin, south in the west to eastern Bolivia, and in Brazil south to Mato Grosso do Sul, western São Paulo and northern Paraná, but absent from pantanal of western Amazonia; apparently isolated in coastal Brazil, from Pernambuco south to northeastern São Paulo. 2. *A. a. tobagensis* more extensive orange bases to s1 to s4. *Range* Trinidad and Tobago. **SIMILAR SPECIES** Yellow-crowned Amazon *A. ochrocephala* (plate 134) no blue lores or superciliary band; red bend of wing and wing-speculum. Blue-fronted Amazon *A. aestiva* (see below) blue forehead; yellow face and throat; yellow or red "shoulders" and red wing-speculum. **LOCALITIES** Asa Wright Nature Center, Spring Hill Estate, Trinidad. El Tuparro National Park, Vichada, and Amacayacu National Park, Amazonas, Colombia. Henri Pittier National Park, Aragua, and Imataca Forest Reserve, Bolívar, Venezuela. La Selva Lodge, Sucumbíos, northeastern Ecuador. Cristalino Jungle Lodge, Cristalino State Park, northern Mato Grosso, and Caiman Lodge Wildlife Refuge, Mato Grosso do Sul, Brazil.

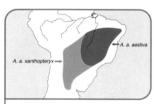

BLUE-FRONTED AMAZON
Amazona aestiva 37cm

Large amazon with yellow face, blue forehead, and red or yellow bend of wing; screeching *krik…kia-krik…krik…krik…kray-o,* melodious *drew-wo…droo-droo-droo…drew-oh…drew-wee…wee,* yelping *help…help.* **DISTRIBUTION** inland eastern Brazil to Paraguay, eastern Bolivia and northern Argentina; up to 1600m; common.

SUBSPECIES two subspecies with broad zone of intergradation. 1. *A. a. aestiva* red bend of wing and red wing-speculum. *Range* inland eastern Brazil, from Pernambuco and southern Pará to western São Paulo and southern Mato Grosso do Sul. 2. *A. a. xanthopteryx* bend of wing and lesser wing-coverts yellow; more extensively yellow on face. *Range* southern Mato Grosso do Sul and formerly Rio Grande do Sul, southern Brazil, northern and eastern Bolivia, and Paraguay to northern Argentina, south to Catamarca, Santiago del Estero, Santa Fé, and Misiones, or extralimitally to northern Buenos Aires. **SIMILAR SPECIES** Orange-winged Amazon *A. amazonica* (see above) yellow, not blue forecrown; no red or yellow "shoulders"; orange wing-speculum. **LOCALITIES** Caiman Lodge Wildlife Refuge, Mato Grosso do Sul, southern Brazil. Noel Kempff National Park, Santa Cruz, northeastern Bolivia. Chancaní Reserve, Córdoba, northern Argentina.

ORANGE-WINGED AMAZON

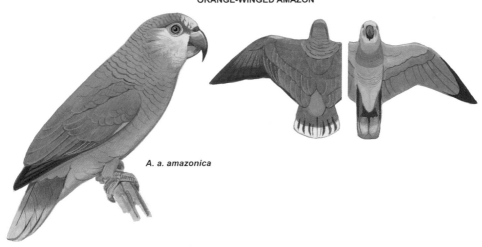

A. a. amazonica

BLUE-FRONTED AMAZON

Juv

A. a. aestiva

A. a. xanthopteryx

PLATE 140 *AMAZONA* PARROTS (in part)

300

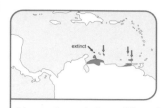

YELLOW-SHOULDERED AMAZON
Amazona barbadensis 33cm

Midsized amazon with white forehead, yellow face, and yellow on bend of wing to lesser wing-coverts; red wing-speculum; yellow thighs; pale bill; JUV yellow restricted to forecrown and upper cheeks; only *Amazona* species in much of range; hoarse, rolling *curr-rak…curr-rak…curr-rak*. **DISTRIBUTION** disjunctly in northern Venezuela, including offshore Islas Margarita and La Blanquilla, and Netherlands Antilles, on Bonaire and formerly Aruba, with feral population on Curaçao; up to 450m; vulnerable; CITES I. **SIMILAR SPECIES** Orange-winged Amazon A. *amazonica* (plate 139) yellow, not white forehead; blue lores and superciliary band; no yellow "shoulders"; orange wing-speculum; bicolored gray/horn-colored bill. Yellow-crowned Amazon A. *ochrocephala* (plate 134) yellow, not white forehead; red "shoulders"; bicolored orange/gray bill. **LOCALITIES** Cuare Wildlife Refuge, Falcón, and Isla Margarita, northern Venezuela. Northern Bonaire Island, Netherlands Antilles.

YELLOW-FACED AMAZON
Amazona xanthops 27cm

Small amazon with green and yellow morphs, but much individual variation within each; broad green edging to feathers producing pronounced scalloping; distinctive pink-red cere and legs; white eye-ring; raucous *gray-o…gray-o…gray-o…to-to-to-to*, and *krew-ee…krew-ee*. Closely associated with *Mauritia* palms in open cerrado woodland. GREEN MORPH forehead and face yellow; sides of breast to flanks deep orange; JUV yellow restricted to forecrown and around eyes to ear-coverts. YELLOW MORPH ♂ head and neck yellow; upper breast green suffused yellow; lower breast and upper abdomen deep yellow, becoming orange-red on sides of breast to flanks; ♀ lower breast and upper abdomen yellowish green; JUV like ♀, but yellow restricted to forecrown, around eyes, and cheeks. **DISTRIBUTION** interior of eastern and southern Brazil, from southern Piauí and southern Maranhão south to Mato Grosso do Sul and western São Paulo, or extralimitally to north-central Bolivia and northernmost Paraguay; up to 300m; near-threatened. **SIMILAR SPECIES** Blue-fronted Amazon A. *aestiva* (plate 139) blue forehead; yellow or red "shoulders," and red wing-speculum; gray bill; larger. Orange-winged Amazon A. *amazonica* (plate 139) blue lores and superciliary band; orange wing-speculum. **LOCALITIES** Caiman Lodge Wildlife Refuge, Mato Grosso do Sul, southern Brazil. Emas National Park, Goiás, and Brasilia National Park, Distrito Federal, Brazil.

YELLOW-SHOULDERED AMAZON

YELLOW-FACED AMAZON

yellow morph ad

green morph ad

green morph juv

PLATE 141 *AMAZONA* PARROTS (in part)

302

TUCUMÁN AMAZON *Amazona tucumana* 32cm

Midsized, all-green amazon with red forecrown and red primary-coverts; pale bill and prominent white eye-ring; JUV red restricted to forehead; only *Amazona* species in range; shrieking *cro-eo...cri-eo...cro-eo*. Closely associated with *Alnus* and mixed *Nothofagus-Podocarpus* montane forests. **DISTRIBUTION** southern Bolivia, in Chuquisaca and Tarija, and northwestern Argentina, in Jujuy, Salta, Tucumán, and Catamarca; mostly 300 to 2200m; locally common but declining, CITES I. **LOCALITY** Finca del Rey National Park, Salta, northwestern Argentina.

RED-SPECTACLED AMAZON
Amazona pretrei 32cm

Midsized, all-green amazon with red forecrown to lores and around eyes; red on bend of wing to carpal edge and red thighs; pale bill and white eye-ring; JUV red restricted to forehead and bend of wing; hoarse *caw...caw...kee-u...kee-u*, repetitive *hee-o...hee-o...hee-o*. Closely associated with *Araucaria* forests. **DISTRIBUTION** southeastern Brazil, in Rio Grande do Sul and neighboring southeastern Santa Catarina, extralimitally reaching easternmost Paraguay and extreme northeastern Argentina in Misiones; mostly 300 to 1000m; vulnerable, CITES I. **SIMILAR SPECIES** Vinaceous Amazon A. *vinacea* (see below) deep mauve-maroon breast; red lores, but green forecrown; dark red bill and dark eye-ring. **LOCALITIES** Aracuri-Esmeralda Ecological Station and Carazinho Municipal Park, Rio Grande do Sul, southeastern Brazil (roosting sites in nonbreeding season).

VINACEOUS AMAZON *Amazona vinacea* 30cm

Only green-headed amazon with deep mauve-maroon breast and dark red bill; red lores and frontal band; feathers of nape and neck broadly edged blue; red wing-speculum; JUV duller mauve breast suffused green, and pale bill with red at base; raucous *tay-o...tayo* or *kray-o...krayo*. Seasonally attracted to *Euterpe* palms for food. **DISTRIBUTION** southeastern Brazil, from southern Bahia and western Espírito Santo south to northeastern Argentina, in Misiones and possibly northeastern Corrientes, and southeastern Paraguay; up to 1200m; vulnerable, CITES I. **SIMILAR SPECIES** Red-spectacled Amazon A. *pretrei* (see above) red forecrown to around eyes and red "shoulders"; green breast; pale bill and prominent white eye-ring. **LOCALITIES** Jacupiranga State Park, São Paulo, and Aparados da Serra National Park, Santa Catarina, southeastern Brazil. Estancia Itabó Nature Reserve, near Puente Kyha, Canendiyú, eastern Paraguay.

TUCUMÁN AMAZON

Juv

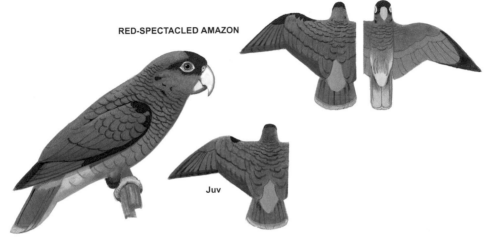

RED-SPECTACLED AMAZON

Juv

VINACEOUS AMAZON

Juv

PLATE 142 *AMAZONA* PARROTS (in part)

304

RED-TAILED AMAZON
Amazona brasiliensis 37cm

Large amazon with distinctive tail pattern and bold head markings; rose-red forecrown and lores; cheeks and ear-coverts mauve-pink; carpal edge red; lateral tail-feathers purple-blue at bases, subterminally banded red, and broadly tipped greenish yellow; pale bill; JUV duller, less extensive head markings; raucous *kraa...kraa...kraa*, also *kli-kli*, *kal-ik*, and *kree-o*. Associated with littoral forests, including mangroves, and estuarine wetlands along mainland coast and on offshore islands. **DISTRIBUTION** coastal plain of southeastern Brazil, including some offshore islands, in southern São Paulo to extreme northeastern Santa Catarina; mostly below 400m; vulnerable, CITES I. **SIMILAR SPECIES** Orange-winged Amazon *A. amazonica* (plate 139) yellow, not red forecrown; yellow face; orange wing-speculum. Vinaceous Amazon *A. vinacea* (plate 141) red lores, but green forecrown and cheeks; mauve-maroon breast; red wing-speculum; red bill. **LOCALITIES** Supergui National Park, Paraná, and Ilha Cardoso State Park, São Paulo, Brazil.

RED-CROWNED AMAZON
Amazona rhodocorytha 35cm

Large amazon with red crown and dusky occiput to nape; forecheeks orange-yellow bordered blue; carpal edge greenish yellow, and red wing-speculum; lateral tail-feathers subterminally banded red and broadly tipped greenish yellow; bicolored pink-gray bill; JUV duller, less extensive head markings; raucous *caa-ua...caa-ua...caa-ua*, also *cheee-ooo* and *nee-it*, loud *koy-ok...koy-ok* or *kow-ow...kow-ow*. **DISTRIBUTION** eastern Brazil, from Alagoas, where possibly extinct, northeastern Bahia and eastern Minas Gerais to Espírito Santo and northern São Paulo; up to 1000m; endangered, CITES I. **SIMILAR SPECIES** Orange-winged Amazon *A. amazonica* (plate 139) see above. Mealy Amazon *A. farinosa* (plate 137) green crown and face; glaucous suffusion on upperparts; little or no red in tail; pale bill and prominent white eye-ring. **LOCALITIES** Monte Pascoal National Park, Bahia, and Sooretama and Linhares Reserves, northern Espírito Santo, eastern Brazil.

BLUE-CHEEKED AMAZON
Amazona dufresniana 34cm

Only amazon with yellow wing-speculum; forehead and lores orange-yellow; cheeks to ear-coverts violet-blue; carpal edge greenish yellow; orange-red on inner webs of lateral tail-feathers; bicolored red/gray bill; JUV forehead dull yellow, and little blue suffusion on cheeks; noisy in flight, loud *queenk-queenk-queenk*, and babbling song when perched. **DISTRIBUTION** range poorly defined; recorded from northern Guyana, northeastern Suriname, northeastern French Guiana, and eastern Venezuela in La Gran Sabana and vicinity of Sierra de Lema, eastern Bolívar; probably also northeastern Brazil, in northern Amapá to northern Roraima; up to 1700m; near-threatened. **SIMILAR SPECIES** Orange-winged Amazon *A. amazonica* (plate 139) darker orange wing-speculum; yellow, not blue cheeks; pale bill. Yellow-crowned Amazon *A. ochrocephala* (plate 134) brighter yellow forecrown; green, not blue cheeks; red "shoulders" and red wing-speculum. Mealy Amazon *A. farinosa* (plate 137) predominantly green forecrown and cheeks; glaucous suffusion on upperparts; red wing-speculum and carpal edge; prominent white eye-ring. **LOCALITIES** recorded only irregularly at most localities. Las Claritas and San Isidro districts, eastern Bolívar, Venezuela. Brownsberg Nature Park, Suriname. Nouragues Field Station, northwest of Ipoucin Crique, French Guiana.

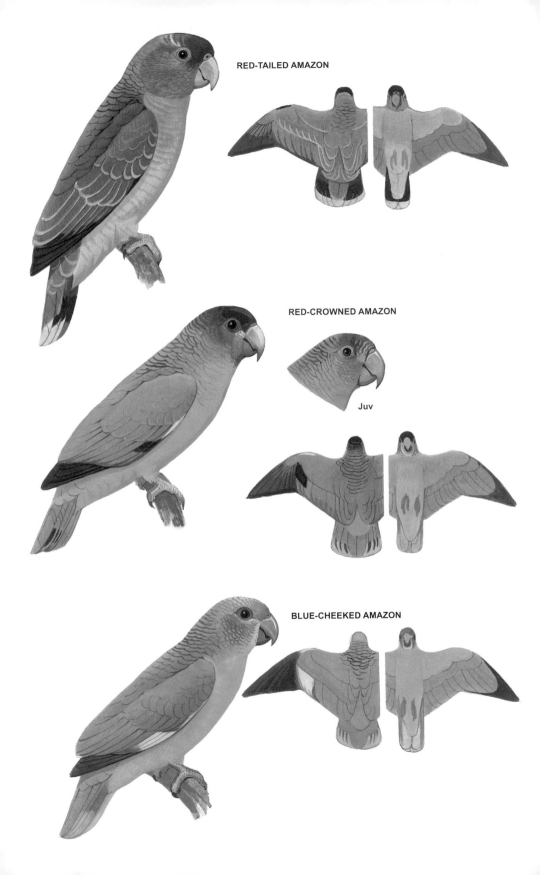

RED-TAILED AMAZON

RED-CROWNED AMAZON

Juv

BLUE-CHEEKED AMAZON

PLATE 143 *AMAZONA* **PARROTS (in part)**

306

ST. VINCENT AMAZON
Amazona guildingii 40cm

Unmistakable, and only *Amazona* species in range; large amazon with highly distinctive coloration featuring white face; orange wing-speculum; two color morphs, but much individual variation; JUV duller than adults, with paler, less clearly defined head markings; loud *quaw...quaw...quaw*, guttural *screee-eee-ah*, shrieking *scree-ree-lee-lee*, grating *draaak* or dry *screeet*. YELLOW-BROWN MORPH predominantly bronze-brown; forecrown and lores to around eyes white; greater underwing-coverts and undersides of flight feathers yellow; tail orange at base, centrally banded violet-blue and broadly tipped yellow. GREEN MORPH predominantly dark green; greater underwing-coverts and undersides of flight feathers green; tail green at base, centrally banded violet-blue and broadly tipped yellow. Favors moist forest; noisy and gregarious. **DISTRIBUTION** confined to St. Vincent, Lesser Antilles, West Indies; mostly below 700m, and recorded mainly in upper reaches of Buccament, Cumberland, and Wallilibou valleys; vulnerable, CITES I. **LOCALITY** walking track through Buccament valley, St. Vincent.

ST. VINCENT AMAZON

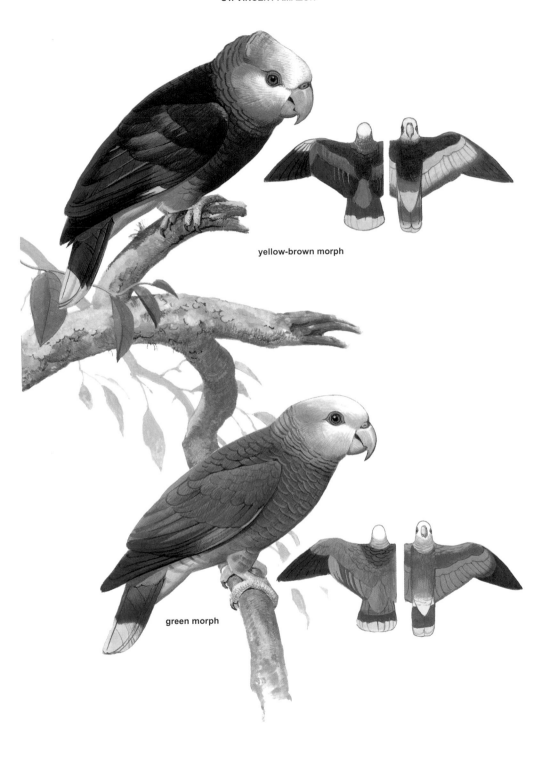

yellow-brown morph

green morph

PLATE 144 *AMAZONA* PARROTS (in part)

308

ST. LUCIA AMAZON *Amazona versicolor* 43cm

Unmistakable, and only *Amazona* species in range; forecrown and lores violet-blue, becoming paler blue on hindcrown to cheeks; red band across foreneck; lower breast to abdomen suffused reddish brown; red wing-speculum; JUV duller with less blue on face, and less red on foreneck; raucous screeching or shrill squawks in flight. Favors moist forest. **DISTRIBUTION** confined to St. Lucia, Lesser Antilles, West Indies; mostly above 300m, and mainly in central and southern mountains; vulnerable, CITES I. **LOCALITIES** Quilesse and Edmond Forest Reserves, St. Lucia.

RED-NECKED AMAZON
Amazona arausiaca 40cm

Smaller of two amazons occurring on Dominica, and differentiated by violet-blue face and green underparts with red on foreneck; red and yellow wing-speculum; pale bill; JUV duller with less blue on face and little or no red on foreneck; drawn-out *rrr-eee*, more high-pitched than calls of A. *imperialis*. Favors evergreen forest in foothills; noisy and gregarious, so contrasting with shy, secretive A. *imperialis*. **DISTRIBUTION** restricted to Dominica, Lesser Antilles, West Indies; mainly 300 to 800m, and mostly below range of A. *imperialis*; vulnerable, CITES I. **SIMILAR SPECIES** Imperial Amazon A. *imperialis* (see below) maroon-purple head and underparts, giving overall dark appearance; larger. **LOCALITIES** Syndicate Estate in Morne Diablotin Forest Reserve, nearby Northern Forest Reserve, and Morne Trois Pitons National Park, Dominica.

IMPERIAL AMAZON *Amazona imperialis* 48cm

Unmistakable; largest amazon with highly distinctive coloration featuring maroon-purple head and underparts; dark maroon wing-speculum and red carpal edge; JUV face dull rufous, but head and underparts green; distinctive trumpeting *eeeee-er*, and harsh screeches, squawks, or whistling notes. Locally dispersed in mountain forests; shy and secretive, so contrasting with noisy, gregarious A. *arausiaca*; overall dark appearance, and raptor-like in flight; **DISTRIBUTION** confined to Dominica, Lesser Antilles, West Indies; mostly 600 to 1300m, and mainly above range of A. *arausiaca*; endangered, CITES I. **SIMILAR SPECIES** Red-necked Amazon A. *arausiaca* (see above) green head and underparts; blue face and red on foreneck; smaller. **LOCALITIES** Northern Forest Reserve, including Morne Diablotin National Park, and Morne Trois Pitons National Park, Dominica.

ST. LUCIA AMAZON

RED-NECKED AMAZON

Juv

IMPERIAL AMAZON

Juv

310   **Illustrated here are extinct or presumed extinct parrots known from museum specimens. Other species from the Mascarene Islands and the West Indies are known only from subfossil remains or drawings and written accounts.**

PARADISE PARROT
Psephotus pulcherrimus 27cm

Midsized black-capped, brown-backed parrot with red "wing-patch" in both sexes; ♂ with red frontal band; JUV like ♀. **DISTRIBUTION** Formerly central-eastern Australia; last recorded 1927.

BLACK-FRONTED PARAKEET
Cyanoramphus zealandicus 25cm

Midsized olive-green parrot with black forehead, scarlet lores and stripe to behind eye, and dark red rump. **DISTRIBUTION** Formerly Tahiti, Society Islands; last recorded 1844.

RAIATEA PARAKEET *Cyanoramphus ulietanus* 25cm

Midsized parrot with brownish-black head, olive-brown upperparts, olive-yellow underparts, and dark red rump. **DISTRIBUTION** Formerly Raiatea, Society Islands; discovered and last recorded 1773

SEYCHELLES PARAKEET *Psittacula wardi* 41cm

Midsized long-tailed green parrot with red "shoulder-patch"; ♂ with black band across lower cheeks and narrow black collar on hindneck, absent in ♀ and JUV; large red bill. **DISTRIBUTION** Formerly Mahé and Silhouette, Seychelles Islands; last recorded June 1881.

NEWTON'S PARAKEET *Psittacula exsul* 40cm

Presumably green and blue morphs existed, but only two specimens blue; midsized long-tailed, dull blue parrot with black band across lower cheeks and narrow black collar on hindneck. **DISTRIBUTION** Formerly Rodrigues, Mascarene Islands; last recorded August 1875.

PARADISE PARROT

♂ ♀

BLACK-FRONTED PARAKEET

RAIATEA PARAKEET

SEYCHELLES PARAKEET

NEWTON'S PARAKEET

PLATE 146   EXTINCT OR PRESUMED EXTINCT PARROTS (in part)

312   Illustrated here are extinct or presumed extinct parrots known from museum
       specimens. Other species from the Mascarene Islands and the West Indies are
       known only from subfossil remains or drawings and written accounts.

MASCARENE PARROT
Mascarinus mascarinus 35cm

Midsized, broad-tailed brownish parrot with lilac-gray head and
black facial mask; red bill. **DISTRIBUTION** Formerly Réunion and
possibly Mauritius, Mascarene Islands; last recorded 1834, in
captivity.

NORFOLK ISLAND KAKA
Nestor productus 38cm

Large, broad-tailed brown parrot with orange cheeks, yellow
abdomen, and orange lower underparts; massive projecting bill; sexes
alike. **DISTRIBUTION** Norfolk Island and adjacent Philip Island; last
recorded about 1851, in captivity.

GLAUCOUS MACAW
Anodorhynchus glaucus 70cm

Large, dull greenish-blue macaw with yellow eye-ring and "tear-
drop" shaped lappets at sides of lower mandible; sexes alike.
DISTRIBUTION Formerly region centered on middle reaches of Río
Paraguay, Río Paraná, and Río Uruguay in southeastern Paraguay,
northeastern Argentina, southeastern Brazil, and possibly
northwestern Uruguay; last recorded early 1930s, in captivity.

CUBAN MACAW *Ara tricolor* 55cm

Large red macaw with golden yellow nape to mantle, blue wings, and
blue lower underparts. **DISTRIBUTION** Cuba, including Isla de Pinos,
and possibly Hispaniola; last recorded 1864.

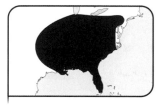

CAROLINA PARAKEET
Conuropsis carolinensis 30cm

Midsized green parrot with yellow head; crown to lores and upper
cheeks orange; sexes alike; JUV forehead to lores tawny-orange,
remainder of head green. **DISTRIBUTION** Formerly eastern U.S.A.;
last captive bird died in February 1918, but wild populations
probably survived until early 1930s.

MASCARENE PARROT

NORFOLK ISLAND KAKA

GLAUCOUS MACAW

CUBAN MACAW

CAROLINA PARAKEET

INDEX OF ENGLISH NAMES

Page numbers are in normal figures;
numbers for plates with accompanying
texts are in *italics*.

Adelaide Rosella 124 *52*
Alexandrine Parakeet 162 *71*
Amazon
  Black-billed 278 *129*
  Blue-cheeked 304 *142*
  Blue-fronted 298 *139*
  Cuban 280 *130*
  Festive 296 *138*
  Green-cheeked 284 *132*
  Hispaniolan 278 *129*
  Imperial 308 *144*
  Lilac-crowned 284 *132*
  Mealy 294 *137*
  Orange-winged 298 *139*
  Puerto Rican 280 *130*
  Red-crowned 304 *142*
  Red-lored 286 *133*
  Red-necked 308 *144*
  Red-spectacled 302 *141*
  Red-tailed 304 *142*
  Scaly-naped 296 *138*
  St Lucia 308 *144*
  St Vincent 306 *143*
  Tucumán 302 *141*
  Vinaceous 302 *141*
  White-chinned 294 *137*
  White-fronted 282 *131*
  Yellow-billed 278 *129*
  Yellow-crowned 288 290 292 *134*
    135 136
  Yellow-faced 300 *140*
  Yellow-headed 292 *136*
  Yellow-lored 282 *131*
  Yellow-naped 290 *135*
  Yellow-shouldered 300 *140*
Amazon Red-fronted Conure 228 *104*
Amazonian Parrotlet 246 *113*
Andean Parakeet 238 *109*
Antipodes Green Parakeet 142 *61*
Austral Conure 212 *96*
Australian King Parrot 112 *46*
Aztec Conure 208 *94*
Azuero Conure 230 *105*

Barred Parakeet 238 *109*
Baudin's Black Cockatoo 22 *1*

Black Lory 36 *8*
Black Parrot 146 *63*
Black-billed Amazon 278 *129*
Black-capped Conure 232 *106*
Black-capped Lory 56 *18*
Black-capped Parrot 260 *120*
Black-cheeked Lovebird 158 *69*
Black-collared Lovebird 158 *69*
Black-fronted Parakeet 310 *145*
Black-lored Parrot 106 *43*
Black-winged Lory 38 *9*
Black-winged Lovebird 154 *67*
Black-winged Parrot 270 *125*
Blaze-winged Conure 214 *97*
Blossom-headed Parakeet 164 *72*
Blue and Yellow Macaw 176 *78*
Blue Lorikeet 54 *17*
Blue-backed Parrot 106 *43*
Bluebonnet 130 *55*
Blue-cheeked Amazon 304 *142*
Blue-collared Parrot 92 *36*
Blue-crowned Conure 188 *84*
Blue-crowned Hanging Parrot 172 *76*
Blue-crowned Lorikeet 54 *17*
Blue-crowned Racquet-tailed Parrot
  100 *40*
Blue-eared Lory 40 *10*
Blue-eyed Cockatoo 28 *4*
Blue-fronted Amazon 298 *139*
Blue-fronted Lorikeet 60 *20*
Blue-fronted Parrotlet 250 *115*
Blue-headed Macaw 182 *81*
Blue-headed Parrot 272 *126*
Blue-headed Racquet-tailed Parrot 100
  40
Blue-naped Parrot 104 *42*
Blue-rumped Parrot 170 *75*
Blue-streaked Lory 38 *9*
Blue-throated Conure 234 *107*
Blue-throated Macaw 176 *78*
Blue-winged Macaw 182 *81*
Blue-winged Parrot 132 *56*
Blue-winged Parrotlet 242 *111*
Blue-winged Racquet-tailed Parrot 102
  41
Bourke's Parrot 134 *57*
Brehm's Tiger Parrot 94 *37*
Bronze-winged Parrot 276 *128*
Brown Lory 36 *8*
Brown-backed Parrotlet 252 *116*

Brown-breasted Conure 222 *101*
Brown-headed Parrot 150 *65*
Brown-hooded Parrot 262 *121*
Brown-necked Parrot 148 *6*
Brown-throated Conure 202 204 *91 92*
Budgerigar 134 57
Buff-faced Pygmy Parrot 78 *29*
Buru Racquet-tailed Parrot 98 *39*

Cactus Conure 206 *93*
Caica Parrot 262 *121*
Cape Parrot 148 *64*
Cardinal Lory 36 *8*
Carnaby's Black Cockatoo 22 *1*
Carolina Parakeet 312 *146*
Chattering Lory 58 *19*
Chestnut-fronted Macaw 182 *81*
Cloncurry Parrot 118 *49*
Cobalt-winged Parakeet 256 *118*
Cockatiel 34 *7*
Cockatoo
  Baudin's Black 22 *1*
  Blue-eyed 28 *4*
  Carnaby's Black 22 *1*
  Gang Gang 26 *3*
  Glossy Black 24 *2*
  Major Mitchell's 26 *3*
  Palm 22 *1*
  Red-tailed Black 24 *2*
  Salmon-crested 30 *5*
  Sulphur-crested 28 *4*
  White-crested 30 *5*
  Yellow-crested 28 *4*
  Yellow-tailed Black 22 *1*
Coconut Lory 42 *11*
Collared Lory 54 *17*
Conure
  Amazon Red-fronted 228 *104*
  Austral 212 *96*
  Aztec 208 *94*
  Azuero 230 *105*
  Black-capped 232 *106*
  Blaze-winged 214 *97*
  Blue-crowned 188 *84*
  Blue-throated 234 *107*
  Brown-breasted 222 *101*
  Brown-throated 202 204 *91 92*
  Cactus 206 *93*
  Cordilleran 196 *88*
  Crimson-bellied 220 *100*

Crimson-fronted 192 *86*
Cuban 192 *86*
Deville's 226 *103*
Dusky-headed 198 *89*
El Oro 232 *106*
Emma's 224 *102*
Fiery-shouldered 222 *101*
Golden 186 *83*
Golden-capped 200 *90*
Golden-plumed 188 *84*
Gray-breasted 224 *102*
Green 190 *85*
Green-cheeked 216 *98*
Hispaniolan 192 *86*
Hocking's 196 *88*
Jandaya 200 *90*
Madeira 230 *105*
Maroon-bellied 214 *97*
Maroon-faced 224 *102*
Maroon-tailed 218 *99*
Mitred 194 *87*
Mountain 196 *88*
Nanday 188 *84*
Olive-throated 208 *94*
Orange-fronted 208 *94*
Pacific 190 *85*
Painted 226 *103*
Patagonian 210 *95*
Peach-fronted 206 *93*
Pearly 220 *100*
Perijá 230 *105*
Red-eared 222 *101*
Red-fronted 198 *89*
Red-masked 198 *89*
Red-throated 190 *85*
Rose-crowned 234 *107*
Rose-fronted 228 *104*
Santa Marta 234 *107*
Santarem 226 *103*
Sinú 230 *105*
Slender-billed 212 *96*
Socorro 190 *85*
Sulphur-winged 234 *107*
Sun 200 *90*
Wavy-breasted 228 *104*
White-eared 224 *102*
White-eyed 194 *87*
White-necked 232 *106*
Yellow-eared 186 *83*
Cordilleran Conure 196 *88*
Corella
  Ducorps's 34 *7*

Goffin's 34 *7*
  Little 32 *6*
  Red-vented 34 *7*
  Slender-billed 32 *6*
  Western 32 *6*
Coxen's Fig Parrot 84 *32*
Crimson Rosella 122 124 *51 52*
Crimson-bellied Conure 220 *100*
Crimson-fronted Conure 192 *86*
Cuban Amazon 280 *130*
Cuban Conure 192 *86*
Cuban Macaw 312 *146*

Derbyan Parakeet 166 *73*
Desmarest's Fig Parrot 86 *33*
Deville's Conure 226 *103*
Double-eyed Fig Parrot 84 *32*
Duchess Lorikeet 66 *23*
Ducorps's Corella 34 *7*
Dusky Lory 38 *9*
Dusky Parrot 276 *128*
Dusky-billed Parrotlet 244 *112*
Dusky-headed Conure 198 *89*

Eastern Rosella 126 *53*
Eclectus Parrot 108 110 *44 45*
Edwards's Fig Parrot 88 *34*
Edwards's Lorikeet 46 *13*
El Oro Conure 232 *106*
Elegant Parrot 132 *56*
Emerald Lorikeet 70 *25*
Emerald-collared Parakeet 162 *71*
Emma's Conure 224 *102*

Fairy Lorikeet 66 *23*
Festive Amazon 296 *138*
Fiery-shouldered Conure 222 *101*
Fig Parrot
  Coxen's 84 *32*
  Desmarest's 86 *33*
  Double-eyed 84 *32*
  Edwards's 88 *34*
  Marshall's 84 *32*
  Orange-breasted 82 *31*
  Red-browed 84 *32*
  Salvadori's 88 *34*
Finsch's Pygmy Parrot 80 *30*
Fischer's Lovebird 156 *68*
Flores Hanging Parrot 72 *26*
Forsten's Lorikeet 46 *13*

Galah 26 *3*

Gang Gang Cockatoo 26 *3*
Geelvink Pygmy Parrot 78 *29*
Glaucous Macaw 312 *146*
Glossy Black Cockatoo 24 *2*
Goffin's Corella 34 *7*
Golden Conure 186 *83*
Golden-capped Conure 200 *90*
Golden-mantled Racquet-tailed Parrot
  98 *39*
Golden-plumed Conure 188 *84*
Golden-shouldered Parrot 130 *55*
Golden-tailed Parrotlet 252 *116*
Golden-winged Parakeet 258 *119*
Goldie's Lorikeet 50 *15*
Grand Eclectus Parrot 108 *44*
Gray Parrot 148 *64*
Gray-breasted Conure 224 *102*
Gray-cheeked Parakeet 256 *118*
Gray-headed Lovebird 154 *67*
Gray-headed Parakeet 166 *73*
Great Green Macaw 180 *80*
Great-billed Parrot 104 *42*
Green Conure 190 *85*
Green Hanging Parrot 72 *26*
Green Racquet-tailed Parrot 100
  40
Green Rosella 120 *50*
Green-cheeked Amazon 284 *132*
Green-cheeked Conure 216 *98*
Green-naped Lorikeet 42 *11*
Green-rumped Hanging Parrot 72 *26*
Green-rumped Parrotlet 240 *110*
Green-winged Macaw 178 *79*
Ground Parrot 136 *58*
Guaiabero 88 *34*

Hanging Parrot
  Blue-crowned 172 *76*
  Flores 72 *26*
  Green 72 *26*
  Green-rumped 72 *26*
  Maroon-rumped 74 *27*
  Moluccan 74 *27*
  Orange-fronted 72 *26*
  Philippine 76 *28*
  Sangihe 74 *27*
  Sri Lankan 172 *76*
  Sula 74 *27*
  Vernal 172 *76*
  Yellow-throated 72 *26*
Hawk-headed Parrot 270 *125*
Hispaniolan Amazon 278 *129*

Hispaniolan Conure 192 *86*
Hocking's Conure 196 *88*
Hooded Parrot 130 *55*
Horned Parakeet 138 *59*
Hyacinth Macaw 174 *77*

Imperial Amazon 308 *144*
Indigo-winged Parrot 268 *124*
Intermediate Parakeet 164 *72*
Iris Lorikeet 50 *15*

Jandaya Conure 200 *90*
Jardine's Parrot 148 *64*
Josephine's Lorikeet 66 *23*

Kaka 144 *62*
Kakapo 144 *62*
Kea 144 *62*
King Parrot
    Australian 112 *46*
    Moluccan 112 *46*
    Papuan 112 *46*
Kuhl's Lorikeet 54 *17*

Lear's Macaw 174 *77*
Lilac-crowned Amazon 284 *132*
Lilac-tailed Parrotlet 248 *114*
Little Corella 32 *6*
Little Lorikeet 52 *16*
Long-tailed Parakeet 170 *75*
Lorikeet
    Blue 54 *17*
    Blue-crowned 54 *17*
    Blue-fronted 60 *20*
    Duchess 66 *23*
    Edwards's 46 *13*
    Emerald 70 *25*
    Fairy 66 *23*
    Forsten's 46 *13*
    Goldie's 50 *15*
    Green-naped 42 *11*
    Iris 50 *15*
    Josephine's 66 *23*
    Kuhl's 54 *17*
    Little 52 *16*
    Meek's 60 *20*
    Mindanao 48 *14*
    Mitchell's 46 *13*
    Musschenbroek's 70 *25*
    Musk 52 *16*
    New Caledonian 60 *20*
    Olive-headed 48 *14*

Ornate 42 *11*
Palm 60 *20*
Papuan 68 *24*
Pohnpei 48 *14*
Purple-crowned 52 *16*
Pygmy 62 *21*
Rainbow 42 44 46 *11 12 13*
Red-chinned 60 *20*
Red-collared 46 *13*
Red-flanked 64 *22*
Red-fronted 64 *22*
Red-throated 62 *21*
Scaly-breasted 48 *14*
Stella's 68 *24*
Stephen's 54 *17*
Striated 62 *21*
Ultramarine 54 *17*
Varied 50 *15*
Weber's 46 *13*
Whiskered 70 *25*
Yellow and Green 48 *14*
Lory
    Black 36 *8*
    Black-capped 56 *18*
    Black-winged 38 *9*
    Blue-eared 40 *10*
    Blue-streaked 38 *9*
    Brown 36 *8*
    Cardinal 36 *8*
    Chattering 58 *19*
    Coconut 42 *11*
    Collared 54 *17*
    Dusky 38 *9*
    Purple-bellied 56 *18*
    Purple-naped 58 *19*
    Rajah 36 *8*
    Red 40 *10*
    Red and Blue 40 *10*
    Violet-necked 38 *9*
    White-naped 58 *19*
    Yellow-bibbed 58 *19*
    Yellow-streaked 36 *8*
Lovebird
    Black-cheeked 158 *69*
    Black-collared 158 *69*
    Black-winged 154 *67*
    Fischer's 156 *68*
    Gray-headed 154 *67*
    Masked 156 *68*
    Nyasa 158 *69*
    Peach-faced 156 *68*
    Red-faced 154 *67*

Macaw
    Blue and Yellow 176 *78*
    Blue-headed 182 *81*
    Blue-throated 176 *78*
    Blue-winged 182 *81*
    Chestnut-fronted 182 *81*
    Cuban 312 *146*
    Glaucous 312 *146*
    Great Green 180 *80*
    Green-winged 178 *79*
    Hyacinth 174 *77*
    Lear's 174 *77*
    Military 180 *80*
    Red-bellied 184 *82*
    Red-fronted 180 *80*
    Red-shouldered 184 *82*
    Scarlet 178 *79*
    Spix's 174 *77*
    Yellow-collared 184 *82*
Madarasz's Tiger Parrot 96 *38*
Madeira Conure 230 *105*
Major Mitchell's Cockatoo 26 *3*
Malabar Parakeet 162 *71*
Mallee Ringneck Parrot 118 *49*
Maroon-bellied Conure 214 *97*
Maroon-faced Conure 224 *102*
Maroon-fronted Parrot 186 *83*
Maroon-rumped Hanging Parrot 74 *27*
Maroon-tailed Conure 218 *99*
Marshall's Fig Parrot 84 *32*
Mascarene Parrot 312 *146*
Masked Lovebird 156 *68*
Masked Shining Parrot 138 *59*
Mauritius Parakeet 160 *70*
Mealy Amazon 294 *137*
Meek's Lorikeet 60 *20*
Meek's Pygmy Parrot 80 *30*
Mexican Parrotlet 240 *110*
Meyer's Parrot 150 *65*
Military Macaw 180 *80*
Mindanao Lorikeet 48 *14*
Mindanao Racquet-tailed Parrot 102 *41*
Mitchell's Lorikeet 46 *13*
Mitred Conure 194 *87*
Modest Tiger Parrot 96 *38*
Moluccan Hanging Parrot 74 *27*
Moluccan King Parrot 112 *46*
Monk Parakeet 210 *95*
Mountain Conure 196 *88*
Mountain Parakeet 236 *108*
Mountain Racquet-tailed Parrot 102 *41*

Mulga Parrot 128 *54*
Musk Lorikeet 52 *16*
Musschenbroek's Lorikeet 70 *25*

Nanday Conure 188 *84*
Naretha Bluebonnet 130 *55*
New Caledonian Lorikeet 60 *20*
Newton's Parakeet 310 *145*
Niam-Niam Parrot 150 *65*
Nicobar Parakeet 166 *73*
Night Parrot 136 *58*
Norfolk Island Kaka 312 *146*
Northern Rosella 126 *53*
Nyasa Lovebird 158 *69*

Olive-headed Lorikeet 48 *14*
Olive-shouldered Parrot 114 *47*
Olive-throated Conure 208 *94*
Orange-bellied Parrot 132 *56*
Orange-breasted Fig Parrot 82 *31*
Orange-cheeked Parrot 264 *122*
Orange-chinned Parakeet 256 *118*
Orange-fronted Conure 208 *94*
Orange-fronted Hanging Parrot 72 *26*
Orange-fronted Parakeet 142 *61*
Orange-headed Parrot 264 *122*
Orange-winged Amazon 298 *139*
Ornate Lorikeet 42 *10*

Pacific Conure 190 *85*
Pacific Parrotlet 246 *113*
Painted Conure 226 *103*
Painted Tiger Parrot 94 *37*
Pale-headed Rosella 126 *53*
Palm Cockatoo 22 *1*
Palm Lorikeet 60 *20*
Papuan King Parrot 112 *46*
Papuan Lorikeet 68 *24*
Paradise Parrot 310 *145*
Parakeet
    Alexandrine 162 *71*
    Andean 238 *109*
    Antipodes Green 142 *61*
    Barred 238 *109*
    Black-fronted 310 *145*
    Blossom-headed 164 *72*
    Carolina 312 *146*
    Cobalt-winged 256 *118*
    Derbyan 166 *73*
    Emerald-collared 162 *71*
    Golden-winged 258 *119*
    Gray-cheeked 256 *118*

    Gray-headed 166 *73*
    Horned 138 *59*
    Intermediate 164 *72*
    Long-tailed 170 *75*
    Malabar 162 *71*
    Mauritius 160 *70*
    Monk 210 *95*
    Mountain 236 *108*
    Newton's 310 *145*
    Nicobar 166 *73*
    Orange-chinned 256 *118*
    Orange-fronted 142 *61*
    Plain 254 *117*
    Plum-headed 164 *72*
    Raiatea 310 *145*
    Red-breasted 168 *74*
    Red-fronted 140 *60*
    Rose-ringed 160 *70*
    Rufous-fronted 238 *109*
    Seychelles 310 *145*
    Sierra 236 *108*
    Slaty-headed 166 *73*
    Tui 258 *119*
    White-winged 254 117
    Yellow-chevroned 254 *117*
    Yellow-fronted 142 *61*
Parrot
    Australian King 112 *46*
    Black 146 *63*
    Black-capped 260 *120*
    Black-lored 106 *43*
    Black-winged 270 *125*
    Blue-backed 106 *43*
    Blue-collared 92 *36*
    Blue-headed 272 *126*
    Blue-naped 104 *42*
    Blue-rumped 170 *75*
    Blue-winged 132 *56*
    Bourke's 134 *57*
    Bronze-winged 276 *128*
    Brown-headed 150 *65*
    Brown-hooded 262 *121*
    Brown-necked 148 *64*
    Caica 262 *121*
    Cape 148 *64*
    Cloncurry 118 *49*
    Dusky 276 *128*
    Eclectus 108 110 *44 45*
    Elegant 132 *56*
    Golden-shouldered 130 *55*
    Grand Eclectus 108 *44*
    Gray 148 *64*

    Great-billed 104 *42*
    Ground 136 *58*
    Hawk-headed 270 *125*
    Hooded 130 *55*
    Indigo-winged 268 *124*
    Jardine's 148 *64*
    Mallee Ringneck 118 *49*
    Maroon-fronted 186 *83*
    Mascarene 312 *146*
    Masked Shining 138 *59*
    Meyer's 150 *65*
    Moluccan King 112 *46*
    Mulga 128 *54*
    Niam-Niam 150 *65*
    Night 136 *58*
    Olive-shouldered 114 *47*
    Orange-bellied 132 *56*
    Orange-cheeked 264 *122*
    Orange-headed 264 *122*
    Papuan King 112 *46*
    Paradise 310 *145*
    Pesquet's 110 *45*
    Pileated 266 *123*
    Plum-headed 274 *127*
    Port Lincoln 118 *49*
    Princess 116 *48*
    Purple-bellied 266 *123*
    Red-bellied 152 *56*
    Red-billed 274 *127*
    Red-capped 120 *50*
    Red-cheeked 90 92 *35 36*
    Red-faced 268 *124*
    Red-rumped 128 *54*
    Red Shining 138 *59*
    Red-sided Eclectus 110 *45*
    Red-winged 114 *47*
    Regent 116 *48*
    Rock 132 *56*
    Rose-faced 262 *121*
    Rüppell's 152 *66*
    Rusty-faced 268 *124*
    Saffron-headed 262 *121*
    Scaly-headed 276 *128*
    Scarlet-chested 134 *57*
    Senegal 152 *66*
    Short-tailed 270 *125*
    Singing 92 *36*
    Superb 116 *48*
    Swift 52 *16*
    Thick-billed 186 *83*
    Turquoise 134 *57*
    Twenty-eight 118 *49*

Parrot *continued*
Vasa 146 *63*
Vulturine 264 *122*
White-bellied 260 *120*
White-capped 274 *127*
White-crowned 272 *126*
Yellow-faced 148 *64*
Parrotlet
Amazonian 246 *113*
Blue-fronted 250 *115*
Blue-winged 242 *111*
Brown-backed 252 *116*
Dusky-billed 244 *112*
Golden-tailed 252 *116*
Green-rumped 240 *110*
Lilac-tailed 248 *114*
Mexican 240 *110*
Pacific 246 *113*
Red-fronted 250 *115*
Sapphire-rumped 252 *116*
Scarlet-shouldered 250 *115*
Spectacled 244 *112*
Spot-winged 248 *114*
Tepui 246 *113*
Yellow-faced 246 *113*
Patagonian Conure 210 *95*
Peach-faced Lovebird 156 *68*
Peach-fronted Conure 206 *93*
Pearly Conure 220 *100*
Perijá Conure 230 *105*
Pesquet's Parrot 110 *45*
Philippine Hanging Parrot 76 *28*
Pileated Parrot 266 *123*
Plain Parakeet 254 *117*
Plum-headed Parakeet 164 *72*
Plum-headed Parrot 274 *127*
Pohnpei Lorikeet 48 *14*
Port Lincoln Parrot 118 *49*
Princess Parrot 116 *48*
Puerto Rican Amazon 280 *130*
Purple-bellied Lory 56 *18*
Purple-bellied Parrot 266 *123*
Purple-crowned Lorikeet 52 *16*
Purple-naped Lory 58 *19*
Pygmy Lorikeet 62 *21*
Pygmy Parrot
Buff-faced 78 *29*
Finsch's 80 *30*
Geelvink 78 *29*
Meek's 80 *30*
Red-breasted 80 *30*
Yellow-capped 78 *29*

Racquet-tailed Parrot
Blue-crowned 100 *40*
Blue-headed 100 *40*
Blue-winged 102 *41*
Buru 98 *39*
Golden-mantled 98 *39*
Green 100 *40*
Mindanao 102 *41*
Mountain 102 *41*
Yellow-breasted 98 *39*
Raiatea Parakeet 310 *145*
Rainbow Lorikeet 42 44 46 *11 12 13*
Rajah Lory 36 *8*
Red and Blue Lory 40 *10*
Red Lory 40 *10*
Red Shining Parrot 138 *59*
Red-bellied Macaw 184 *82*
Red-bellied Parrot 152 *56*
Red-billed Parrot 274 *127*
Red-breasted Parakeet 168 *74*
Red-breasted Pygmy Parrot 80 *30*
Red-browed Fig Parrot 84 *32*
Red-capped Parrot 120 *50*
Red-cheeked Parrot 90 92 *35 36*
Red-chinned Lorikeet 60 *20*
Red-collared Lorikeet 46 *13*
Red-crowned Amazon 304 *142*
Red-eared Conure 222 *101*
Red-faced Lovebird 154 *67*
Red-faced Parrot 268 *124*
Red-flanked Lorikeet 64 *22*
Red-fronted Conure 198 *89*
Red-fronted Lorikeet 64 *22*
Red-fronted Macaw 180 *80*
Red-fronted Parakeet 140 *60*
Red-fronted Parrotlet 250 *115*
Red-lored Amazon 286 *133*
Red-masked Conure 198 *89*
Red-necked Amazon 308 *144*
Red-rumped Parrot 128 *54*
Red-shouldered Macaw 184 *82*
Red-sided Eclectus Parrot 110 *45*
Red-spectacled Amazon 302 *141*
Red-tailed Amazon 304 *142*
Red-tailed Black Cockatoo 24 *2*
Red-throated Conure 190 *85*
Red-throated Lorikeet 62 *21*
Red-vented Corella 34 *7*
Red-winged Parrot 114 *47*
Regent Parrot 116 *48*
Rock Parrot 132 *56*
Rose-crowned Conure 234 *107*

Rose-faced Parrot 262 *121*
Rose-fronted Conure 228 *104*
Rosella
Adelaide 124 *52*
Crimson 122 124 *51 52*
Eastern 126 *53*
Green 120 *50*
Northern 126 *53*
Pale-headed 126 *53*
Western 126 *53*
Yellow 124 *52*
Rose-ringed Parakeet 160 *70*
Rufous-fronted Parakeet 238 *109*
Rüppell's Parrot 152 *66*
Rusty-faced Parrot 268 *124*

Saffron-headed Parrot 262 *121*
Salmon-crested Cockatoo 30 *5*
Salvadori's Fig Parrot 58 *34*
Sangihe Hanging Parrot 74 *27*
Santa Marta Conure 234 *107*
Santarem Conure 226 *103*
Sapphire-rumped Parrotlet 252 *116*
Scaly-breasted Lorikeet 48 *14*
Scaly-headed Parrot 276 *128*
Scaly-naped Amazon 296 *138*
Scarlet Macaw 178 *79*
Scarlet-chested Parrot 134 *57*
Scarlet-shouldered Parrotlet 250
115
Senegal Parrot 152 66
Seychelles Parakeet 310 *145*
Shining Parrot
Masked 138 *59*
Red 138 *59*
Short-tailed Parrot 270 *125*
Sierra Parakeet 236 *108*
Singing Parrot 92 *36*
Sinú Conure 230 *105*
Slaty-headed Parakeet 166 *73*
Slender-billed Conure 212 *96*
Slender-billed Corella 32 *6*
Socorro Conure 190 *85*
Spectacled Parrotlet 244 *112*
Spix's Macaw 174 *77*
Spot-winged Parrotlet 248 *114*
Sri Lankan Hanging Parrot 172 *76*
St Lucia Amazon 308 *144*
St Vincent Amazon 306 *143*
Stella's Lorikeet 68 *24*
Stephen's Lorikeet 54 *17*
Striated Lorikeet 62 *21*

Sula Hanging Parrot 74 *27*
Sulphur-crested Cockatoo 28 *4*
Sulphur-winged Conure 234 *107*
Sun Conure 200 *90*
Superb Parrot 116 *48*
Swift Parrot 52 *16*

Tepui Parrotlet 246 *113*
Thick-billed Parrot 186 *83*
Tiger Parrot
  Brehm's 94 *37*
  Madarasz's 96 *38*
  Modest 96 *38*
  Painted 94 *37*
Tucumán Amazon 302 *141*
Tui Parakeet 258 *119*
Turquoise Parrot 134 *57*
Twenty-eight Parrot 118 *49*

Ultramarine Lorikeet 54 *17*

Varied Lorikeet 50 *15*
Vasa Parrot 146 *63*

Vernal Hanging Parrot 172 *76*
Vinaceous Amazon 302 *141*
Violet-necked Lory 38 *9*
Vulturine Parrot 264 *122*

Wavy-breasted Conure 228 *104*
Weber's Lorikeet 46 *13*
Western Corella 32 *6*
Western Rosella 126 *53*
Whiskered Lorikeet 70 *25*
White-bellied Parrot 260 *120*
White-capped Parrot 274 *127*
White-chinned Amazon 294 *137*
White-crested Cockatoo 30 *15*
White-crowned Parrot 272 *126*
White-eared Conure 224 *102*
White-eyed Conure 194 *87*
White-fronted Amazon 282 *131*
White-naped Lory 58 *19*
White-necked Conure 232 *106*
White-winged Parakeet 254 *117*

Yellow and Green Lorikeet 48 *14*

Yellow Rosella 124 *52*
Yellow-bibbed Lory 58 *19*
Yellow-billed Amazon 278 129
Yellow-breasted Racquet-tailed Parrot
  98 *39*
Yellow-capped Pygmy Parrot 78 *29*
Yellow-chevroned Parakeet 254 *117*
Yellow-collared Macaw 184 *82*
Yellow-crested Cockatoo 28 *4*
Yellow-crowned Amazon 288 290 292
  134 135 136
Yellow-eared Conure 186 *83*
Yellow-faced Amazon 300 *140*
Yellow-faced Parrot 148 *64*
Yellow-faced Parrotlet 246 *113*
Yellow-fronted Parakeet 142 *61*
Yellow-headed Amazon 292 *136*
Yellow-lored Amazon 282 *131*
Yellow-naped Amazon 290 *135*
Yellow-shouldered Amazon 300 *140*
Yellow-streaked Lory 36 *8*
Yellow-tailed Black Cockatoo 22 *1*
Yellow-throated Hanging Parrot 72 *26*

INDEX OF SCIENTIFIC NAMES

Page numbers are in normal figures, and numbers for colored plates with accompanying texts are in *italics*.

abbotti, Cacatua sulphurea 28
abbotti, Psittacula alexandri 168
abbotti, Psittinus cyanurus 170 *75*
ablectaneus, Agapornis canus 154 *67*
accipitrinus, Deroptyus 270 *125*
acuticaudata, Aratinga 188 *84*
adscitus, Platycercus 126 *53*
aeruginosa, Aratinga pertinax 202 *91*
aestiva, Amazona 298 *139*
affinis, Tanygnathus megalorhynchus 104 *42*
Agapornis canus 154 *67*
Agapornis fischeri 156 *68*
Agapornis lilianae 158 *69*
Agapornis nigrigenis 158 *69*
Agapornis personatus 156 *68*
Agapornis pullarius 154 *67*
Agapornis roseicollis 156 *68*
Agapornis swindernianus 158 *69*
Agapornis taranta 154 *67*
agilis, Amazona 278 *129*
alba, Cacatua 30 *5*
albiceps, Eolophus roseicapilla 26 *3*
albidinucha, Lorius 58 *19*
albifrons, Amazona 282 *131*
albipectus, Pyrrhura 232 *106*
alexandrae, Polytelis 116 *48*
alexandri, Psittacula 168 *74*
Alisterus amboinensis 112 *46*
Alisterus chloropterus 112 *46*
Alisterus scapularis 112 *46*
alpinus, Neopsittacus pullicauda 70 *25*
alticola, Aratinga 196 *88*
amabilis, Charmosyna 62 *21*
amabilis, Cyclopsitta gulielmitertii 82 *31*
amabilis, Loriculus 74 *27*
Amazona aestiva 298 *139*
Amazona agilis 278 *129*
Amazona albifrons 282 *131*
Amazona amazonica 298 *139*
Amazona arausiaca 308 *144*
Amazona autumnalis 286 *133*
Amazona barbadensis 300 *140*
Amazona brasiliensis 304 *142*

Amazona collaria 278 *129*
Amazona dufresniana 304 *142*
Amazona farinosa 294 *137*
Amazona festiva 296 *138*
Amazona finschi 284 *132*
Amazona guildingii 306 *143*
Amazona imperialis 308 *144*
Amazona kawalli 294 *137*
Amazona leucocephala 280 *130*
Amazona mercenaria 296 *138*
Amazona ochrocephala 288 290 292 *134 135 136*
Amazona pretrei 302 *141*
Amazona rhodocorytha 304 *142*
Amazona tucumana 302 *141*
Amazona ventralis 278 *129*
Amazona versicolor 308 *144*
Amazona vinacea 302 *141*
Amazona viridigenalis 284 *132*
Amazona vittata 280 *130*
Amazona xantholora 282 *131*
Amazona xanthops 300 *140*
amazonica, Amazona 298 *139*
amazonina, Hapalopsittaca 268 *124*
amazonum, Pyrrhura 226 *103*
ambiguus, Ara 180 *80*
amboinensis, Alisterus 112 *46*
andinus, Cyanoliseus patagonus 210 *95*
anerythra, Pyrrhura lepida 220 *100*
Anodorhynchus glaucus 312 *146*
Anodorhynchus hyacinthinus 174 *77*
Anodorhynchus leari 174 *77*
antelius, Pionus sordidus 274
anthopeplus, Polytelis 116 *48*
aolae, Micropsitta finschii 80
apicalis, Loriculus philippensis 76 *28*
Aprosmictus erythropterus 114 *47*
Aprosmictus jonguillaceus 114 *47*
Ara ambiguus 189 *80*
Ara ararauna 176 *78*
Ara chloropterus 178 *79*
Ara glaucogularis 176 *78*
Ara macao 178 *79*
Ara militaris 180 *80*
Ara rubrogenys 180 *80*
Ara severus 182 *81*
Ara tricolor 312 *146*
ararauna, Ara 176 *78*
Aratinga acuticaudata 188 *84*

Aratinga alticola 196 *88*
Aratinga aurea 206 *93*
Aratinga auricapillus 200 *90*
Aratinga branickii 188 *84*
Aratinga brevipes 190 *85*
Aratinga cactorum 206 *93*
Aratinga canicularis 208 *94*
Aratinga chloroptera 192 *86*
Aratinga erythrogenys 198 *89*
Aratinga euops 192 *86*
Aratinga finschi 192 *86*
Aratinga frontata 196 *88*
Aratinga hockingi 196 *88*
Aratinga holochlora 190 *85*
Aratinga jandaya 200 *90*
Aratinga leucophthalma 194 *87*
Aratinga mitrata 194 *87*
Aratinga nana 208 *94*
Aratinga nenday 188 *84*
Aratinga pertinax 202 204 *91 92*
Aratinga rubritorquis 190 *85*
Aratinga solstitialis 200 *90*
Aratinga strenua 190 *85*
Aratinga wagleri 198 *89*
Aratinga weddellii 198 *89*
arausiaca Amazona 308 *144*
arfaki, Oreopsittacus 70 *25*
arubensis, Aratinga pertinax 202 *91*
aruensis, Cyclopsitta diophthalma 84 *32*
aruensis, Eclectus roratus 110
aruensis, Geoffroyus geoffroyi 90 *35*
astec, Aratinga nana 208 *94*
aterrimus, Probosciger 22 *1*
atra, Chalcopsitta 36 *8*
aurantiiceps, Poicephalus flavifrons 148 *64*
aurantiifrons, Loriculus 72 *26*
aurantiigena, Gypopsitta barrabandi 264 *122*
aurantiocephala, Gypopsitta 264 *122*
aurea, Aratinga 206 *93*
auricapillus, Aratinga 200 *90*
auriceps, Cyanoramphus 142 *61*
auricollis, Primolius 184 *82*
auricularis, Pyrrhura emma 224 *102*
aurifrons, Aratinga auricapillus 200 *90*
aurifrons, Bolborhynchus 236 *108*
auropalliata, Amazona ochrocephala 290 *135*

australis, Pyrrhura molinae 216 *98*
australis, Vini 54 *17*
autumnalis, Amazona 286 *133*
avensis, Psittacula eupatria 162 *71*
aymara, Bolborhynchus 236 *108*

bahamensis, Amazona leucocephala
   280 *130*
banksii, Calyptorhynchus 24 *2*
barbadensis, Amazona 300 *140*
barklyi, Corocopsis nigra 146 *63*
barnardi, Barnardius 118 *49*
Barnardius barnardi 118 *49*
Barnardius zonarius 118 *49*
barrabandi, Gypopsitta 264 *122*
batavicus, Touit 248 *114*
batavorum, Loriculus aurantiifrons 72
   26
baudinii, Calyptorhynchus 22 *1*
beccarii, Micropsitta pusio 78 *29*
behni, Brotogeris chiriri 254
belizensis, Amazona ochrocephala 292
   136
beniensis, Brotogeris cyanoptera 256
berlepschi, Pyrrhura melanura 218 *99*
bernsteini, Chalcopsitta atra 36
beryllinus, Loriculus 172 *76*
biaki, Eclectus roratus 110
bloxami, Cyanoliseus patagonus 210 *95*
blythii, Psittaculirostris desmarestii 86
   33
bodini, Amazona festiva 296 *138*
Bolbopsittacus lunulatus 88 *34*
Bolborhynchus aurifrons 236 *108*
Bolborhynchus aymara 236 *108*
Bolborhynchus ferrugineifrons 238 *109*
Bolborhynchus lineola 238 *109*
Bolborhynchus orbygnesius 238 *109*
bolivianus, Ara militaris 180
bonapartei, Loriculus philippensis 76
   28
borealis, Psittacula krameri 160 *70*
bornea, Eos 40 *10*
bourkii, Neopsephotus 134 *57*
bournsi, Loriculus philippensis 76
brachyurus, Graydidascalus 270 *125*
branickii, Aratinga 188 *84*
brasiliensis, Amazona 304 *142*
brehmii, Psittacella 94 *37*
brevipes, Aratinga 190 *85*
brewsteri, Aratinga holochlora 190
Brotogeris chiriri 254 *117*

Brotogeris chrysoptera 258 *119*
Brotogeris cyanoptera 256 *118*
Brotogeris jugularis 256 *118*
Brotogeris pyrrhoptera 256 *118*
Brotogeris sanctithomae 258 *119*
Brotogeris tirica 254 *117*
Brotogeris versicolorus 254 *117*
bruijnii, Micropsitta 80 *30*
buergersi, Geoffroyus simplex 92 *36*
burbidgii, Tanygnathus sumatranus 106
buruensis, Alisterus amboinensis 112
   46

Cacatua alba 30 *5*
Cacatua ducorpsii 34 *7*
Cacatua galerita 28 *4*
Cacatua goffiniana 34 *7*
Cacatua haematuropygia 34 *7*
Cacatua moluccensis 30 *5*
Cacatua ophthalmica 28 *4*
Cacatua pastinator 32 *6*
Cacatua sanguinea 32 *6*
Cacatua sulphurea 28 *4*
Cacatua tenuirostris 32 *6*
cactorum, Aratinga 206 *93*
caeruleiceps, Pyrrhura 230 *105*
caeruleiceps, Trichoglossus
   haematodus 42 *11*
caeruleus, Psephotus haematonotus
   128 *54*
caica, Gypopsitta 262 *121*
caixana, Aratinga cactorum 206 *93*
cala, Psittacula alexandri 168 *74*
caledonicus, Platycercus 120 *50*
calita, Myiopsitta monachus 210 *95*
callainipictus, Bolbopsittacus lunulatus
   88
calliptera, Pyrrhura 222 *101*
Callocephalon fimbriatum 26 *3*
callogenys, Aratinga leucophthalma
   194
callopterus, Alisterus chloropterus 112
calthorpae, Psittacula 162 *71*
Calyptorhynchus banksii 24 *2*
Calyptorhynchus baudinii 22 *1*
Calyptorhynchus funcrcus 22 *1*
Calyptorhynchus lathami 24 *2*
Calyptorhynchus latirostris 22 *1*
camiguinensis, Loriculus philippensis
   76 *28*
caniceps, Psittacula 166 *73*
canicularis, Aratinga 208 *94*

canipalliata, Amazona mercenaria 296
   138
canus, Agapornis 154 *67*
capistratus, Trichoglossus haematodus
   46 *13*
cardinalis, Chalcopsitta 36 *8*
caribaea, Amazona ochrocephala 290
   135
carolinensis, Conuropsis 312 *146*
catamene, Loriculus 74 *27*
catumbella, Agapornis roseicollis 156
caucae, Forpus conspicillatus 244
caymanensis, Amazona leucocephala
   280
cervicalis, Psittaculirostris desmarestii
   86 *33*
Chalcopsitta atra 36 *8*
Chalcopsitta cardinalis 36 *8*
Chalcopsitta duivenbodei 36 *8*
Chalcopsitta sintillata 36 *8*
chalcopterus, Pionus 276 *128*
challengeri, Eos histrio 40 *10*
chapmani, Amazona farinosa 294
chapmani, Pyrrhura melanura 218 *99*
Charmosyna amabilis 62 *21*
Charmosyna diadema 60 *20*
Charmosyna josefinae 66 *23*
Charmosyna margarethae 66 *23*
Charmosyna meeki 60 *20*
Charmosyna multistriata 62 *21*
Charmosyna palmarum 60 *20*
Charmosyna papou 68 *24*
Charmosyna placentis 64 *22*
Charmosyna pulchella 66 *23*
Charmosyna rubrigularis 60 *20*
Charmosyna rubronotata 64 *22*
Charmosyna toxopei 60 *20*
Charmosyna wilhelminae 62 *21*
chathamensis, Cyanoramphus
   novaezelandiae 140
chiripepe, Pyrrhura frontalis 214 *97*
chiriri, Brotogeris 254 *117*
chlorocercus, Lorius 58 *19*
chlorogenys, Aratinga mitrata 194 *87*
chlorolepidotus, Trichoglossus 48 *14*
chloroptera, Aratinga 192 *86*
chloroptera, Chalcopsitta sintillata 36
   8
chloropterus, Alisterus 112 *46*
chloropterus, Ara 178 *79*
chloroxantha, Micropsitta keiensis 78
   29

chryseura, Touit surdus 252

chrysogaster, Neophema 132 *56*

chrysogenys, Aratinga pertinax 204

chrysonotus, Loriculus philippensis 76 *28*

chrysophrys, Aratinga pertinax 204

chrysoptera, Brotogeris 258 *119*

chrysopterygius, Psephotus 130 *55*

chrysosema, Brotogeris chrysoptera 258 *119*

chrysostoma, Neophema 132 *56*

citrinocristata, Cacatua sulphurea 28 *4*

clarae, Aratinga canicularis 208 *94*

coccineifrons, Cyclopsitta diophthalma 84 *32*

coccineopterus, Aprosmictus erythropterus 114 *47*

coccinicollaris, Gypopsitta haematotis 262 *121*

coelestis, Forpus 246 *113*

coerulescens, Pyrrhura lepida 220

collaria, Amazona 278 *129*

collaris, Psittacella modesta 96

columboides, Psittacula 162 *71*

comorensis, Coracopsis vasa 146 *63*

concinna, Glossopsitta 52 *16*

conlara, Cyanoliseus patagonus 210

conspicillatus, Forpus 244 *112*

Conuropsis carolinensis 312 *146*

cooki, Cyanoramphus novaezelandiae 140 *60*

corallinus, Pionus sordidus 274 *127*

cornelia, Eclectus roratus 108 *44*

cornutus, Eunymphicus 138 *59*

Corocopsis nigra 146 *63*

Corocopsis vasa 146 *63*

costaricensis, Touit 250 *115*

cotorra, Myiopsitta monachus 210

couloni, Primolius 182 *81*

coxeni, Cyclopsitta diophthalma 84 *32*

crassirostris, Forpus xanthopterygius 242

crassus, Poicephalus 150 *65*

cruentata, Pyrrhura 234 *107*

cryptoxanthus, Poicephalus 150 *65*

cumanensis, Diopsittaca nobilis 184 *82*

cyanicarpus, Geoffroyus geoffroyi 90 *35*

cyanicollis, Geoffroyus geoffroyi 90 *35*

cyanocephala, Psittacula 164 *72*

cyanochlorus, Forpus passerinus 240

cyanogenia, Eos 38 *9*

Cyanoliseus patagonus 210 *95*

cyanonothus, Eos bornea 40 *10*

cyanophanes, Forpus passerinus 240 *110*

Cyanopsitta spixii 174 *77*

cyanopterus, Ara macao 178 *79*

cyanoptera, Brotogeris 256 *118*

cyanopygius, Forpus 240 *110*

Cyanoramphus auriceps 142 *61*

Cyanoramphus malherbi 142 *61*

Cyanoramphus novaezelandiae 140 *60*

Cyanoramphus ulietanus 310 *145*

Cyanoramphus unicolor 142 *61*

Cyanoramphus zealandicus 310 *145*

cyanuchen, Lorius lory 56 *18*

cyanurus, Cyanoramphus novaezelandiae 140 *60*

cyanurus, Psittinus 170 *75*

cycloporum, Charmosyna josefinae 66 *23*

Cyclopsitta diophthalma 84 *32*

Cyclopsitta gulielmitertii 82 *31*

dachilleae, Nannopsittaca 246 *113*

damarensis, Poicephalus meyeri 150

dammermani, Psittacula alexandri 168

defontainei, Psittacula longicauda 170

deliciosus, Forpus passerinus 240

deplanchii, Trichoglossus haematodus 44 *12*

derbiana, Psittacula 166 *73*

derbyi, Cacatua pastinator 32

Deroptyus accipitrinus 270 *125*

desmarestii, Psittaculirostris 86 *33*

devillei, Pyrrhura 214 *97*

devittatus, Lorius hypoinochrous 56 *18*

diadema, Amazona autumnalis 286 *133*

diadema, Charmosyna 60 *20*

didimus, Glossopsitta concinna 52

diemenensis, Platycercus eximius 126 *53*

dilectissimus, Touit 250 *115*

dilutissima, Pyrrhura peruviana 228

diophthalma, Cyclopsitta 84 *32*

Diopsittaca nobilis 184 *82*

discolor, Lathamus 52 *16*

discurus, Prioniturus 100 *40*

dissimilis, Psephotus 130 *55*

djampeanus, Trichoglossus haematodus 46

dohertyi, Loriculus philippensis 76

domicella, Lorius 58 *19*

dorsalis, Alisterus amboinensis 112

drouhardi, Coracopsis vasa 146 *63*

ducorpsii, Cacatua 34 *7*

dufresniana, Amazona 304 *142*

duivenbodei, Chalcopsitta 36 *8*

duponti, Tanygnathus sumatranus 106

eburnirostrum, Aratinga canicularis 208 *94*

echo, Psittacula 160 *70*

Eclectus roratus 108 110 *44 45*

edwardsii, Psittaculirostris 88 *34*

egregia, Pyrrhura 222 *101*

eidos, Forpus sclateri 244

eisenmanni, Pyrrhura 230 *105*

elecica, Platycercus eximius 126 *53*

elegans, Neophema 132 *56*

elegans, Platycercus 122 124 *51 52*

eleonora, Cacatua galerita 28

emini, Agapornis swindernianus 158 *69*

emma, Pyrrhura 224 *102*

Enicognathus ferrugineus 212 *96*

Enicognathus leptorhynchus 212 *96*

Eolophus roseicapilla 26 *3*

Eos bornea 40 *10*

Eos cyanogenia 38 *9*

Eos histrio 40 *10*

Eos reticulata 38 *9*

Eos semilarvata 40 *10*

Eos squamata 38 *9*

erebus, Calyptorhynchus lathami 24

erithacus, Psittacus 148 *64*

erythrogenys, Aratinga 198 *89*

erythropterus, Aprosmictus 114 *47*

erythrothorax, Lorius lory 56 *18*

Eunymphicus cornutus 138 *59*

euops, Aratinga 192 *86*

eupatria, Psittacula 162 *71*

euteles, Trichoglossus 48 *14*

everetti, Tanygnathus sumatranus 106 *43*

excelsa, Psittacella picta 94 *37*

exilis, Loriculus 72 *26*

eximius, Platycercus 126 *53*

exsul, Brotogeris jugularis 256

exsul, Psittacula 310 *145*

fantiensis, Poicephalus gulielmi 148 *64*

farinosa, Amazona 294 *137*

fasciata, Psittacula alexandri 168 *74*

ferrugineifrons, Bolborhynchus 238 *109*

ferrugineus, Enicognathus 212 *96*

festiva, Amazona 296 *138*

fimbriatum, Callocephalon 26 *3*

finschi, Amazona 284 *132*

finschi, Aratinga 192 *86*

finschii, Micropsitta 80 *30*

finschii, Psittacula 166 *73*

fischeri, Agapornis 156 *68*

fitzroyi, Cacatua galerita 28

flaveolus, Platycercus elegans 124 *52*

flavescens, Forpus xanthopterygius 242

flavicans, Prioniturus 98 *39*

flavicans, Trichoglossus haematodus 44 *12*

flavifrons, Poicephalus 148 *64*

flavissimus, Forpus xanthopterygius 242 *111*

flaviventris, Pezoporus wallicus 136 *58*

flavopalliatus, Lorius garrulus 58 *19*

flavoptera, Pyrrhura molinae 216 *98*

flavotectus, Trichoglossus haematodus 46

flavoviridis, Trichoglossus 48 *14*

fleurieuensis, Platycercus elegans 124 *52*

floresianus, Geoffroyus geoffroyi 90

flosculus, Loriculus 72 *26*

forbesi, Cyanoramphus auriceps 142 *61*

Forpus coelestis 246 *113*

Forpus conspicillatus 244 *112*

Forpus cyanopygius 240 *110*

Forpus passerinus 240 *110*

Forpus sclateri 244 *112*

Forpus xanthops 246 *113*

Forpus xanthopterygius 242 *111*

forsteni, Trichoglossus haematodus 46 *13*

fortis, Trichoglossus haematodus 46 *13*

freeri, Tanygnathus sumatranus 106

frontalis, Pyrrhura 214 *97*

frontata, Aratinga 196 *88*

fuertesi, Hapalopsittaca 268 *124*

fulgidus, Psittrichas 110 *45*

funereus, Calyptorhynchus 22 *1*

fuscata, Pseudeos 38 *9*

fuscicollis, Poicephalus robustus 148 *64*

fuscifrons, Cyclopsitta gulielmitertii 82 *31*

fuscifrons, Deroptyus accipitrinus 270 *125*

fuscus, Pionus 276 *128*

galerita, Cacatua 28 *4*

galgulus, Loriculus 172 *76*

garrulus, Lorius 58 *19*

gaudens, Pyrrhura hoffmanni 234 *107*

geelvinkiana, Micropsitta 78 *29*

geoffroyi, Geoffroyus 90 92 *35 36*

Geoffroyus geoffroyi 90 92 *35 36*

Geoffroyus heteroclitus 92 *36*

Geoffroyus simplex 92 *36*

glaucogularis, Ara 176 *78*

glaucus, Anodorhynchus 312 *146*

Glossopsitta concinna 52 *16*

Glossopsitta porphyrocephala 52 *16*

Glossopsitta pusilla 52 *16*

godmani, Psittaculirostris desmarestii 86 *33*

goffiniana, Cacatua 34 *7*

goldiei, Psitteuteles 50 *15*

goliath, Probosciger aterrimus 22 *1*

goliathina, Charmosyna papou 68 *24*

gramineus, Tanygnathus 106 *43*

grandis, Oreopsittacus arfaki 70 *25*

graptogyne, Calyptorhynchus banksii 24 *2*

Graydidascalus brachyurus 270 *125*

griseipecta, Aratinga pertinax 202 *91*

griseipectus, Pyrrhura 224 *102*

guarouba, Guaruba 186 *83*

Guaruba guarouba 186 *83*

guatemalae, Amazona farinosa 294 *137*

guayaquilensis, Ara ambiguous 180

guildingii, Amazona 306 *143*

gulielmi, Poicephalus 148 *64*

gulielmitertii, Cyclopsitta 82 *31*

gustavi, Brotogeris cyanoptera 256 *118*

gymnopis, Cacatua sanguinea 32 *6*

Gypopsitta aurantiocephala 264 *122*

Gypopsitta barrabandi 264 *122*

Gypopsitta caica 262 *121*

Gypopsitta haematotis 262 *121*

Gypopsitta pulchra 262 *121*

Gypopsitta pyrilia 262 *121*

Gypopsitta vulturina 264 *122*

habroptila, Strigops 144 *62*

haematodus, Trichoglossus 42 44 46 *11 12 13*

haematogaster, Northiella 130 *55*

haematonotus, Psephotus 128 *54*

haematorrhous, Northiella haematogaster 130 *55*

haematotis, Gypopsitta 262 *121*

haematotis, Pyrrhura 222 *101*

haematuropygia, Cacatua 34 *7*

haemorrhous, Aratinga acuticaudata 188 *84*

hallstromi, Psittacella madaraszi 96

halmaturinus, Calyptorhynchus lathami 24

Hapalopsittaca amazonina 268 *124*

Hapalopsittaca fuertesi 268 *124*

Hapalopsittaca melanotis 270 *125*

Hapalopsittaca pyrrhops 268 *124*

harterti, Micropsitta pusio 78

harterti, Psittacella brehmii 94

hellmayri, Tanygnathus megalorhynchos 104

hesterna, Amazona leucocephala 280 *130*

heteroclitus, Geoffroyus 92 *36*

hilli, Platycercus venustus 126 *53*

himalayana, Psittacula 166 *73*

histrio, Eos 40 *10*

hochstetteri, Cyanoramphus novaezelandiae 140 *60*

hockingi, Aratinga 196 *88*

hoffmanni, Pyrrhura 234 *107*

hollandicus, Nymphicus 34 *7*

holochlora, Aratinga 190 *85*

hondurensis, Amazona ochrocephala 292 *136*

huetii, Touit 250 *115*

huonensis, Psittacella madaraszi 96

hyacinthinus, Anodorhynchus 174 *77*

hyacinthinus, Geoffroyus heteroclitus 92 *36*

hybridus, Tanygnathus lucionensis 104

hypoinochrous, Lorius 56 *18*

hypophonius, Alisterus amboinensis 112 *46*

icterotis, Ognorhynchus 186 *83*

icterotis, Platycercus 126 *53*

immarginata, Pyrrhura haematotis 222
101
imperialis, Amazona 308 144
inornata, Amazona farinosa 294 137
inseparabilis, Cyclopsitta diophthalma
84 32
insignis, Chalcopsitta atra 36 8
insularis, Forpus cyanopygius 240
intensior, Charmosyna placentis 64 22
intermedia, Psittacula 164 72
intermedia, Psittaculirostris
desmarestii 86
intermedius, Bolbopsittacus lunulatus
88 34
intermixta, Psittacella brehmii 94
iris, Psitteuteles 50 15

jandaya, Aratinga 200 90
jobiensis, Geoffroyus geoffroyi 92
jobiensis, Lorius lory 56 18
johnstoniae, Trichoglossus 48 14
jonquillaceus, Aprosmictus 114 47
josefinae, Charmosyna 66 23
jugularis, Brotogeris 256 118
juneae, Psittacula roseata 164

kangeanensis, Psittacula alexandri 168
kawalli, Amazona 294 137
keiensis, Micropsitta 78 29
keyensis, Geoffroyus geoffroyi 90 35
koenigi, Aratinga acuticaudata 188
kordoana, Charmosyna rubronotata 64
22
krameri, Psittacula 160 70
kriegi, Pyrrhura frontalis 214
kuhli, Eolophus roseicapilla 26
kuhlii, Vini 54 17

lacerus, Pionus maximiliani 276
lathami, Calyptorhynchus 24 2
Lathamus discolor 52 16
latirostris, Calyptorhynchus 22 1
leadbeateri, Lophocroa 26 3
leari, Anodorhynchus 174 77
lehmanni, Aratinga pertinax 202
lepida, Pyrrhura 220 100
leptorhynchus, Enicognathus 212 96
leucocephala, Amazona 280 130
leucogaster, Pionites 260 120
leucophthalma, Aratinga 194 87
leucotis, Pyrrhura 224 102
libs, Corocopsis nigra 146

lilacina, Amazona autumnalis 286 133
lilianae, Agapornis 158 69
lineola, Bolborhynchus 238 109
longicauda, Psittacula 170 75
longipennis, Diopsittaca nobilis 184
Lophocroa leadbeateri 26 3
lorentzi, Psittacella picta 94 37
Loriculus amabilis 74 27
Loriculus aurantiifrons 72 26
Loriculus beryllinus 172 76
Loriculus catamene 74 27
Loriculus exilis 72 26
Loriculus flosculus 72 26
Loriculus galgulus 172 76
Loriculus philippensis 76 28
Loriculus pusillus 72 26
Loriculus sclateri 74 27
Loriculus stigmatus 74 27
Loriculus tener 72 26
Loriculus vernalis 172 76
Lorius albidinucha 58 19
Lorius chlorocercus 58 19
Lorius domicella 58 19
Lorius garrulus 58 19
Lorius hypoinochrous 56 18
Lorius lory 56 18
lory, Lorius 56 18
luchsi, Myiopsitta monachus 210 95
lucianii, Pyrrhura 226 103
lucionensis, Tanygnathus 104 42
luconensis, Prioniturus 100 40
lunulatus, Bolbopsittacus 88 34

macao, Ara 178 79
macgillivrayi, Barnardius barnardi 118
49
macgillivrayi, Eclectus roratus 110
macgillivrayi, Probosciger aterrimus 22
macleayana, Cyclopsitta diophthalma
84 32
maclennani, Geoffroyus geoffroyi 90
35
macrorhynchus, Calyptorhynchus
banksii 24 2
maculata, Aratinga solstitialis 200
mada, Prioniturus 98 39
madaraszi, Psittacella 96 38
magnirostris, Psittacula eupatria 162
major, Aratinga aurea 206
major, Neopsittacus musschenbroekii
70 25
major, Oreopsittacus arfaki 70 25

major, Psittacella madaraszi 96
major, Psittacula alexandri 168
malachitacea, Triclaria 266 123
malherbi, Cyanoramphus 142 61
malindangensis, Prioniturus
waterstradti 102 41
manilata, Orthopsittaca 184 82
manillensis, Psittacula krameri 160
maracana, Primolius 182 81
margaretae, Bolborhynchus aurifrons
236
margarethae, Charmosyna 66 23
margaritensis, Aratinga pertinax 202
91
marshalli, Cyclopsitta diophthalma 84
32
Mascarinus mascarinus 312 146
mascarinus, Mascarinus 312 146
massaicus, Poicephalus gulielmi 148
64
massena, Trichoglossus haematodus 44
12
matschiei, Poicephalus meyeri 150 65
maugei, Aratinga chloroptera 192
maximiliani, Pionus 276 128
medius, Neopsittacus musschenbroekii
70 25
meeki, Charmosyna 60 20
meeki, Loriculus aurantiifrons 72 26
meeki, Micropsitta 80 30
megalorhynchos, Tanygnathus 104 42
melanoblepharus, Pionus maximiliani
276
melanocephalus, Pionites 260 120
melanogenia, Cyclopsitta gulielmitertii
82
melanonotus, Touit 252 116
melanopterus, Platycercus elegans 122
51
melanotis, Hapalopsittaca 270 125
melanura, Pyrrhura 218 99
Melopsittacus undulatus 134 57
menstruus, Pionus 272 126
mercenaria, Amazona 296 138
meridionalis, Nestor 144 62
mesotypus, Poicephalus senegalus 152
metae, Forpus conspicillatus 244 112
mexicanus, Ara militaris 180
meyeri, Poicephalus 150 65
meyeri, Trichoglossus flavoviridis 48
14
Micropsitta bruijnii 80 30

Micropsitta finschii 80 *30*
Micropsitta geelvinkiana 78 *29*
Micropsitta keiensis 78 *29*
Micropsitta meeki 80 *30*
Micropsitta pusio 78 *29*
micropteryx, Trichoglossus haematodus 42
microtera, Pyrrhura amazonum 226
militaris, Ara 180 *80*
mindanensis, Bolbopsittacus lunulatus 88 *34*
mindorensis, Loriculus philippensis 76 *28*
mindorensis, Prioniturus discurus 100 *40*
minor, Alisterus scapularis 112
minor, Aratinga frontata 196 *88*
minor, Enicognathus ferrugineus 212
minor, Geoffroyus geoffroyi 92 *36*
misoriensis, Micropsitta geelvinkiana 78 *29*
mitchelli, Trichoglossus haematodus 46 *13*
mitrata, Aratinga 194 *87*
modesta, Psittacella 96 *38*
modesta, Psittacula longicauda 170 *75*
molinae, Pyrrhura 216 *98*
mollis, Lophochroa leadbeateri 26 *3*
moluccanus, Trichoglossus haematodus 44 *12*
moluccensis, Cacatua 30 *5*
monarchoides, Polytelis anthopeplus 116 *48*
monachus, Myiopsitta 210 *95*
montanus, Prioniturus 102 *41*
morotaianus, Lorius garrulus 58
moszkowskii, Alisterus chloropterus 112 *46*
multistriata, Charmosyna 62 *21*
musschenbroekii, Neopsittacus 70 *25*
Myiopsitta monachus 210 *95*
mysoriensis, Geoffroyus geoffroyi 92 *36*

nana, Amazona albifrons 282
nana, Aratinga 208 *94*
nanina, Micropsitta finschii 80 *30*
Nannopsittaca dachilleae 246 *113*
Nannopsittaca panychlora 246 *113*
narethae, Northiella haematogaster 130 *55*
naso, Calyptorhynchus banksii 24 *2*

nattereri, Amazona ochrocephala 288 *134*
necopinata, Micropsitta bruijnii 80 *30*
nenday, Aratinga 188 *84*
Neophema chrystogaster 132 *56*
Neophema chrysostoma 132 *56*
Neophema elegans 132 *56*
Neophema petrophila 132 *56*
Neophema pulchella 134 *57*
Neophema splendida 134 *57*
Neopsephotus bourkii 134 *57*
Neopsittacus musschenbroekii 70 *25*
Neopsittacus pullicauda 70 *25*
neoxena, Aratinga acuticaudata 188
nesophilus, Trichoglossus haematodus 42
Nestor meridionalis 144 *62*
Nestor notabilis 144 *62*
Nestor productus 312 *146*
neumanni, Aratinga acuticaudata 188
nicefori, Aratinga leucophthalmus 194
nicobarica, Psittacula longicauda 170 *75*
nigra, Coracopsis 146 *63*
nigrescens, Platycercus elegans 122 *51*
nigrifrons, Cyclopsitta gulielmitertii 82 *31*
nigrigenis, Agapornis 158 *69*
nigrogularis, Trichoglossus haematodus 42
nipalensis, Psittacula eupatria 162
nobilis, Diopsittaca 184 *82*
normantoni, Cacatua sanguinea 32
Northiella haematogaster 130 *55*
notabilis, Nestor 144 *62*
novaezelandiae, Cyanoramphus 140 *60*
Nymphicus hollandicus 34 *7*

obiensis, Eos squamata 38
obiensis, Geoffroyus geoffroyi 92 *36*
obscura, Pyrrhura egregia 222
occidentalis, Barnardius zonarius 118
occidentalis, Pezoporus 136 *58*
occidentalis, Psittaculirostris desmarestii 86 *33*
ochrocephala, Amazona 288 290 292 *134 135 136*
ocularis, Aratinga pertinax 204 *92*
Ognorhynchus icterotis 186 *83*
olallae, Forpus xanthopterygius 242

ophthalmica, Cacatua 28 *4*
oratrix, Amazona ochrocephala 292 *136*
orbygnesius, Bolborhynchus 238 *109*
orcesi, Pyrrhura 232 *106*
Oreopsittacus arfaki 70 *25*
ornata, Charmosyna placentis 64 *22*
ornatus, Trichoglossus 42 *11*
Orthopsittaca manilata 184 *82*

pachyrhyncha, Rhynchopsitta 186 *83*
pacifica, Pyrrhura melanura 218 *99*
pallescens, Northiella haematogaster 130 *55*
palliceps, Platycercus adscitus 126 *53*
pallidus, Pionites melanocephalus 260 *120*
pallida, Psittacella brehmii 94 *37*
pallidior, Charmosyna placentis 64 *22*
pallidus, Forpus cyanopygius 240
palmarum, Amazona leucocephala 280
palmarum, Charmosyna 60 *20*
panamensis, Amazona ochrocephala 288 *134*
panychlora, Nannopsittaca 246 *113*
papou, Charmosyna 68 *24*
paraensis, Aratinga pertinax 204 *92*
parvifrons, Pyrrhura 228 *104*
parvipes, Amazona ochrocephala 290 *135*
parvirostris, Psittacula krameri 160
parvula, Cacatua sulphurea 28 *4*
passerinus, Forpus 240 *110*
pastinator, Cacatua 32 *6*
patagonus, Cyanoliseus 210 *95*
perlata, Pyrrhura 220 *100*
peronica, Psittacula alexandri 168
personata, Prosopeia 138 *59*
personatus, Agapornis 156 *68*
pertinax, Aratinga 202 204 *91 92*
peruviana, Hapalopsittaca melanotis 270 *125*
peruviana, Pyrrhura 228 *104*
peruviana, Vini 54 *17*
petrophila, Neophema 132 *56*
Pezoporus occidentalis 136 *58*
Pezoporus wallicus 136 *58*
pfrimeri, Pyrrhura 224 *102*
Phigys solitarius 54 *17*
philippensis, Loriculus 76 *28*
phoenicura, Pyrrhura molinae 216
picta, Psittacella 94 *37*

picta, Pyrrhura 226 *103*

pileata, Micropsitta bruijnii 80

pileata, Pionopsitta 266 *123*

Pionites leucogaster 260 *120*

Pionites melanocephalus 260 *120*

Pionopsitta pileata 266 *123*

Pionus chalcopterus 276 *128*

Pionus fuscus 276 *128*

Pionus maximiliani 276 *128*

Pionus menstruus 272 *126*

Pionus senilis 272 *126*

Pionus seniloides 274 *127*

Pionus sordidus 274 *127*

Pionus tumultuosus 274 *127*

pistra, Trichoglossus johnstoniae 48 *14*

placentis, Charmosyna 64 *22*

platenae, Prioniturus 100 *40*

platurus, Prioniturus 98 *39*

Platycercus adscitus 126 *53*

Platycercus caledonicus 120 *50*

Platycercus elegans 122 124 *51 52*

Platycercus eximius 126 *53*

Platycercus icterotis 126 *53*

Platycercus venustus 126 *53*

Poicephalus crassus 150 *65*

Poicephalus cryptoxanthus 150 *65*

Poicephalus flavifrons 148 *64*

Poicephalus gulielmi 148 *64*

Poicephalus meyeri 150 *65*

Poicephalus robustus 148 *64*

Poicephalus rueppellii 152 *66*

Poicephalus rufiventris 152 *66*

Poicephalus senegalus 152 *66*

polychloros, Eclectus roratus 110 *45*

Polytelis alexandrae 116 *48*

Polytelis anthopeplus 116 *48*

Polytelis swainsonii 116 *48*

ponsi, Pionus sordidus 274

pontius, Psittinus cyanurus 170

porphyrocephala, Glossopsitta 52 *16*

pretrei, Amazona 302 *141*

Primolius auricollis 184 *82*

Primolius couloni 182 *81*

Primolius maracana 182 *81*

Prioniturus discurus 100 *40*

Prioniturus flavicans 98 *39*

Prioniturus luconensis 100 *40*

Prioniturus mada 98 *39*

Prioniturus montanus 102 *41*

Prioniturus platenae 100 *40*

Prioniturus platurus 98 *39*

Prioniturus verticalis 102 *41*

Prioniturus waterstradti 102 *41*

Probosciger aterrimus 22 *1*

productus, Nestor 312 *146*

Prosopeia personata 138 *59*

Prosopeia tabuensis 138 *59*

proxima, Micropsitta meeki 80 *30*

Psephotus chrysopterygius 130 *55*

Psephotus dissimilis 130 *55*

Psephotus haematonotus 128 *54*

Psephotus pulcherrimus 310 *145*

Psephotus varius 128 *54*

Pseudeos, fuscata 38 *9*

Psittacella brehmii 94 *37*

Psittacella madaraszi 96 *38*

Psittacella modesta 96 *38*

Psittacella picta 94 *37*

Psittacula alexandri 168 *74*

Psittacula calthorpae 162 *71*

Psittacula caniceps 166 *73*

Psittacula columboides 162 *71*

Psittacula cyanocephala 164 *72*

Psittacula derbiana 166 *73*

Psittacula echo 160 *70*

Psittacula eupatria 162 *71*

Psittacula exsul 310 *145*

Psittacula finschii 166 *73*

Psittacula himalayana 166 *73*

Psittacula intermedia 164 *72*

Psittacula krameri 160 *70*

Psittacula longicauda 170 *75*

Psittacula roseata 164 *72*

Psittacula wardi 310 *145*

Psittaculirostris desmarestii 86 *33*

Psittaculirostris edwardsii 88 *34*

Psittaculirostris salvadorii 88 *34*

Psittacus erithacus 148 *64*

Psitteuteles goldiei 50 *15*

Psitteuteles iris 50 *15*

Psitteuteles versicolor 50 *15*

Psittinus cyanurus 170 *75*

Psittrichas fulgidus 110 *45*

pucherani, Geoffroyus geoffroyi 92 *36*

pulchella, Charmosyna 66 *23*

pulchella, Neophema 134 *57*

pulcherrimus, Psephotus 310 *145*

pulchra, Gypopsitta 262 *121*

pullarius, Agapornis 154 *67*

pullicauda, Neopsittacus 70 *25*

purpuratus, Touit 252 *116*

Purpureicephalus spurius 120 *50*

pusilla, Glossopsitta 52 *16*

pusillus, Loriculus 72 *26*

pusio, Micropsitta 78 *29*

pyrilia, Gypopsitta 262 *121*

pyrrhops, Hapalopsittaca 268 *124*

pyrrhoptera, Brotogeris 256 *118*

Pyrrhura albipectus 232 *106*

Pyrrhura amazonum 226 *103*

Pyrrhura caeruleiceps 230 *105*

Pyrrhura calliptera 222 *101*

Pyrrhura cruentata 234 *107*

Pyrrhura devillei 214 *97*

Pyrrhura egregia 222 *101*

Pyrrhura eisenmanni 230 *105*

Pyrrhura emma 224 *102*

Pyrrhura frontalis 214 *97*

Pyrrhura griseipectus 224 *102*

Pyrrhura haematotis 222 *101*

Pyrrhura hoffmanni 234 *107*

Pyrrhura lepida 220 *100*

Pyrrhura leucotis 224 *102*

Pyrrhura lucianii 226 *103*

Pyrrhura melanura 218 *99*

Pyrrhura molinae 216 *98*

Pyrrhura orcesi 232 *106*

Pyrrhura parvifrons 228 *104*

Pyrrhura perlata 220 *100*

Pyrrhura peruviana 228 *104*

Pyrrhura pfrimeri 224 *102*

Pyrrhura picta 226 *103*

Pyrrhura rhodocephala 234 *107*

Pyrrhura roseifrons 228 *104*

Pyrrhura rupicola 232 *106*

Pyrrhura snethlageae 230 *105*

Pyrrhura subandina 230 *105*

Pyrrhura viridicata 234 *107*

queenslandica, Cacatua galerita 28

ramuensis, Cyclopsitta gulielmitertii 82

regulus, Loriculus philippensis 76 *28*

reichenowi, Pionus menstruus 272 *126*

reichenowi, Poicephalus meyeri 150 *65*

restricta, Pyrrhura molinae 216

reticulata, Eos 38 *9*

rhodocephala, Pyrrhura 234 *107*

rhodocorytha, Amazona 304 *142*

rhodops, Geoffroyus geoffroyi 90 *35*

Rhynchopsitta pachyrhyncha 186 *83*

Rhynchopsitta terresi 186 *83*

riciniata, Eos squamata 38 *9*

riedeli, Eclectus roratus 108 *44*

robertsi, Bolborhynchus aurifrons 236
108

robustus, Poicephalus 148 *64*

roratus, Eclectus 108 110 *44 45*

rosea, Micropsitta bruijnii 80 *30*

roseata, Psittacula 164 *72*

roseicapilla, Eolophus 26 *3*

roseicollis, Agapornis 156 *68*

roseifrons, Pyrrhura 228 *104*

rosenbergii, Trichoglossus haematodus
42 *11*

rosselianus, Lorius hypoinochrous 56
18

rothschildi, Charmosyna pulchella 66
23

ruber, Loriculus sclateri 27

rubiginosus, Trichoglossus 48 *14*

rubrifrons, Chalcopsitta sintillata 36 *8*

rubrigularis, Charmosyna 60 *20*

rubrigularis, Pionus menstruus 272

rubrirostris, Bolborhynchus aurifrons
236 *108*

rubritorquis, Aratinga 190 *85*

rubritorquis, Trichoglossus haematodus
46 *13*

rubrogenys, Ara 180 *80*

rubronotata, Charmosyna 64 *22*

rueppellii, Poicephalus 152 *66*

rufiventris, Poicephalus 152 *66*

rupicola, Pyrrhura 232 *106*

saissetti, Cyanoramphus
novaezelandiae 140 *60*

saltuensis, Amazona albifrons 282
131

salvadorii, Lorius lory 56 *18*

salvadorii, Psittaculirostris 88 *34*

salvini, Amazona autumnalis 286 *133*

samueli, Calyptorhynchus banksii 24 *2*

sanctithomae, Brotogeris 258 *119*

sandiae, Pyrrhura rupicola 232

sangirensis, Tanygnathus sumatranus
106

sanguinea, Cacatua 32 *6*

saturatus, Pionus sordidus 274 *127*

saturatus, Poicephalus meyeri 150

sauvissima, Cyclopsitta gulielmitertii
82 *31*

scapularis, Alisterus 112 *46*

sclateri, Forpus 244 *112*

sclateri, Loriculus 74 *27*

semilarvata, Eos 40 *10*

semitorquatus, Barnardius zonarius 118
49

senegalus, Poicephalus 152 *66*

senilis, Pionus 272 *126*

seniloides, Pionus 274 *127*

sepikiana, Charmosyna josefinae 66
23

septentrionalis, Nestor meridionalis
144 *62*

septentrionalis, Trichoglossus
haematodus 44

severus, Ara 182 *81*

siamensis, Psittacula eupatria 162

sibilans, Coracopsis nigra 146

simplex, Geoffroyus 92 *36*

sinerubris, Prioniturus platurus 98 *39*

sintillata, Chalcopsitta 36 *8*

siquijorensis, Loriculus philippensis 76
28

siy, Pionus maximiliani 276 *128*

snethlageae, Pyrrhura 230 *105*

socialis, Neopsittacus pullicauda 70
25

solimoensis, Brotogeris chrysoptera
258 *119*

solitarius, Phigys 54 *17*

solomonensis, Eclectus roratus 110 *45*

solstitialis, Aratinga 200 *90*

somu, Lorius lory 56 *18*

sordida, Pyrrhura molinae 216 *98*

sordidus, Pionus 274 *127*

souancei, Pyrrhura melanura 218 *99*

spengeli, Forpus xanthopterygius 242
111

spixii, Cyanopsitta 174 *77*

splendens, Prosopeia tabuensis 138
59

splendida, Neophema 134 *57*

spurius, Purpureicephalus 120 *50*

squamata, Eos 38 *9*

stellae, Charmosyna papou 68 *24*

stenolophus, Probosciger aterrimus 22
1

stepheni, Vini 54 *17*

stictopterus, Touit 248 *114*

stigmatus, Loriculus 74 *27*

strenua, Aratinga 190 *85*

stresemanni, Micropsitta pusio 78

stresemanni, Trichoglossus haematodus
46

Strigops, habroptila 144 *62*

suahelicus, Poicephalus robustus 148

suavissima, Cyclopsitta gulielmitertii
82 *31*

subadelaidae, Platycercus elegans 174
52

subaffinis, Tanygnathus
megalorhynchos 104

subandina, Pyrrhura 230 *105*

subcollaris, Psittacella modesta 96 *38*

subplacens, Charmosyna placentis 64
22

sudestiensis, Geoffroyus geoffroyi 90
35

sulaensis, Alisterus amboinensis 112

sulphurea, Cacatua 28 *4*

sumatranus, Tanygnathus 106 *43*

sumbensis, Tanygnathus
megalorhynchos 104

surdus, Touit 252 *116*

surinama, Aratinga pertinax 204 *92*

swainsonii, Polytelis 116 *48*

swindernianus, Agapornis 158 *69*

syringanuchalis, Chalcopsitta
duivenbodei 36

tabuensis, Prosopeia 138 *59*

takatsukasae, Brotogeris sanctithomae
258 *119*

talautensis, Eos histrio 40

talautensis, Prioniturus platurus 98

talautensis, Tanygnathus lucionensis
104 *42*

tanganyikae, Poicephalus cryptoxanthus
150

Tanygnathus gramineus 106 *43*

Tanygnathus lucionensis 104 *42*

Tanygnathus megalorhynchos 104 *42*

Tanygnathus sumatranus 106 *43*

taranta, Agapornis 154 *67*

taviuensis, Prosopeia tabuensis 138

tener, Loriculus 72 *26*

tenuifrons, Brotogeris chrysoptera 258
119

tenuirostris, Cacatua 32 *6*

terresi, Rhynchopsitta 186 *83*

theresae, Hapalopsittaca amazonina
268 *124*

tigrinus, Bolborhynchus lineola 238
109

timneh, Psittacus erithacus 148 *64*

timorlaoensis, Geoffroyus geoffroyi 90

tirica, Brotogeris 254 *117*

tobagensis, Amazona amazonica 298

tortugensis, Aratinga pertinax 202
Touit batavicus 248 *114*
Touit costaricensis 250 *115*
Touit dilectissimus 250 *115*
Touit huetii 250 *115*
Touit melanonotus 252 *116*
Touit purpuratus 252 *116*
Touit stictopterus 248 *114*
Tout surdus 252 *116*
toxopei, Charmosyna 60 *20*
transfreta, Cacatua sanguinea 32
transilis, Aratinga wagleri 198
transvaalensis, Poicephalus meyeri
  150
tresmariae, Amazona ochrocephala 292
  136
Trichoglossus chlorolepidotus 48 *14*
Trichoglossus euteles 48 *14*
Trichoglossus flavoviridis 48 *14*
Trichoglossus haematodus 42 44 46 *11*
  12 13
Trichoglossus johnstoniae 48 *14*
Trichoglossus ornatus 42 *11*
Trichoglossus rubiginosus 48 *14*
Triclaria malachitacea 266 *123*
tricolor, Ara 312 *146*
tristami, Micropsitta finschii 80
triton, Cacatua galerita 28 *4*
tucumana, Amazona 302 *141*
tucumana, Aratinga mitrata 194 *87*
tuipara, Brotogeris chrysoptera 258
  119
tumultuosus, Pionus 274 *127*
tytleri, Psittacula longicauda 170 *75*

ugandae, Agapornis pullarius 154
ulietanus, Cyanoramphus 310 *145*
ultramarina, Vini 54 *17*
undulatus, Melopsittacus 134 *57*
unicolor, Cyanoramphus 142 *61*

uvaeensis, Eunymphicus cornutus 138
  59

varius, Psephotus 128 *54*
vasa, Coracopsis 146 *63*
velezi, Hapalopsittaca amazonina 268
  124
venezuelae, Aratinga pertinax 204 *92*
ventralis, Amazona 278 *129*
venustus, Platycercus 126 *53*
vernalis, Loriculus 172 *76*
versicolor, Alisterus amboinensis 112
versicolor, Amazona 308 *144*
versicolor, Psitteuteles 50 *15*
versicolorus, Brotogeris 254 *117*
versteri, Poicephalus senegalus 152
  66
verticalis, Prioniturus 102 *41*
vicinalis, Aratinga nana 208
vinacea, Amazona 302 *141*
Vini australis 54 *17*
Vini kuhlii 54 *17*
Vini peruviana 54 *17*
Vini stepheni 54 *17*
Vini ultramarina 54 *17*
virago, Cyclopsitta diophthalma 84 *32*
virenticeps, Amazona farinosa 294
  137
viridicata, Pyrrhura 234 *107*
viridiceps, Touit purpuratus 252 *116*
viridicrissalis, Lorius lory 56 *18*
viridifrons, Micropsitta finschii 80
  30
viridigenalis, Amazona 284 *132*
viridipectus, Micropsitta keiensis 78
  29
viridissimus, Forpus passerinus 240
vittata, Amazona 280 *130*
vosmaeri, Eclectus roratus 108 *44*
vulturina, Gypopsitta 264 *122*

wagleri, Aratinga 198 *89*
wahnesi, Charmosyna papou 68 *24*
wallicus, Pezoporus 136 *58*
wardi, Psittacula 310 *145*
waterstradti, Prioniturus 102 *41*
weberi, Trichoglossus haematodus 46
  13
weddellii, Aratinga 198 *89*
westralensis, Cacatua sanguinea 32 *6*
wetterensis, Aprosmictus jonquillaceus
  114
wetterensis, Psitteuteles iris 50 *15*
whiteheadi, Prioniturus discurus 100
wilhelminae, Charmosyna 62 *21*
woodi, Amazona finschi 284 *132*
worcesteri, Loriculus philippensis 76
  28

xanthanotus, Calyptorhynchus funereus
  22 *1*
xanthogenia, Aratinga pertinax 202
  91
xanthogenys, Platycercus icterotis 126
  53
xantholaema, Amazona ochrocephala
  288 *134*
xantholora, Amazona 282 *131*
xanthomerius, Pionites leucogaster 260
  120
xanthops, Amazona 300 *140*
xanthops, Forpus 246 *113*
xanthopterygius, Forpus 242 *111*
xanthopteryx, Amazona aestiva 298
  139
xanthurus, Pionites leucogaster 260
  120

zealandicus, Cyanoramphus 310 *145*
zenkeri, Agapornis swindernianus 158
zonarius, Barnardius 118 *49*